Chemistry of the Atmosphere

Chemistry of the Atmosphere

MURRAY J. McEWAN and
LEON F. PHILLIPS
Chemistry Department,
University of Canterbury,
Christchurch, New Zealand.

A HALSTED PRESS BOOK

JOHN WILEY & SONS
New York

© 1975 by M. J. McEwan and L. F. Phillips

First published 1975
by Edward Arnold (Publishers) Ltd.,
London

Published in the USA
by Halsted Press, a Division
of John Wiley & Sons, Inc.
New York

Library of Congress Cataloging in Publication Data

McEwan, Murray, J.
    The chemistry of the atmosphere.

    "A Halsted Press book."
    Includes bibliographies.
    1. Atmospheric chemistry.   I. Phillips, Leon
F., joint author.   II. Title.
QC879.6.M3   1975   551.5'11   75-14487
ISBN 0-470-58393-2

Printed in Great Britain

Preface

The subject of atmospheric chemistry has undergone a remarkable expansion during the last few years, spectacular growth having occurred both in the amount of interest attracted to the field and in the total volume of accumulated knowledge. Some of this growth can be associated with the recent upsurge of concern about matters relating to the environment, but most has resulted from the cumulative effect of a succession of experimental and theoretical advances in the investigation of the chemistry of planetary atmospheres, and in the study of atmospheric phenomena under laboratory conditions. A list of outstanding examples of such advances would include the improvements in rocketry and rocket-borne instrumentation which have enabled detailed information to be obtained about the composition of the atmospheres of Venus and Mars, the use of computer simulation to elucidate the behaviour of complex photochemical systems, and the development of the flowing afterglow method for measuring rates of fast ion–molecule reactions.

In view of this expansion, it seemed to us unsatisfactory that the subject of atmospheric chemistry appeared in textbooks either in the form of a chapter of a book devoted mainly to some other field such as upper atmosphere physics, or, in a single one of its aspects, as a monograph on a topic such as air pollution. The present book is intended to remedy this situation by providing a reasonably compact review of the whole field, with emphasis on matters that are likely to be of interest to chemists. The book is intended primarily as an introduction to the chemistry of the atmosphere for persons with a background in chemistry or chemical physics, and in this capacity might serve as the basis of a course at graduate or advanced undergraduate level. However, we have endeavoured, particularly in the last few chapters, to point out limitations of current knowledge and to summarize a considerable body of current experimental data, so that the book should also be useful to research workers in the field. In giving references to sources of experimental data we have preferred, where possible, to cite key review articles rather than individual papers; thus the lists of references at the ends of the chapters should offer a number of suitable stepping-off points for an investigation of the literature at first hand.

It is a pleasure to record our gratitude to Professor Harold Schiff of York University, Toronto, who initiated our interest in atmospheric chemistry, and to both Professor Schiff and Dr. Eldon Ferguson of the Environmental

Research Laboratories, Boulder, for their helpful criticism of the manuscript. We must also thank Dr. Amos Horney and his colleages at the Directorate of Chemical Sciences of the U.S. Air Force Office of Scientific Research for their support of our own research in this field. Finally, we are greatly indebted to our Publishers, who first pointed out the need for such a book, and to Mrs. Jill Dolby who patiently typed the several versions of the manuscript.

Christchurch, N.Z.   Leon F. Phillips
1974   Murray J. McEwan

Contents

1
General Characteristics of the Atmosphere

1.1 Introduction

The thin film of gas that clings to the surface of the earth performs many functions, and has a correspondingly large number of important properties. To a biologist its most outstanding attribute is the ability to support life-forms which derive their energy from the interaction of sunlight with water, oxygen and carbon dioxide. A less obvious biological function is that of shielding life on the surface from harsh, short wavelength radiation emitted by the sun. To a meteorologist the atmosphere is basically a sink for solar energy, and its most interesting characteristic is the ability to transform that energy into wind systems and rainfall averages. To a radiophysicist the most important feature of the atmosphere is the extremely tenuous region above an altitude of fifty kilometres, where the concentration of ions and free electrons is sufficient to justify the name ionosphere, and from which radio waves are bent back towards the surface in such a way as to permit long distance radio communication. To a poet the atmosphere is a medium for sunsets; to an astronomer it is a regrettable necessity.

To a chemist the atmosphere appears as a continuous, large scale photo-chemical experiment. As an experimental system it is unusual in being agreeably free from wall effects, and in that the nature of the radiation being absorbed, the species undergoing photolysis, and the type of primary process involved—photoionization, photodissociation, or photo-excitation—are all dependent upon the altitude of observation rather than on the will of the observer. Interpretation of observations on the atmosphere is further complicated by the effects of winds, vertical mass transport, and the alternation of night and day; factors such as these take the place of the experimental variables which one would alter in a laboratory. Since this is a book about the chemistry of the atmosphere it is natural that the outcome of the photochemical experiment should be our main concern. Nevertheless it is important to remember that this is just one aspect of the situation, albeit a fundamental one, and that the results of investigations by meteor-ologists, astronomers and radiophysicists, not to mention biologists and poets, are often not only relevant but also illuminating.

For purposes of study the atmosphere is usually divided into two main regions. The study of the dynamics of the lower atmosphere, or troposphere, is considered to be part of meteorology, whereas the study of the physics and chemistry of the upper atmosphere forms the subject of aeronomy. In a normal (i.e. clean) atmosphere, photochemistry begins at the top of the troposphere and becomes more complex and interesting as the altitude increases. Thus the chemistry of the atmosphere might fairly be regarded as part of aeronomy. However, a clean atmosphere cannot always be taken for granted, and the photochemical reactions which may occur at low altitudes in a polluted atmosphere are of great scientific interest and considerable practical importance. In general, workers in the field of atmospheric photochemistry seldom limit themselves even to reactions which involve normal atmospheric constituents and common pollutants; for those reactions of interest which cannot be accommodated under the heading of pollution studies there is the alternative plea of relevance to the atmospheres of planets other than the earth. The atmospheres of other planets have appeared, until recently, to be solely the concern of astronomers, but with increasing knowledge, derived in part from interplanetary probes and partly from improved astronomical techniques, has come the realization that many exciting and fundamental photochemical problems in this area remain to be solved. Consequently our present account of the chemistry of the atmosphere will include a discussion of topics which commonly appear in texts on astronomy, aeronomy, and air pollution, as well as photochemistry. The remainder of this present chapter comprises a description of the gross physical structure of the atmosphere, and of the factors which govern this structure. It should also serve as an introduction to the terminology employed in upper atmosphere work.

1.2 Atmospheric structure and terminology

The earth's atmosphere is divided into regions primarily on the basis of temperature gradients, as shown in Fig. 1.1. Typically, at middle latitudes, the temperature falls with increasing altitude (has a 'positive lapse rate') in the *troposphere*, passes through a minimum at the tropopause, and then rises with increasing altitude in the *stratosphere*. A region in which temperature increases with altitude is termed an *inversion*, or inversion layer. Because cold air is denser than warm air an inversion is normally very stable against vertical mixing effects; the stratosphere is no exception to this. At the stratopause the temperature passes through a more or less well-defined maximum, then decreases through the *mesosphere* to a second minimum at the mesopause. Above the mesopause, in the *thermosphere*, the kinetic temperature again increases with altitude, reaching a maximum value in the region of 1500–2000 K at a height of several thousand kilometres. Above the thermosphere, and demarcated from it by the thermopause, is an essentially isothermal region, the *exosphere*. At any particular time and place the actual structure of the atmospheric temperature profile may differ appreciably from Fig. 1.1, for example as a consequence of winds or

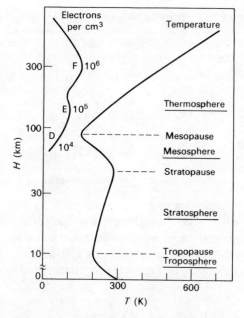

Fig. 1.1 Atmospheric regions and nomenclature.

of local temperature inversions, but in general the curve shown is at least qualitatively correct.

Various regions or layers of the atmosphere have properties which are sufficiently characteristic to justify their being given special names, most notably the *ionosphere*. The ionosphere is itself divided into D, E and F regions according to electron density and altitude. The region below 90 km is called the D region, that between 90 and 140 km the E region, and that above 140 km the F region. Between the different regions there is not necessarily any minimum of electron density (though a small minimum may be present between the E and F regions as indicated in Fig. 1.1) nor even a marked change in gradient of the electron concentration profile. The F region is commonly sub-divided into an F1 'ledge', with an F2 layer above. The outermost layer of the ionosphere, in which the motions of the ions (mainly H^+) are largely governed by the earth's magnetic field, is termed the magnetosphere. The nature of the positive ions which are present varies markedly with altitude, as shown in Table 1.1, and provides an alternative basis for distinguishing between regions. The interesting and varied chemistry which results from the interactions of both positive and negative ions with neutral constituents of the atmosphere forms the subject of Chapter 6.

The presence of an ionosphere is a consequence of the absorption of photoionizing radiation from the sun in the outer layers of the atmosphere.

Table 1.1   Composition of the ionosphere (quiet, daytime, mid-solar cycle)

Region	Altitude (km)	T (Kelvin)	Log[M]*	Log[A]*	Principal Ions
D	60	250	15	10	NO^+, $H(H_2O)_2^+$, H_3O^+, $NO_3^-(H_2O)_n$
E	110	250	13	11	NO^+, O_2^+, e^-
Sporadic E	110	250	13	11	Mg^+, Fe^+, Si^+, Ca^+, e^-
F_1	170	700	10	10	NO^+, O_2^+, O^+, e^-
F_2	300	1500	8	9	O^+, N^+, O_2^+, NO^+, e^-
Magnetosphere	1000	1700		5	H^+, He^+ (?)

* [M] = concentration of molecular species, [A] = concentration of atomic species, in particles cm^{-3}.

Absorption of somewhat longer wavelength vacuum ultraviolet radiation causes molecular oxygen, and minor constituents such as water and carbon dioxide, to be largely dissociated in the daytime above about 100 km. Thus at this altitude there is a marked change in the composition of the atmosphere from what it is at sea level. In addition, the composition of the atmosphere varies with altitude above 100 km, with atomic hydrogen and to a lesser extent helium becoming predominant at very high levels. The region below about 90 km, where the bulk composition of the atmosphere is essentially the same as at sea level, is termed the *homosphere*, and the region above is termed the *heterosphere*. In the homosphere diffusion is relatively slow and the composition of the air is governed by mixing. In contrast, diffusion is very important in the heterosphere. Thus, for example, during the day molecular oxygen diffuses upwards until photodissociation occurs, then the resulting oxygen atoms diffuse downwards until they reach a region where the pressure is high enough for termolecular recombination to occur. At the boundary of the homosphere and heterosphere, in a belt between about 80 and 110 km, there is a region which contains relatively large concentrations of atomic oxygen and hydrogen, together with reactive molecular species such as OH, NO and O_3. The reactive species take part in a variety of chemical processes, many of which lead to the formation of products in excited states, with consequent emission of weak but detectable luminescence. Luminescence emitted between 80 and 110 km is a major component of the night-time *airglow*, i.e. of the light emitted by the atmosphere as a result of photochemical processes. The airglow is not to be confused with the *aurora*, which is a very much more intense emission that results from bombardment of the atmosphere by electrons and protons from the sun. When these charged particles enter the earth's magnetic field they are obliged to spiral along the magnetic lines of force and thus to enter the atmosphere in the vicinity of one of the magnetic poles. (In exceptional circumstances auroral displays may extend a long way from the magnetic poles; observations have been reported from the Mediterranean and from the British Isles.) Although the aurora is a fascinating phenomenon, and the spectroscopic study of auroral emissions has given useful information about the composition and temperature of the upper atmosphere, it is considered to be outside the scope of this book. The airglow, and the reactions involved in its production, are considered in detail in Chapter 5.

The major photodissociation process occurring near 100 km involves absorption in the Schumann-Runge continuum of molecular oxygen, at wavelengths below 175 nm. At lower altitudes and higher pressures, absorption in the weaker Herzberg continuum below 242 nm also leads to dissociation of O_2. Dissociation in the Herzberg continuum is important in a region which extends down to the tropopause, with the maximum rate of oxygen atom production occurring at an altitude of about 30 km. As a result of the two important dissociation processes, chemical reactions can occur in a band which includes the whole of the stratosphere, together with the mesosphere and the lower part of the thermosphere. This band is usually referred to as the *chemosphere*. The composition of the chemosphere, and the reactions of ground state and excited neutral species which are believed to occur there, form the subject of Chapter 4.

The most obvious outcome of photodissociation of O_2 at the relatively high pressures characteristic of the stratosphere is the formation of a layer which contains ozone at a peak concentration of 10^{12}–10^{13} molecule cm^{-3}. This layer is occasionally termed the *ozonosphere*. The main processes involved are

$$O_2 + h\nu \rightarrow O + O \tag{1.1}$$

$$O + O_2 + M \rightarrow O_3 + M + 100\,kJ \tag{1.2}$$

$$O + O_3 \rightarrow 2O_2 + 390\,kJ \tag{1.3}$$

$$O_3 + h\nu \rightarrow O_2 + O \tag{1.4}$$

where M is any available third body (usually N_2 or O_2). Ozone is itself readily photolysed, both by visible light, which it absorbs weakly, and by ultraviolet light in a band below 290 nm, where it absorbs very strongly. A very important consequence of the presence of ozone in the stratosphere is that sunlight of wavelength below 290 nm is prevented from reaching ground level, where its arrival would be extremely inimical to life such as has developed on the surface. The photodissociation of ozone (eqn. 1.4) regenerates atomic oxygen, which is able to reform ozone by reaction 1.2, or alternatively to destroy more ozone by reaction 1.3. We may note that for every mole of ozone that is formed by reaction 1.2, 100 kJ of energy is liberated, and for every mole that is destroyed by reaction 1.3 a further 390 kJ is evolved. Consequently the cycle of reactions 1.1–1.4 releases a great deal of heat into the atmosphere, and is in fact responsible for the observed increase of temperature with altitude in the stratosphere. This photochemical heating of the atmosphere is naturally of considerable interest to meteorologists. The reaction sequence 1.1–1.4 has other important consequences: for example, when ozone is photolysed at wavelengths below 310 nm the O_2 molecule appearing on the right hand side of eqn. 1.4 can be the metastable $O_2(^1\Delta_g)$, which is one of the most important neutral constituents of the D region. This subject will be taken up again in Chapters 4, 5 and 6.

1.3 Composition, temperature and scale height

The major components of air at ground level, and the proportion of each, are given in Table 1.2. These components govern the bulk properties of air throughout the homosphere. The typical amounts of minor molecular

Table 1.2   Major constituents of dry air at sea level

Species	Molecular Weight	Percentage	Concentration (molecules cm^{-3} at NTP)
N_2	28.02	78.08	2.098×10^{19}
O_2	32.00	20.95	5.629×10^{18}
Ar	39.95	0.934	2.510×10^{17}
CO_2	44.01	0.33	8.87×10^{15}

constituents at ground level are given in Table 1.3. Although they are present only in small amounts the minor constituents play a disproportionate role in the photochemical processes which occur in the chemosphere.

Height profiles of the concentration of the more important neutral constituents are shown qualitatively in Fig. 1.2. The main virtue of this figure is that, like Fig. 1.1, it is a useful summary of the trends to be expected in a normal atmosphere. However, the concentration values shown for the reactive constituents are not to be taken too literally, and in practice the detailed structure of the atmosphere, as summarized in Figs. 1.1 and 1.2, varies in a predictable way with latitude, time of day, season, and position in the eleven year cycle of solar activity. Important features to be noted in Fig. 1.2 include the virtually steady exponential fall-off of molecular nitrogen concentration up to very high altitudes, in contrast to the molecular oxygen profile which shows a marked change of slope near 100 km where dissociation to oxygen atoms becomes important, the peak of ozone concentration in the lower stratosphere, and the presence of reactive species such as nitric oxide and atomic hydrogen, in addition to atomic oxygen, at significant concentrations in the neighbourhood of the mesopause.

The variation of atmospheric pressure with altitude is governed by the *hydrostatic equation*, which can be derived with the aid of Fig. 1.3. The pressure is P at an altitude z, and $P - \delta P$ at altitude $z + \delta z$. The difference in pressure is due to the weight of gas in a column of unit cross section and

Table 1.3 Minor constituents of air in the troposphere

Species	Typical Mole Fraction*
H_2O	$10^{-5}-10^{-2}$
Ne	1.82×10^{-5}
CH_4	1.5×10^{-6}
Kr	1.14×10^{-6}
O_3	10^{-8}
N_2O	2×10^{-7}
H_2	5×10^{-7}
CO	$6 \times 10^{-8}-2 \times 10^{-7}$
Xe	8.7×10^{-8}
$NO + NO_2$	$5 \times 10^{-10}-2 \times 10^{-8}$

* Mole fraction = mixing ratio, or 'volume mixing ratio'; chemists generally prefer the former term, physicists the latter.

length δz. Hence

$$-\delta P = \delta z \cdot \rho \cdot g$$

or in the limit $\delta z \to 0$

$$dP/dz = -\rho g \qquad (1.5)$$

where g is the acceleration due to gravity and ρ is the gas density. For air we can use the perfect gas equation

$$PV = nRT \qquad (1.6)$$

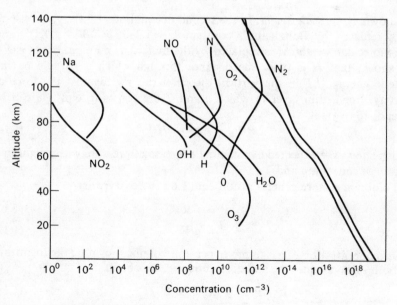

Fig. 1.2 Representative concentration profiles of neutral species in the atmosphere.

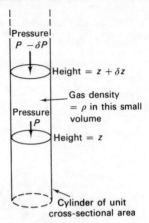

Fig. 1.3 Model for calculating the variation of pressure with altitude.

together with the expression for the density

$$\rho = nM/V \tag{1.7}$$

(where n is the number of moles of gas of mean molecular weight M in the volume V at temperature T) to obtain

$$dP/P = -dz/H \tag{1.8}$$

The important quantity

$$H = RT/Mg = kT/mg \tag{1.9}$$

is known as the *scale height*. In the second of equations 1.9 the gas constant R is replaced by Boltzmann's constant $k = 1.3805 \times 10^{-23}$ J K^{-1}, and the molecular weight M is replaced by the molecular mass m. It can easily be shown that H is twice the distance through which a particle having kinetic energy $\frac{1}{2}kT$ in the vertical direction can rise against the force of gravity. For a thin layer in which g and T are constant, eqn. 1.8 can be integrated to yield

$$P/P_0 = \exp(-z/H) \tag{1.10}$$

which shows the expected exponential decrease of pressure with height in a region of constant g and T.

To obtain a more general result, eqn. 1.6 can be rewritten

$$P = CRT \tag{1.11}$$

where

$$C = n/V \tag{1.12}$$

is the concentration of gas in moles per unit volume. Forming the logarithm of both sides of eqn. 1.11 and differentiating, we obtain

$$dP/P = dC/C + dT/T \tag{1.13}$$

i.e.

$$dC/C + dT/T = -dz/H \tag{1.14}$$

From eqn. 1.9

$$dH/dz = (R/Mg)\,dT/dz$$

or $\qquad dH/dz = \beta = $ a constant $\qquad\qquad$ (1.15)

in a region which is thin enough for both g and the temperature gradient dT/dz to be constant. Hence in this case, from eqn. 1.8,

$$dP/P = -dH(\beta H)^{-1}$$

or $\qquad P/P_0 = (H/H_0)^{-1/\beta}$ $\qquad\qquad$ (1.16)

Similarly, by integrating eqn. 1.14 for such a region, we obtain

$$C/C_0 = (H/H_0)^{-(1+1/\beta)} \qquad\qquad (1.17)$$

The value of g varies only slowly with height,* so that in the homosphere, where the mean molecular weight is constant, a measurement of the variation of pressure with altitude, i.e. of the scale height, amounts to a determination of temperature. In this way temperature profiles for the homosphere have been determined on numerous occasions from pressure measurements made during rocket flights.

Where diffusive separation occurs, as in the heterosphere, there is a separate eqn. 1.14 for each component, with, in the simplest situation, the scale height appropriate to the component's molecular weight. More complicated situations arise, for example, in the case of a minor constituent undergoing fast chemical reactions with major constituents having differing scale heights. Temperatures in the thermosphere cannot be determined from pressure or density measurements without making assumptions about composition and mean molecular weight. Useful independent temperature estimates can be obtained spectroscopically for this region, the main difficulty being the determination of the precise altitude at which a particular spectroscopic emission feature originates.

In the next chapter we consider the properties of the earth's atmosphere as a photochemical system, and of the sun as a light source for photolysis. Experimental methods for studying the chemistry of the atmosphere are described in Chapter 3, and the results of studies relating to the chemosphere, the airglow and the ionosphere, are discussed in Chapters 4, 5 and 6, respectively. In Chapter 7 we consider the chemical processes which can occur in the troposphere and stratosphere as a consequence of air pollution. In Chapter 8 we discuss the information which is currently available about the atmospheres of other planets, especially Venus, Mars and Jupiter.

---

* Typically $g = 980$ cm s^{-2} at sea level, 950 at 100 km, 922 at 200 km, 895 at 300 km, 869 at 400 km, and 844 at 500 km.

General references

1 Kuiper, G. P., Ed., 'The Earth as a Planet', University of Chicago Press, 1954.
2 Ratcliffe, J. A., Ed., 'Physics of the Upper Atmosphere', Academic Press, New York, 1960.
3 Whitten, R. C. and Poppoff, I. G. 'Fundamentals of Aeronomy', John Wiley and Sons, Inc., New York, 1971.
4 Chamberlain, J. W. 'Physics of the Aurora and Airglow', Academic Press, New York, 1961.
5 Ratcliffe, J. A. 'Introduction to the Ionosphere and Magnetosphere', Cambridge University Press, 1972.
6 Landel, R. F. and Rembaum, A., Eds., 'Chemistry in Space Research', American Elsevier, New York, 1972.

2
The Atmosphere as a Photochemical System

2.1 Introduction: basic photochemical principles

Our aim in this chapter is to set the stage for the various atmospheric phenomena which will be the focus of interest in the remainder of the book. We begin by discussing the elementary principles of photochemistry[1] and the laws of light absorption, with particular reference to factors which cause the formation of layers of photolysis products in the upper atmosphere. Next we consider the characteristics of the sun as a photochemical light source, and the distribution of intensity as a function of wavelength in the sunlight which strikes the outer layers of the earth's atmosphere. In the final section of the chapter we discuss experimental data concerning the ability of atmospheric gases to absorb the sun's ultraviolet radiation.

2.1.1 THE LAWS OF PHOTOCHEMISTRY

There are two fundamental laws of photochemistry, the first being that a molecule must absorb the incident light before it can become activated, and the second that the absorption of one photon results in the activation of one molecule.* The absorption and activation step is termed the *primary process*. Secondary chemical reactions occur as a result of the activated species produced in the primary process interacting with other molecules. The final outcome of the photochemical experiment depends on both the nature of the primary process, and the availability of other molecules to take part in secondary reactions. The *quantum yield* of a photochemical reaction is the ratio of the number of product molecules generated by the overall reaction, to the number of photons absorbed in the primary process. If secondary reactions leading to product formation are inefficient the quantum yield will be low, as in the photochemical production of HBr from H_2 and Br_2; if the secondary reactions constitute a chain with relatively slow termination steps the quantum yield may be very large, as in the photochemical production of HCl from H_2 and Cl_2. Many reactions are known which have quantum yields close to unity; an example is the production

---

* Under conditions of extremely intense illumination, as in a focused laser beam, processes can occur which are initiated by the simultaneous absorption of two or more photons. Such non-linear processes (so called because the probability of the process is proportional to the square, or higher power, of the light intensity) are not significant in the atmosphere.

of CO during the photolysis of CO_2 with 147 nm xenon resonance radiation.

It is possible also to define a quantum yield for the reverse process of light emission, as in the emission of fluorescence from an optically excited species, or the emission of chemiluminescence by an excited molecule formed in an atom recombination process. In such a case the quantum yield, or quantum efficiency, is equal to the probability that the excited molecule will emit its energy as a quantum of electromagnetic radiation, rather than lose it by some non-radiative mechanism.

If a species A becomes electronically excited as a result of light absorption (eqn. 2.1)

$$A + h\nu \rightarrow A^* \tag{2.1}$$

the excitation energy may subsequently be lost by fluorescence,

$$A^* \rightarrow A + h\nu' \tag{2.2}$$

where the emission frequency ν' is not necessarily identical with the excitation frequency ν, or by reaction with a 'quencher' Q:

$$A^* + Q \rightarrow products \tag{2.3}$$

Other modes of deactivation of an excited molecular species include internal conversion to high vibrational levels of lower electronic states,

$$A^* \rightarrow A^\dagger \tag{2.4}$$

intersystem crossing to an excited state of different multiplicity,

$$A^* \rightarrow A^{**} \tag{2.5}$$

and dissociation:

$$A^* \rightarrow B + C \tag{2.6}$$

Included under quenching are such processes as collision induced intersystem crossing, and excimer formation (eqn. 2.3a)

$$A^* + Q \rightarrow [AQ^*] \rightarrow A + Q + h\nu'' \tag{2.3a}$$

which prevent the species A^* from radiating light of frequency ν'. In the absence of the quencher Q, and at pressures low enough for deactivation of A^* by collision with the parent species A to be negligible the photodissociation of the parent species occurs with a *quantum yield* Φ_d which is given by

$$\Phi_d = k_{2.6}/(k_{2.2} + k_{2.4} + k_{2.5} + k_{2.6}) \tag{2.7}$$

Under these conditions the *fluorescent efficiency* Φ_f and the *mean lifetime* of the excited state, τ, are given by

$$\Phi_f = k_{2.2}/(k_{2.2} + k_{2.4} + k_{2.5} + k_{2.6}) \tag{2.8}$$

and

$$\tau = 1/(k_{2.2} + k_{2.4} + k_{2.5} + k_{2.6}) \tag{2.9}$$

respectively. In the presence of Q the fluorescent intensity I is less than the initial value I_0, i.e. is quenched. The intensity of fluorescence is proportional to the steady state concentration of A*, which is given by

$$[A^*] = I_{abs}/(k_{2.2} + k_{2.3}[Q] + k_{2.4} + k_{2.5} + k_{2.6})$$

where I_{abs} is the rate of absorption of light by the species A. (For simplicity the absorption of light by Q is assumed negligible.) Hence

$$I_0/I = 1 + k_{2.3}[Q]/(k_{2.2} + k_{2.4} + k_{2.5} + k_{2.6})$$

or $\qquad I_0/I = 1 + k_{2.3}\tau[Q]$ \hfill (2.10)

which is the well known *Stern-Volmer equation* governing the quenching of fluorescence. Deviations from the predicted linear dependence of I_0/I on [Q] are commonly observed during the quenching of fluorescence in solution, but are seldom encountered in gas phase studies. For a molecule undergoing an allowed transition the mean lifetime τ is typically in the range 10^{-7}–10^{-9} s. The quenching process 2.3 commonly occurs on almost every collision between A* and Q, so that with uncharged species the rate constant $k_{2.3}$ is in the range 10^{-10}–10^{-11} cm^3 molecule^{-1} s^{-1}. With $k = 5 \times 10^{-11}$ cm^3 molecule^{-1} s^{-1} and $\tau = 10^{-8}$ s the intensity of fluorescence is reduced by a factor of 2 when $[Q] = 2 \times 10^{18}$ molecule cm^{-3}, i.e. about 0.1 atmosphere at a temperature of 300 K.

The subject of photochemistry is concerned with the determination of quantum yields, with the identification of primary processes, and with the elucidation of the reactions which intervene between the primary activation step and the formation of final products.

2.1.2 LIGHT ABSORPTION: FORMS OF THE BEER–LAMBERT LAW

In the atmosphere we are concerned with processes which are initiated by light absorption; therefore, the laws governing the absorption of light are of fundamental importance. Chemists customarily consider light absorption in terms of the familiar Beer–Lambert law, written in the form

$$\text{Log}_{10}(I_0/I)_\lambda = \varepsilon_\lambda C l \hfill (2.11)$$

Here l is the length in centimetres of the light path through an absorbing substance whose concentration is C moles per litre, and whose molar extinction coefficient at wavelength λ is ε_λ. The ratio of incident to transmitted radiation at the wavelength λ is $(I_0/I)_\lambda$. Often the subscript λ is omitted and the formula 2.11 is assumed to apply in a sufficiently narrow band of wavelengths. If the natural logarithm of I_0/I is used in eqn. 2.11 instead of the logarithm to base ten, the extinction coefficient ε is replaced by an absorption coefficient α. Spectroscopists who study the absorption by gases of vacuum ultraviolet radiation commonly work in terms of an absorption coefficient

κ defined by

$$I = I_0 \exp(-\kappa d) \tag{2.12}$$

where d is the path length in centimetres reduced to NTP. The units of κ are thus cm^{-1}. If P is the pressure of the absorbing gas in atmospheres, and T is the absolute temperature, eqn. 2.12 can be written in terms of the actual path length l, as

$$I = I_0 \exp(-\kappa P l \times 273/T) \tag{2.12a}$$

Occasionally the base 10 is used in place of e in eqn. 2.12, with an absorption coefficient K cm^{-1}, i.e.

$$I = I_0 \, 10^{-Kd} \tag{2.12b}$$

To complete our inventory of different forms of the Beer–Lambert law we may note that physicists usually prefer to work in terms of absorption cross sections rather than absorption or extinction coefficients. The absorption cross section, which may be designated by either q or σ, is defined by the relation

$$I = I_0 \exp(-qnl) \tag{2.13}$$

where n is the number of absorbing molecules per cm^3 in the light path whose length is l cm. In aeronomy it is usual to quote concentrations in units of particles per cm^3, which makes eqn. 2.13 the most convenient form of Beer's law for numerical calculations. The different quantities governing the absorption process are related by

$$\kappa = 2.303K; \quad \alpha = 2.303\varepsilon; \quad \kappa = Lq; \quad \alpha = 273.2R\kappa \tag{2.14}$$

where Loschmidt's number L is 2.687×10^{19} cm^{-3} (the number of molecules per cm^3 at NTP), and the gas constant R is 0.0820 litre atm mol^{-1} deg^{-1}.

2.1.3   RATE COEFFICIENTS FOR PHOTON ABSORPTION PROCESSES

If a species A is photolysed to give primary products B and C according to the equation

$$A + h\nu \rightarrow B + C \tag{2.15}$$

the rate of the primary process is customarily written in terms of a rate coefficient J, where

$$d[B]/dt = d[C]/dt = J_{2.15}[A] \tag{2.16}$$

The quantity $J_{2.15}$ takes account of the wavelength variation of the absorption cross section q_λ of species A, of the wavelength distribution of intensity I_λ from the light source, and of the wavelength variation of the quantum efficiency Φ_λ of the primary process. In an optically thin absorbing layer the amount of light absorbed is given by

$$-\delta I_\lambda = I_\lambda q_\lambda [A] \, \delta l \tag{2.17}$$

where [A] is expressed in molecule cm^{-3} and the path length δl is in centimetres. In a volume of 1 cm^3 the rate of the primary process due to light of wavelength λ is, therefore,

$$(dB/dt)_\lambda = \Phi_\lambda I_\lambda q_\lambda [A] \tag{2.18}$$

Hence the rate coefficient is given by

$$J_{2.15} = \int_0^\infty \Phi_\lambda I_\lambda q_\lambda \, d\lambda \tag{2.19}$$

In practice this integration would be carried out numerically, and would be limited to the wavelength range of the absorption bands in which the primary process of interest was known to occur.

2.1.4 THE FORMATION OF LAYERS IN THE ATMOSPHERE

In Chapter 1 we noted that the pressure of gas in the atmosphere diminishes with height according to the formula

$$P = P_0 \exp(-h/H) \tag{2.20}$$

where h is the height above the point where the pressure is P_0, and the scale height H is given by

$$H = RT/Mg \tag{2.21}$$

The result 2.20 applies strictly only to a region in which the temperature T, the mean molecular weight M, and the gravitational acceleration g are constant. In such a region the combination of the exponential *decrease* of transmitted intensity with decreasing altitude due to the changing value of the product Pl in eqn. 2.12a, and the exponential *increase* of P with decreasing altitude given by eqn. 2.20, produces a well-defined maximum in the rate of loss of energy by a beam of sunlight being absorbed in the atmosphere, as shown qualitatively in Fig. 2.1. The maximum rate of loss of energy from the beam corresponds to a maximum rate of production of primary photolysis products—ions, neutral fragments or excited species. Thus on the basis of this picture we can conclude that photolysis products should be produced in layers, the height of the peak concentration in a particular layer being a function of the product of the partial pressure of the absorbing species and its absorption coefficient for the radiation of interest. The formation of layers in this way was first discussed by Chapman,[2] and the mathematical function which describes the shape of such a layer is called a Chapman function. The mathematical form of a Chapman layer can be derived as follows:

The *solar zenith angle* χ is defined in Fig. 2.2. From this diagram it can be seen that the increment in path length dl is related to the increment in

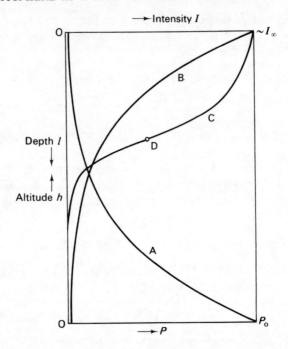

Fig. 2.1   Light absorption in the atmosphere. Curve A shows the variation of gas pressure with altitude; curve B shows the variation of light intensity with depth at constant gas pressure; curve C shows the variation of intensity with depth when pressure increases exponentially, as in curve A. The maximum slope of curve C, corresponding to maximum rate of absorption of light, occurs near the altitude of point D.

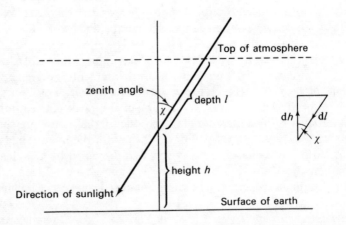

Fig. 2.2   Definition of the solar zenith angle χ. The inset shows that $dl = -dh\sec\chi$.

height dh by the expression

$$dl = -dh \sec \chi \tag{2.22}$$

The differential form of eqn. 2.12*a*, taking the standard pressure as one atmosphere and putting $T = 273$ K for simplicity, is

$$dI = -IP\kappa \, dl$$
$$= IP\kappa \sec \chi \, dh \tag{2.23}$$

Substituting for P from eqn. 2.20 gives

$$d(\mathrm{Ln}\, I) = P_0 \kappa \sec \chi \, e^{-h/H} \, dh \tag{2.24}$$

Integrating over light intensity from I_∞ to I, and over the light path from infinity to h, we obtain

$$\mathrm{Ln}(I/I_\infty) = -P_0 \kappa H \sec \chi \, e^{-h/H}$$

or

$$I = I_\infty \exp(-P_0 \kappa H \sec \chi \, e^{-h/H}) \tag{2.25}$$

The rate of deposition of energy from the beam is given by

$$r = -dI/dl$$
$$= \cos \chi \, (dI/dh)$$

so

$$r = I_\infty P_0 \kappa \cos \chi \exp(-h/H - P_0 \kappa H \sec \chi \, e^{-h/H}) \tag{2.26}$$

The condition for r to be a maximum is $dr/dh = 0$, which reduces to

$$P_0 \kappa H \sec \chi = \exp(h_\mathrm{m}/H) \tag{2.27}$$

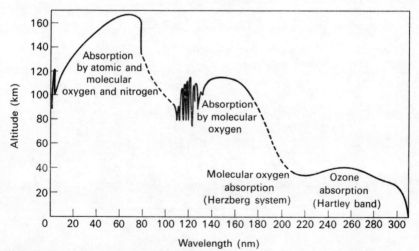

Fig. 2.3 Altitude of maximum light absorption with the sun overhead, as a function of wavelength. (After H. Friedman, *Physics of the Upper Atmosphere*, J. A. Ratcliffe Ed., Academic Press, New York, 1960.)

where h_m, the height of the maximum, is independent of I_0. Comparing eqns. 2.25 and 2.27, we find that h_m corresponds to the height at which the intensity is down by a factor of e from I_∞. The value of h_m in the earth's atmosphere at zero solar zenith angle ($\chi = 0$) is plotted as a function of wavelength in Fig. 2.3.

The value of r at the maximum is, from eqn. 2.26,

$$r_m = I_\infty P_0 \kappa \cos \chi \exp(-1 - h_m/H)$$

so that

$$r/r_m = \exp(1 - z - e^{-z}) \tag{2.28}$$

where the dimensionless quantity z is equal to $(h - h_m)/H$. The dependence of r/r_m on z is shown by the curve for $\chi = 0$ in Fig. 2.4. The other curves, which are for different zenith angles as indicated, are plotted against z as defined for the curve with $\chi = 0$, so that the shift of h_m to higher altitude with increasing χ can be seen.

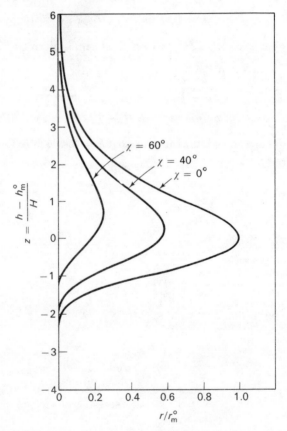

Fig. 2.4 Rate of light absorption as a function of reduced height $Z = (h - h_m)/H$ for different values of the zenith angle χ; $r_m{}^0$ and $h_m{}^0$ are the maximum rate and the altitude for maximum rate, respectively, for $\chi = 0$.

The simple theory given above has been modified to take account of the curvature of the earth, the difference being significant for zenith angles greater than about 70°. The theory can be further elaborated by allowing for the variation of composition and temperature with height, and for the variation of κ with wavelength across an absorption band. However, in the process much of the virtue of the simple theory is lost, and in any event there are other factors, such as the vertical diffusion of absorbing species and photolysis products, which are much more difficult to take into account.

2.2 The incoming radiation

The most important source of incoming radiation is, of course, the sun, but radiation from the rest of the galaxy may also be significant. The sun emits over the whole spectral range, from radio waves down to gamma rays. Of this radiation, that below 290 nm is most relevant to the photochemistry of the upper atmosphere; in the troposphere important photochemical processes may be initiated by light of wavelengths up to about 400 nm as a result of absorption by common pollutants such as NO_2 and SO_2.

2.2.1 PROPERTIES OF THE SUN AS A LIGHT SOURCE

For the purpose of considering its characteristics as a light source we can divide the sun into four regions, namely the interior, the photosphere, the chromosphere, and the corona.[3] The interior of the sun is the part which is at a sufficient depth (actually only a few hundred kilometres) for radiation to be unable to escape into space without undergoing repeated absorption and re-emission. This absorption and re-emission process causes the light to have a wavelength distribution which corresponds to the temperature of the region being traversed. Light from the interior ultimately diffuses into the thin photosphere, from which it escapes with an intensity distribution in the visible and infrared that corresponds to a black-body temperature of about 6000 K. In the outer photosphere the temperature is about 4200 K; light absorption in this region produces the many dark Fraunhofer lines which are a characteristic feature of the visible solar spectrum. In the ultraviolet the absorption by the outer photosphere causes the intensity distribution to fall off at a faster rate than would correspond to a black body at 6000 K, and in the vacuum ultraviolet at ~ 130 nm the emission temperature is close to 4200 K. The chromosphere is a rather inhomogeneous region, about 12 000 km in thickness, between the outer photosphere, at 4200 K, and the corona, where the temperature is of the order of 10^6 K. The chromosphere consists largely of neutral hydrogen atoms, and in consequence is optically dense to light of wavelengths below the Lyman limit (the onset of the ionization continuum of H) at 91 nm. The transition from the chromosphere to the corona is marked by the change from neutral to ionized hydrogen. The corona is visible during a total eclipse, or at other times with the aid of a coronagraph, as a white halo spreading many millions of miles from the sun's disc. Both the huge extent of the corona (described by the scale height RT/Mg, where g is the gravitational acceleration of the sun) and the observation of emission lines from very highly ionized atoms such as Ni(XVI) $(= Ni^{+15})$, Fe(XV) and Ar(XIV), require the coronal temperature to be around 10^6 K. The high temperature of the corona is attributed to heating by shock waves which arise from turbulence in the photosphere. The coronal emission is the major component of sunlight in the extreme ultraviolet and X-ray region, where it consists of a weak continuum, to which radiation from $H^+ + e^-$ recombination contributes intensity below 91 nm, and that from $He^+ + e^-$ contributes below 20.4 nm. Superimposed on this continuum are numerous X-ray lines resulting from electron impact excitation of heavy atoms.

At times the coronal X-ray emission undergoes a sudden, marked increase as a result of strong heating by solar flares. Solar flares, and related disturbances such as outbursts of radio noise, coronal hot spots, and certain types of prominences, are associated with relatively small active centres on the sun's surface, usually in the vicinity of sunspots. During the course of a solar flare, which may last minutes or hours, the X-ray emission from the flare area increases by a factor of the order of 10^8, and a variety of ionospheric effects become apparent as a result of increased ionization in the 60–100 km region of the earth's sunlit hemisphere. An active group of sunspots may give rise to as many as fifty flares during its approximately 13 day passage across the visible disc of the sun. The sunspot number,* and the frequency of solar flares, show an eleven year periodicity, with sunspot number values varying from an average of two or three during quiet years to one hundred or more in periods of intense activity. As an indicator of the amount of solar activity at any particular time it is customary to quote the average intensity of the flux received at a wavelength of 10.7 cm in the microwave region. This flux typically varies from 7–25×10^{-21} W m^{-2} Hz^{-1} between solar minimum and solar maximum.

2.2.2   THE SOLAR SPECTRUM

The total amount of light at all wavelengths received per unit area normal to the sun's direction, as measured outside the earth's atmosphere, is termed the solar constant, H_0. The value of H_0 is close to 1400 W m^{-2}, and on the basis of palaeontological and geological records would appear to have changed very little in the last billion years. The irradiance at a particular

---

* The sunspot number, obtained by averaging records from a number of observatories, is equal to ten times the number of sunspot groups plus the number of individual spots visible at a given time.

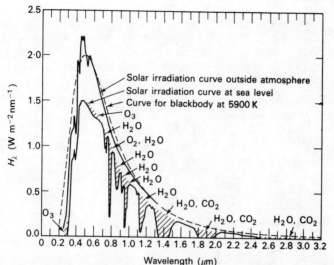

Fig. 2.5 Irradiance H_λ as a function of wavelength. The shaded areas show the light which is absent at sea level because of absorption by the species named. (From *Handbook of Geophysics and Space Environments*, McGraw-Hill, New York, 1965.)

wavelength, H, is shown as a function of wavelength in Fig. 2.5, both for the light incident on the upper atmosphere and for the light arriving at sea level. It is clear from Fig. 2.5 that light of wavelength less than 200 nm, which largely governs the structure and composition of the upper atmosphere, makes up an exceedingly small fraction of H_0.

Table 2.1   Intensity distribution in sunlight below 500 nm
($H_0 = 1390$ W m^{-2}, $F_{10.7 \text{cm}} \sim 13{-}17 \times 10^{-21}$ W m^{-2} Hz^{-1}.)

Wavelength Interval $\Delta\lambda$ (nm)	$H_{\Delta\lambda}$ (W m^{-2} $\Delta\lambda^{-1}$)	$H_{\Delta\lambda}/H_0$ (%)
0.1–0.8	$< 10^{-9}$	$< 10^{-11}$*
0.8–4.0	2×10^{-6}	1.4×10^{-7}†
4.0–10.0	0.16×10^{-3}	1.1×10^{-5}‡
10.0–30.3	1.6×10^{-3}	1.1×10^{-4}
30.38	0.35×10^{-3}	2.5×10^{-5}
30.4–46.0	0.16×10^{-3}	1.1×10^{-5}
46.0–121.5	0.91×10^{-3}	6.3×10^{-5}
121.6	5.0×10^{-3}	3.5×10^{-4}
121.6–180.0	3.30×10^{-2}	2.3×10^{-3}
180–225	0.90	6.5×10^{-2}
225–300	17	1.3
300–400	110	7.9
400–500	200	14.4

* Increased by factors of 10^3 and 10^5 for disturbed sun and class 3 flare respectively.
† Increased by at least a factor of 50 for disturbed sun.
‡ Increased by a factor of ~ 7 for disturbed sun.

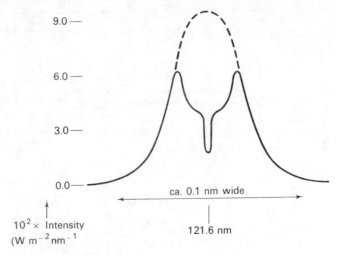

Fig. 2.6   Line profile of solar Lyman-α at 120 km.

The solar emission spectrum at wavelengths below 500 nm is summarized in Table 2.1. Below 190 nm the emission is almost entirely in the form of atomic lines; the black-body continuum is undetectable below 150 nm. The strongest individual feature is the Lyman–α line of atomic hydrogen at 121.6 nm in the vacuum ultraviolet, whose total intensity is typically 5×10^{-3} W m^{-2}. The profile of this line, as received at an altitude of 120 km,[4] is shown diagrammatically in Fig. 2.6. The line is very broad, with a Doppler width appropriate to the temperature of the photosphere, and has a wide reversal 'crater' due to absorption by hot hydrogen in the photosphere and chromosphere. At the centre of this broad crater is a much narrower well due to absorption by relatively cool hydrogen atoms in the earth's upper atmosphere and in the intervening space. The peculiar importance of Lyman–α is that it happens to coincide in wavelength with a deep window in the absorption spectrum of O_2, and so is able to penetrate to an altitude of about 75 km in the D region (cf. Fig. 2.3). The other line which is listed separately in Table 2.1 is the He(II) line corresponding to Lyman–α of atomic hydrogen, which is emitted with relatively high intensity at 30.38 nm in the extreme ultraviolet. Unfortunately, at the time of writing there is considerable uncertainty, perhaps amounting to a factor of two, in listed values of solar flux below 180 nm. For precise calculations in the future the reader is recommended to select values from the current literature.

2.3   Absorption coefficients of atmospheric gases

The photochemistry of the upper atmosphere is governed by the photolysis of the species O_2, N_2 and O_3. H_2O, H_2, NO and CO_2 are important as minor constituents of the earth's upper atmosphere, and CO_2 as a major component of the atmospheres of Venus and Mars. Absorption by SO_2 and NO_2 can also lead to significant effects in the earth's troposphere. We now consider these substances individually. We shall also give brief consideration to the absorption of light by methane, ammonia, nitrous oxide, hydrogen peroxide, and carbon monoxide.

2.3.1 MOLECULAR OXYGEN

Potential energy curves for the lower electronic states of O_2 are shown qualitatively in Fig. 2.7. In relation to the photochemistry of the atmosphere, the most important electronic transitions in absorption from the ground state are the forbidden Herzberg system ($A^3\Sigma_u^+ \leftarrow X^3\Sigma_g^-$), which has a weak dissociation continuum extending from 185 nm to at least 242 nm, and the fully-allowed Schumann-Runge system ($B^3\Sigma_u^- \leftarrow X^3\Sigma_g^-$), which has sharp band structure from 175 nm to 200 nm, with an intense dissociation continuum between about 137 and 175 nm. Between the photoionization limit of O_2 at 102.7 nm and the Schumann-Runge continuum are a number of overlapping diffuse bands and continua, with a series of deep windows between the bands below 130 nm. One of these windows coincides with Lyman-α at 121.6 nm. Below the limit at 102.7 nm is a complex region of

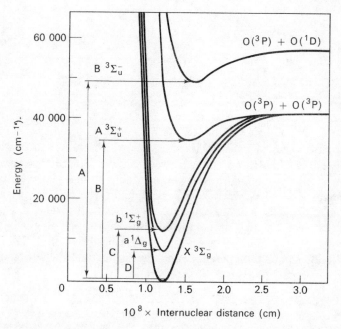

Fig. 2.7 Potential energy curves of O_2. The transitions indicated correspond to band systems as follows: A. Schumann-Runge, B. Herzberg, C. Atmospheric, D. Infrared Atmospheric.

bands and continua, which has been studied with a continuum light source down to 60 nm, and with line sources, which permit measurements to be made only at discrete wavelengths, below 60 nm. Below 60 nm the photo-ionization yield is unity; between 60 and 100 nm it varies between unity and about 0.2, passing through a shallow minimum in the vicinity of 80 nm. This behaviour is interpreted in terms of competition between predissociation and preionization.*

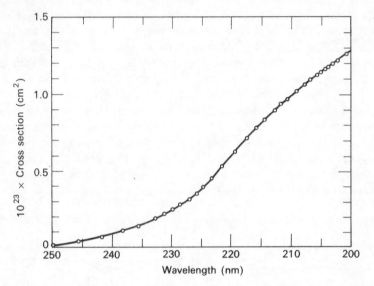

Fig. 2.8   Absorption by O_2 in the region of the Herzberg continuum.

Measured absorption cross sections of O_2, as obtained by Ditchburn and Young[5] in the region of the Herzberg continuum, are shown in Fig. 2.8. Hasson and Nicholls[6] have obtained results which are in good agreement with those of Ditchburn and Young between 200 and 230 nm. It was necessary to work at moderately high resolution in order to obtain cross sections at wavelengths located between known Schumann–Runge absorption lines. Nevertheless Hudson,[7] in an extremely useful critical review, has pointed out that between 200 and 210 nm the published cross sections can be almost entirely accounted for by absorption in pressure-broadened Schumann–Runge bands, so that both the shape and intensity of the Herzberg continuum must still be regarded as somewhat uncertain.

---

* In predissociation the potential energy curve of the upper electronic state crosses a repulsive curve derived from neutral fragments, and one can say that the molecule dissociates as a result of undergoing a non-radiative transition to the repulsive curve. In preionization the same kind of process occurs, but the crossing involves a repulsive curve which corresponds, at infinite separation, to ionized fragments which are usually just a positive ion and an electron.

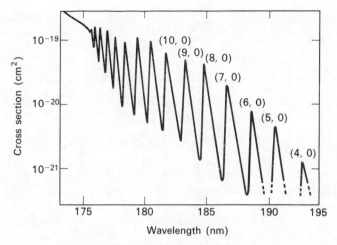

Fig. 2.9 The Schumann-Runge bands of O_2 below 195 nm.

Figs. 2.9, 2.10 and 2.11 show the absorption cross sections measured by Watanabe, Inn and Zelikoff[8] in the regions of the Schumann–Runge bands, the Schumann–Runge continuum, and between this continuum and the photoionization limit, respectively. In Hudson's review these results are compared with data obtained by other groups of workers. In general the agreement is within the combined limits of experimental error, but there are

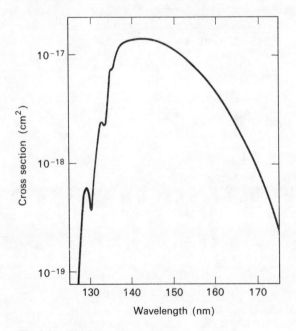

Fig. 2.10 The Schumann-Runge continuum of O_2.

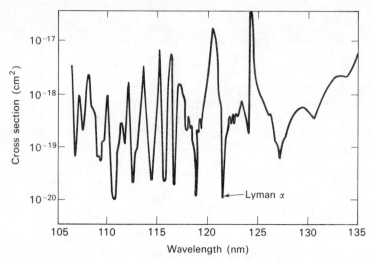

Fig. 2.11   Absorption cross sections of O_2 between 105 and 135 nm.

particular instances of disagreement that are hard to account for. In the important case of the Schumann–Runge continuum the cross section is known probably to within $\pm 5\%$ (excluding systematic errors) over the range 137–180 nm, but the actual shape of the continuum is not so well known.

For the important metastable species $O_2(^1\Delta_g)$ absorption cross sections do not appear to have been determined at the time of writing. Wayne[9] has measured the photoionization cross section of $O_2(^1\Delta_g)$ at the wavelength

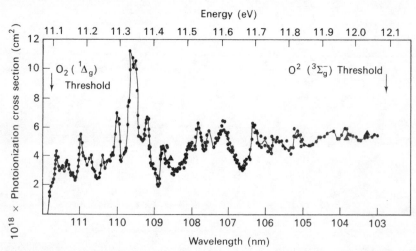

Fig. 2.12   Photoionization cross sections of $O_2(^1\Delta_g)$ between 111.8 nm and the photo-ionization threshold of $O_2(^3\Sigma_g^-)$. The triangles indicate points obtained at 103.2, 103.8 and 108.5 nm using a sharp line source.

of the 106.7 nm argon resonance line, obtaining a value of $3.2 \pm 0.8 \times 10^{-18}$ cm^2. On the basis of this value, and relative photoionization efficiency data of Cairns and Samson,[10] Wayne calculated photoionization cross sections for $O_2(^1\Delta_g)$ between 103.5 and 111.8 nm. Huffman and coworkers[11] have determined relative photoionization cross sections for $O_2(^1\Delta_g)$ with higher resolution than was used by Cairns and Samson, and have used Wayne's value at 106.7 nm to place them on an absolute basis. The results are shown in Fig. 2.12.

2.3.2 MOLECULAR NITROGEN

Molecular nitrogen absorbs only very weakly at wavelengths above 100 nm, the most important absorption feature being the $(v', 0)$ progression of the forbidden Lyman–Birge–Hopfield system $(a^1\Pi_g \leftarrow X^1\Sigma_g^+)$, which lies between 110 and 145 nm.[12] The $a^1\Pi_g$ state is weakly predissociated above $v = 6$, but the cross section for photodissociation is too small for this absorption to produce a significant percentage dissociation of N_2 in the atmosphere. Between 60 and 100 nm there are a number of strong bands with well resolved rotational structure. Bands at 97.2 and 82.8 nm show evidence of predissociation in the form of anomalously broad rotational lines, and the presence of an underlying dissociation continuum with a cross section of $2–4 \times 10^{-19}$ cm^2 between 85 and 100 nm has been suggested. Because of preionization the rotational structure normally appears diffuse below the ionization limit at 79.58 nm. Below 60 nm absorption by N_2 is almost entirely continuous. The ionization yield of N_2 is essentially constant at 1.0 at wavelengths below 70 nm, with a value that fluctuates between 0.4 and 1.0, especially in the vicinity of band maxima, above 70 nm. At wavelengths below 66 nm the photoionization process gives rise to fluorescence from electronically excited N_2^+.

2.3.3 OZONE

The most important feature of the absorption spectrum of O_3 consists of the Hartley band and continuum between 200 and 300 nm. This system, which is shown in Fig. 2.13, adjoins the Huggins bands, located between 300 and 350 nm, which are shown in Fig. 2.14. Ozone also absorbs in the visible region, in the weak Chappuis band which is illustrated in Fig. 2.15. These Figures are taken from a paper by Griggs,[13] in which comparison is made with earlier work of Inn and Tanaka.[14] Below 200 nm the absorption consists of broad bands, probably overlying a dissociation continuum. It is energetically possible to photodissociate ozone at wavelengths up to 1.14 μm in the near infrared, and at shorter wavelengths a considerable variety of dissociation products could in principle be obtained. Wavelength thresholds

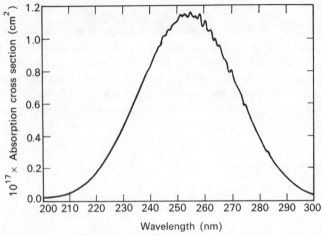

Fig. 2.13 Hartley band and continuum of ozone.

for the production of different states of O and O_2 are given in Table 2.2.[1] Dissociation processes which lead to O and O_2 in states of differing multiplicity correspond to spin-forbidden transitions and should be unimportant in comparison with allowed transitions which occur in the same wavelength region.

2.3.4 WATER VAPOUR

The ultraviolet absorption spectrum of H_2O consists of a region of continuous absorption between 145 and 186 nm, diffuse bands between 69

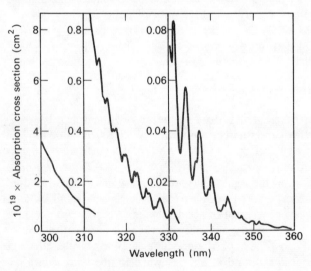

Fig. 2.14 Huggins bands of ozone.

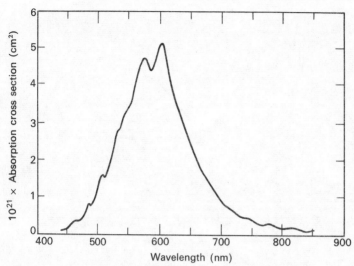

Fig. 2.15 Chappuis band of ozone.

and 145 nm, and continuous absorption again below 69 nm. The first ionization limit occurs at 98.4 nm, so that bands below this wavelength are superimposed upon an ionization continuum. Fig. 2.16 shows the data of Watanabe and Zelikoff[15] for the region between 120 and 180 nm. Other data are discussed by Hudson.[7]

Absorption in the long wavelength continuum is expected to result in the dissociation of water molecules which diffuse upwards into the mesosphere. Nicolet[16] has calculated the rate of photodissociation of water vapour in the mesosphere and lower thermosphere, at various solar zenith angles. The results, shown in Fig. 2.17, indicate that Lyman–α is responsible for most of the photodissociation occurring between 70 and 100 km.

2.3.5 MOLECULAR HYDROGEN

The absorption spectrum of H_2 consists of sharp absorption bands between 84 nm and about 115 nm, bands superimposed on a dissociation continuum between 80 and 84 nm, bands superimposed on an ionization

Table 2.2 Wavelength thresholds (nm) in the photodissociation of ozone

	$O_2(^3\Sigma_g^-)$	$O_2(^1\Delta_g)$	$O_2(^1\Sigma_g^+)$	$O_2(^3\Sigma_u^+)$	$O_2(^3\Sigma_u^-)$
$O(^3P)$	1140	590	460	230	170
$O(^1D)$	410	310	260	167	150
$O(^1S)$	234	196	179	129	108

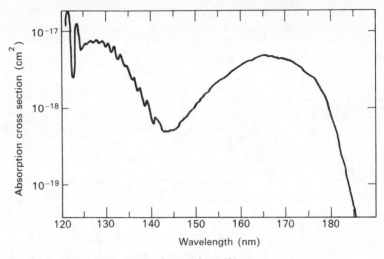

Fig. 2.16 Absorption spectrum of water above 120 nm.

continuum between 74 and 80 nm, and continuous absorption below 74 nm. Fig. 2.18, derived from Hudson's review, shows the wavelength dependence of the absorption cross section below 70 nm. Below 70 nm the ionization yield is 1.0; above this wavelength it fluctuates, being different even for different rotational lines within the individual bands.

2.3.6 NITRIC OXIDE

Nitric oxide is of particular interest because its first ionization limit occurs at the relatively long wavelength of 135 nm. This makes it a prime candidate for ionization by Lyman–α radiation in the D region. Above the ionization limit the absorption consists of the β, δ, ε and γ band systems, possibly with overlapping dissociation continua in limited regions, the banded absorption extending up to about 235 nm (see Fig. 5.3). Between 58.4 nm (the wavelength of the He(I) resonance line) and the ionization limit the absorption spectrum consists of many more or less sharp bands with overlapping ionization and dissociation continua. The most reliable data for absorption coefficients and ionization yields in this region are those of Watanabe, Matsunaga and Sakai,[17] who present their results in the form of an extensive tabulation. At the important Lyman–α wavelength (121.6 nm) the absorption coefficient is 62.5 cm^{-1} (corresponding to a cross section of 2.3×10^{-18} cm^2) and the measured ionization yield is 0.81.

2.3.7 CARBON DIOXIDE

The absorption spectrum of CO_2 between 90 and 180 nm has been described in terms of four separate dissociation continua, with superimposed

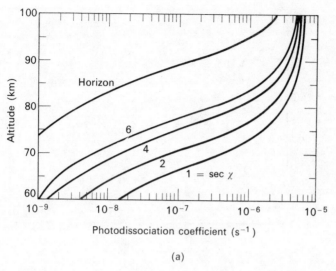

(a)

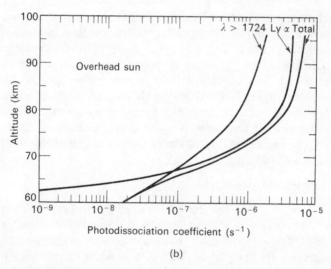

(b)

Fig. 2.17 Rate of photodissociation of water vapour by Lyman-α and by radiation of wavelength greater than 172.4 nm, calculated by Nicolet[16] as a function of altitude: (a) for total radiation, as a function of zenith angle χ; (b) showing importance of Lyman-α, with overhead sun.

bands, some of which appear sharp and others diffuse. Hudson[7] points out that the evidence for the presence of the four underlying continua is not conclusive. The data of Inn, Watanabe and Zelikoff[18] are given in Fig. 2.19. The cross sections obtained by Nakata, Watanabe and Matsunaga[19] are in good agreement with the earlier data of Inn *et al.*, except in the vicinity of sharp bands where the results of Nakata *et al.* are to be preferred because

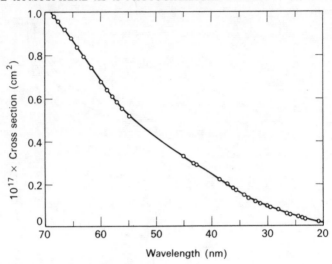

Fig. 2.18 Absorption cross sections for H_2 below 70 nm.

they used a narrower instrumental bandwidth. The first ionization limit of CO_2 is at 98.6 nm, and the associated ionization continuum appears smooth from this limit down to 84 nm, after which there is a banded region extending to 68 nm, followed by a further smooth continuum which has been observed as far as 10 nm. The ionization yield is essentially unity below 70 nm; between 70 and 90 nm it fluctuates slowly between limits of about 50 and 90%. Photoionization is accompanied at short wavelengths by dissociation to ionic and neutral fragments, and by the production of electronically excited CO_2^+, which gives rise to strong fluorescence in the near ultraviolet between 280 and 400 nm.

2.3.8 SULPHUR DIOXIDE

For SO_2 and NO_2 in the troposphere the spectral region of interest is the near ultraviolet. Fig. 2.20 shows absorption cross sections for SO_2 in the region 220–340 nm. Hudson[7] has emphasized that the measured values of cross sections are likely to be strongly dependent on instrumental bandwidth. The band at 290 nm is assigned[20] to the transition $\tilde{A}^1B_1 - \tilde{X}^1A_1$; the corresponding $\tilde{a}^3B_1 - \tilde{X}^1A_1$ transition occurs in absorption between 340 and 390 nm, with a total integrated intensity about 10^{-4} times that of the 290 nm band. The long wavelength limit for dissociation of SO_2 to SO and O is about 218 nm. The first ionization limit of SO_2 is just below 100 nm; measured cross sections are available down to 105 nm.[7]

2.3.9 NITROGEN DIOXIDE

The very complex absorption spectrum of NO_2 in the visible and near ultraviolet, which appears to be due to several overlapping transitions,

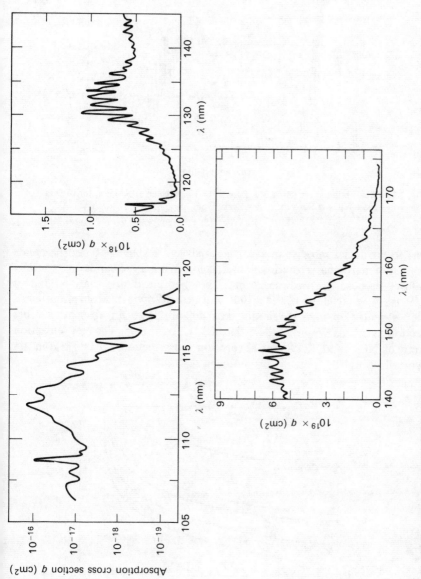

Fig. 2.19 Absorption cross section of CO_2.

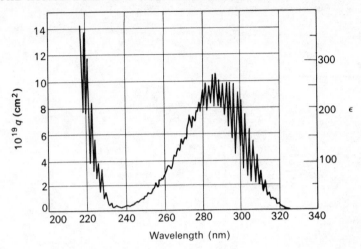

Fig. 2.20　Absorption spectrum of SO_2 (after J. G. Calvert and J. N. Pitts, Jnr., *Photochemistry*, John Wiley & Sons Inc., New York, 1966).

has so far defied more than a partial analysis.[20] Above 400 nm the bands consist of many sharp rotational lines; below 370 nm the bands are generally diffuse because of predissociation. The theoretical dissociation limit is 399 nm. The envelope of the visible and near ultraviolet absorption bands slopes gently up to a maximum near 400 nm, with an absorption cross section at the maximum of the order of 3×10^{-19} cm^2. The first ionization limit of NO_2 is at 128.8 nm; absorption cross sections in this region are typically 2×10^{-17} cm^2.

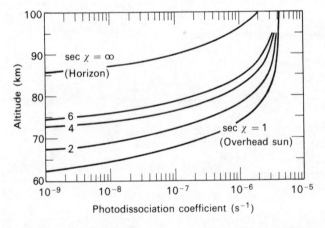

Fig. 2.21　Rate of photodissociation of methane by Lyman-α, calculated by Nicolet,[16] as a function of altitude and solar zenith angle χ.

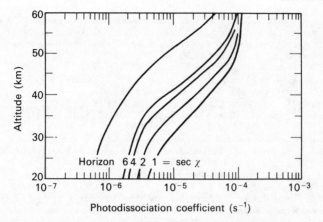

Fig. 2.22 Rate of photodissociation of H_2O_2, calculated by Nicolet,[16] as a function of altitude and solar zenith angle.

2.3.10 OTHER GASES

Other gases whose absorption cross sections may be significant in relation to the photochemistry of planetary atmospheres include methane, ammonia, nitrous oxide, hydrogen peroxide and carbon monoxide. The first three of these substances absorb strongly in the vacuum ultraviolet, with absorption cross sections of the order of 10^{-17} cm² below 150 nm. Fig. 2.21 shows the rate of photodissociation of methane by Lyman–α, calculated by Nicolet[16] as a function of altitude and solar zenith angle. The role of H_2O_2 in the upper atmosphere is uncertain at this time; the possibility exists that it may be an important constituent of the stratosphere. The absorption spectrum of H_2O_2 is continuous in the near ultraviolet (the long wavelength limit for photodissociation to OH + OH is 562 nm) with an absorption cross section of the order of 10^{-20} cm². Below about 185 nm the cross section increases sharply to more than 10^{-18} cm².[21] Nicolet's[16] estimate of the rate of photodissociation of H_2O_2 in the stratosphere and mesosphere is shown in Fig. 2.22. Carbon monoxide resembles N_2, with which it is isoelectronic, in absorbing relatively weakly above 100 nm. The first ionization limits of these five gases (CH_4, NH_3, N_2O, H_2O_2, CO) are at 98, 122, 96, 110 and 88.5 nm, respectively. For further information and references to original papers the reader is advised to consult Hudson's review article[7] and Herzberg's book.[20]

References

1 For a full treatment of this subject see, for example, Calvert, J. G. and Pitts, J. N., Jnr., '*Photochemistry*', John Wiley and Sons, New York, 1966.
2 Chapman, S. *Proc. Phys. Soc.*, **43**, 26, 484 (1931).

3 Evans, J. W., Gast, P. R. and Jursa, A. S., in '*Handbook of Geophysics and Space Environments*', Valley, S. L., Ed., McGraw-Hill, New York, 1965; Friedman, H., in '*Physics of the Upper Atmosphere*', Ratcliffe, J. A., Ed., Academic Press, New York, 1960; Hinteregger, H. E., *Annls Géophys.*, **26,** 547 (1970).

4 Purcell, J. D. and Tousey, R., *J. geophys. Res.*, **65,** 370 (1960).

5 Ditchburn, R. W. and Young, P. A., *J. atmos. terr. Phys.* **24,** 127 (1962).

6 Hasson, V. and Nicholls, R. W., *J. Phys.*, B, **4,** 1789 (1971).

7 Hudson, R. D., *Rev. Geophys. and Space Phys.*, **9,** 305 (1971).

8 Watanabe, K., Inn, E. C. Y. and Zelikoff, M., *J. chem. Phys.*, **21,** 1026 (1953).

9 Wayne, R. P., *Ann. N.Y. Acad. Sci.*, **171,** 199 (1970).

10 Cairns, R. B. and Samson, J. A. R., *Phys. Rev.*, **139,** A1403 (1965).

11 Huffman, R. E., Paulsen, D. E., Larrabee, J. C. and Cairns, R. B., *J. geophys. Res.*, **76,** 1028 (1971).

12 Lofthus, A., '*The Molecular Spectrum of Nitrogen*', University of Oslo, 1960.

13 Griggs, M., *J. chem. Phys.*, **49,** 857 (1968).

14 Inn, E. C. Y. and Tanaka, Y., *J. opt. Soc. Am.*, **43,** 870 (1953).

15 Watanabe, K. and Zelikoff, M., *J. opt. Soc. Am.*, **43,** 753 (1953).

16 Nicolet, M., *Annls Géophys.*, **26,** 531 (1970).

17 Watanabe, K., Matsunaga, F. M. and Sakai, H., *Appl. Opt.*, **6,** 391 (1967).

18 Inn, E. C. Y., Watanabe, K. and Zelikoff, M., *J. chem. Phys.*, **21,** 1648 (1953).

19 Nakata, R. S., Watanabe, K. and Matsunaga, F. M., *Sci. Lt.*, Tokyo, **14,** 54 (1965).

20 Herzberg, G., '*Molecular Spectra and Molecular Structure, III. Electronic Spectra of Polyatomic Molecules*'. D. Van Nostrand Company, Princeton, N.J., 1966.

21 Holt, R. B., McLane, C. K. and Oldenberg, O., *J. chem. Phys.*, **16,** 225, 638 (1948).

3
Experimental Methods

3.1 Introduction

In this chapter we consider the various techniques that can be used to obtain experimental information about the chemistry of the atmosphere. Investigation of the troposphere presents relatively few problems because the whole region is accessible to the experimenter, and the photochemical processes that occur are fairly easy to simulate in the laboratory. Thus our main concern will be the experimental procedures used to obtain stratospheric, mesospheric and ionospheric data.

Many different approaches have been used in investigations of chemical processes in the upper atmosphere. Among the more important are:

(1) ground based observations of airglow emission,
(2) the use of rocket-borne photometers and mass spectrometers,
(3) ground observations of seeding experiments, in which chemicals are released from rockets at high altitudes, and
(4) laboratory studies of the atomic and molecular species which have been shown to be present by field experiments.

Detailed descriptions of the types of rockets, photometers and mass spectrometers used in aeronomic studies are considered to be outside the scope of this book. Our viewpoint is that of the laboratory chemist, consequently our discussion will emphasize laboratory techniques. Nevertheless, since the current understanding of chemical processes occurring at high altitudes results from pooling information gained from both field and laboratory observations, we shall describe the salient features of both kinds of experiment.

3.2 Ground-based observations of airglow emission

The term airglow was introduced by Elvey in 1950 to describe non-auroral light emission from the upper atmosphere. The term nightglow was subsequently used to describe night-time airglow, and twilight glow to describe the emission observed when the sun appears to be below the horizon at ground level but is still in view from an altitude of 50–150 km. Most airglow features are extremely weak; consequently common laboratory prism and grating spectrographs and spectrometers are not sensitive enough to be used in airglow studies. It is necessary to work with instruments having a large aperture (typically f/1 or better) combined with reasonably high dispersion.

High speed spectrographs commonly use a form of flat-field Schmidt camera based on a design by Meinel.[1] The disadvantage of a spectrograph is that an exposure of two minutes or longer may be required to record a weak nightglow feature for which a photoelectric spectrometer could give an intensity reading in about ten seconds. Spectrographs are also limited by the lack of photographic emulsions sensitive in the infrared above about 1 μm. Spectrographs in which the output of an image orthicon detector is photographed are significantly faster than those with direct photographic detection; with such an instrument measurements of the red atomic oxygen line in the nightglow have been made with an exposure time of only ten seconds.

A spectrograph covers a wide spectral range in one exposure, in contrast to a scanning monochromator which detects light in a narrow bandwidth, and is particularly well adapted to quantitative intensity measurements. Probably the best type of scanning monochromator for airglow studies, as for many other purposes, is the Fastie–Ebert spectrometer whose main features are shown diagrammatically in Fig. 3.1. The resolution of such an instrument in first order is proportional to the total number of lines ruled on the grating; the throughput of light at a fixed resolution (controlled by the slit width) is proportional to the sine of the grating angle. Thus it is preferable to use a grating with a large blaze angle. For optimum sensitivity the photodetector is normally cooled, and, in the case of a photomultiplier detector, pulses due to individual photons may be counted. A variety of electronic techniques, including on-line time-averaging computers, have been used to improve the signal-to-noise ratio in recorded spectra. For the study of individual spectral features at high resolution in a narrow wavelength region the Fabry–Perot spectrometer coupled with a photoelectric detector can be used, with light outside the band-pass of interest excluded by an interference filter.

There are considerable experimental difficulties associated with the

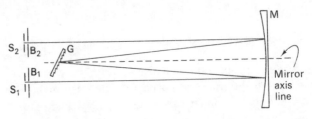

Fig. 3.1 Arrangement of optical components in a Fastie-Ebert grating monochromator. M is a spherical concave mirror, G is a plane grating, B_1 and B_2 are light baffles and S_1 and S_2 are entrance and exit slits in the focal plane of the concave mirror. For optimum resolution slits S_1 and S_2 must be curved.

ground level detection of emission lines originating at high altitudes, even at night when there is only starlight, moonlight, and perhaps scattered radiation from town lighting to contend with. The major problem during observation of twilight and dayglow emissions is the need to isolate the emission feature of interest from the strong background of sunlight scattered by air molecules and by aerosol and dust particles; the recent growth of knowledge of the dayglow is largely a consequence of technical advances made in overcoming this particular problem. The stringency of the require-ments for an instrument which is to be capable of detecting a dayglow line from the ground can be seen if we compare the intensity of a representative dayglow feature with that of the superimposed background. Thus, for example, the strong atomic oxygen line in the green at 557.7 nm may, in a bright aurora, have an intensity of 50 kilorayleighs* (i.e. $4\pi I = 50$ kR), and must be seen above a daytime sky background that may amount to 10 000 kR in a 0.1 nm bandwidth. If the sky spectrum is recorded at 0.01 nm resolution the background is still 20 times the line signal. Furthermore the sky background is never a flat continuum but invariably possesses structure due to terrestrial and solar (Fraunhofer) absorption lines. A direct solar spectrum which does not contain the dayglow emission is therefore required for comparison.

The most successful device for subtracting the solar background emission from the total light intensity at ground level appears to be the scanning

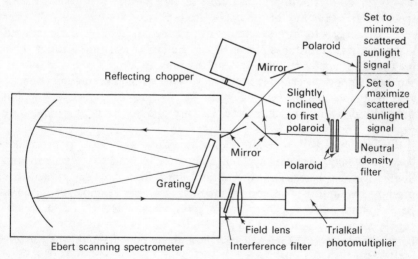

Fig. 3.2   Scanning polarimeter used by Noxon and co-workers to observe dayglow emission from the ground.

* The brightness of an airglow emission feature is usually measured in *rayleigh* units (R), where 1 rayleigh is the brightness of an isotropic source emitting 10^6 photons per cm^2 column per second in all directions. The intensity I of this source is therefore $10^6/4\pi$ photon cm^{-2} s^{-1} steradian^{-1}.

polarimeter used by Noxon and co-workers[2] in the environs of Boston. This instrument capitalizes on the fact that sunlight which has been scattered through a right-angle is strongly polarized, whereas dayglow emissions due to resonance scattering and chemiluminescent reactions possess very small and zero polarization respectively. If the two 90° polarized components of scattered sunlight are presented alternately to a photoelectric detector it is possible to reduce the intensity of the stronger component with a shutter or diaphragm so that the detector gives no a.c. output signal. Then if the instrument is made to scan over an airglow feature the polarization ratio changes during the scan and an a.c. output signal is produced. A diagram of an early version of the instrument is given in Fig. 3.2. The reader is referred to articles by Noxon[2] and Hunten[3] for further discussion of methods of detecting weak airglow emissions from the ground during both day and night.

3.3 Experiments with rocket-borne photometers and mass spectrometers

Rocket-borne instruments can provide 'in situ' measurements of the airglow and of molecular or ionic species present in the atmosphere, giving information on both the species present and their height distribution. The ceiling for most jet aircraft is below 14 km, and that of balloons around 40–50 km, so that both are unsuitable as high altitude platforms for mass spectrometric sampling of the upper atmosphere. However, they do offer some advantages for airglow studies during the day because of the reduced sky brightness at altitudes above 10 km. For some emission features there may also be greatly reduced atmospheric absorption above about 14 km; the most important example is the (0, 0) band of the infrared atmospheric system of molecular oxygen, which can be observed from an aeroplane, but only with difficulty at ground level because of absorption by ground state O_2.[4]

Rocket flights with scanning monochromators and mass spectrometers have been made from many launching sites, in both the Northern and Southern hemispheres. The information gained from these experiments has prompted a number of laboratory investigations. For example, Narcisi and Bailey[5] monitored the positive ion composition of the D region with a quadrupole mass spectrometer during a Nike Cajun rocket flight at altitudes between 50 km and 112 km. Because the ambient pressure of the mesosphere is relatively high, (varying from 0.6 to 10^{-3} torr* between 50 and 90 km) and the operating pressure of the mass spectrometer was below 5×10^{-4} torr it was necessary to include in the payload a liquid N_2 cooled zeolite sorption pump, as shown in Fig. 3.3. The mass spectrometer is shown included in the rocket payload in Fig. 3.4. Among the positive ions detected were some

---

* 1 torr = $\frac{1}{760}$ atmosphere = 133.32 N m^{-2}.

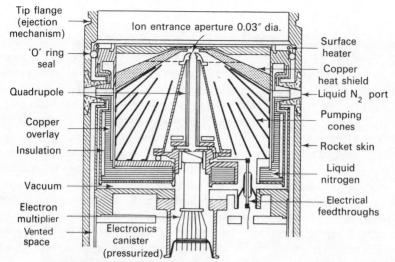

Fig. 3.3 Cross section of the quadrupole mass spectrometer and liquid nitrogen cooled sorption pump used in the rocket flights of Narcisi.[5]

Fig. 3.4 Photograph of the rocket nose cone of Narcisi[5] showing the mass spectrometer and pump assembly incorporated in the rocket payload. (Kindly provided by Dr R. S. Narcisi, Air Force Cambridge Research Laboratories, Bedford, Mass., U.S.A.)

unexpected peaks corresponding to mass numbers 19 and 37. The assignment of these peaks to H_3O^+ (19), and $H_3O^+.H_2O$ (37) led to laboratory studies of the rates and equilibria of clustering reactions for water molecules and positive ions. A theory to account for the presence of these ion clusters was later proposed by Fehsenfeld and Ferguson[6], as discussed in Chapter 6.

As another example, Evans *et al.*[7] included a photometer in the 166 kg payload of an Aerobee 150 rocket fired from the White Sands missile range, and determined an altitude profile for emission of the infrared atmospheric system $(a^1\Delta_g—X^3\Sigma_g^-)$ of O_2 at 1.27 μm in the dayglow. From the intensity of this emission an estimate of the concentration of $O_2(^1\Delta_g)$ was made, and the result turned out to be unexpectedly large. Subsequently the high concentration of $O_2(^1\Delta_g)$ was used to account for the observed O_2^+ concentration in the D region[8], by attributing formation of O_2^+ to photo-ionization of $O_2(^1\Delta_g)$. Interest was thereby attracted to other reactions of $O_2(^1\Delta_g)$, which have since been the subject of laboratory studies.

3.4 Ground observations of seeding experiments

The majority of the chemical releases from rockets have been intended to gather information about physical rather than chemical processes occurring at high altitudes. The procedure is to release the chemical above a predetermined altitude, either continuously or in bursts, by escape of a vapour or by explosion of a grenade filled with a suitable charge. Among the chemicals which have been released are alkali and alkaline earth metals, and compounds such as barium oxide, aluminium oxide, nitric oxide, nitrogen dioxide and trimethylaluminium. The latter undergoes flash vaporization and eventually produces AlO. The chemical cloud is observed photometrically, and from its displacement with time, wind velocities and diffusion coefficients can be determined. A measure of temperature can be obtained from the band profile of the AlO emission spectrum.[9] Useful information about winds and temperature gradients can also be obtained by measuring the rate of propagation through the atmosphere of the sound of a grenade explosion.

A series of rocket flights, in which a laboratory method for monitoring oxygen atom concentrations was used in the E region, was reported by Spindler.[10] Nitric oxide was released at night over a 75–125 km altitude range. The reaction between the nitric oxide seeding gas and atomic oxygen in the upper atmosphere:

$$NO + O \rightarrow [NO_2^*] \rightarrow NO_2 + h\nu \tag{3.1}$$

followed by
$$O + NO_2 \rightarrow O_2 + NO \tag{3.2}$$

is chemiluminescent, the intensity I of the resulting NO_2 emission at low pressures being related to the atomic oxygen concentration by the formula

$$I = k_{3.1}[NO][O] \tag{3.3}$$

Reaction 3.2 is fast enough to maintain the concentration of NO at an essentially constant value. Thus it was possible to determine a relative oxygen atom concentration profile between 78 and 112 km and also to observe some differences in the spectrum of the NO_2 continuum between upper atmosphere conditions and the conditions of previous laboratory studies.

3.5 Laboratory studies

Many different kinds of spectroscopic, photolytic and kinetic experiments have been performed with constituents of the atmosphere, including both normal constituents and pollutants such as hydrocarbons, SO_2, and NO_2. Certain of the experimental techniques can yield information of more than one kind. For example, flash photolysis can be used to study both the spectroscopy and kinetics of transient species, as well as the photochemistry of the parent molecule, and mass spectrometry can yield both ionization potentials and kinetic data. The possibility of studying processes which might occur in non-terrestrial atmospheres makes this a very large field indeed. Therefore in this section we shall be considering the general subject of current experimental methods for determining rates and mechanisms of reactions of gaseous radicals, small molecules, and ions in their ground and excited states.

3.5.1 FLOW SYSTEM STUDIES OF NEUTRAL SPECIES

Many reactions of interest to the aeronomist involve highly reactive neutral species, such as atoms and free radicals, which have short kinetic lifetimes. The experimental techniques used to measure reaction rates therefore require fast time resolution, operation at low pressures to reduce the collision frequency, or both. One of the most successful methods of studying the reactions of these short-lived species is to observe their behaviour in a low pressure *discharge flow system*. A flow tube, typically of 2.5 cm internal diameter, is operated at pressures in the range 0.1–10 torr, with a flow speed of the order of 10 metres per second. Atoms (hydrogen, oxygen, nitrogen, chlorine and bromine have all been used) together with both vibrationally and electronically excited molecules of the parent gas, are generated by passing the gas stream through an electric discharge before it enters the reaction tube. A few centimetres downstream from the discharge the more reactive ionized species and short-lived excited states have disappeared, leaving a steady flow of ground state atoms and/or metastable excited species in the carrier gas. By poisoning the walls, i.e. applying a coating which inhibits heterogeneous recombination processes,[11] atom recombination can be made slow at the low pressure within the tube. The reaction of the atoms or excited metastable species with a reactant gas added downstream from the discharge region can then be followed with the aid of a suitably chosen analytical technique. An approximately linear time scale is provided by the distance between the mixing region and the region where the analysis is performed.

An example of the use of a discharge flow system with mass spectrometric analysis is the study by Phillips and Schiff[12] of the reaction of hydrogen atoms with ozone (Fig. 3.5). They produced the H atoms by passing H_2 gas,

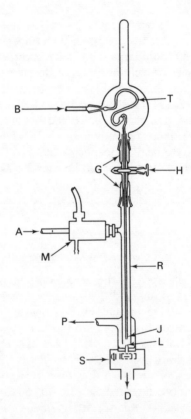

Fig. 3.5   Discharge flow system used by Phillips and Schiff for studies of atom reactions. Gas entering at A is partially dissociated by a discharge inside the microwave cavity M, before flowing down the reaction tube R and out to the pump at P. The stable reactant enters at B, flowing through the flexible tube T to enter the reaction tube through the jet J. The jet J can be moved along the axis of the reaction tube by turning the handle H. Guide tubes G keep it positioned in the centre of the reaction tube. The small sampling leak L leads, via the mass spectrometer ion source S, to a diffusion pump D. In later versions of the apparatus jet J is replaced by a small sphere with six 0.1 mm holes, and the jet is moved by an external magnet operating on an iron slug.

mixed with a large excess of argon or helium, through a microwave discharge, and then observed the ensuing reaction of the atoms with O_3. The progress of the reaction was followed by monitoring the decrease in the O_3 concentration, as indicated by the height of the peak at mass 48, as a function of reaction time. The reaction time was varied by altering the distance from the O_3 inlet to the sampling leak of the mass spectrometer. They found the rate constant of the reaction

$$H + O_3 \rightarrow OH + O_2 \tag{3.4}$$

to be 2.5×10^{-11} cm^3 molecule^{-1} s^{-1} at 300 K. Reaction 3.4 is important in the atmosphere in that it provides the vibrationally excited OH radicals whose emission spectrum in the near infrared is the strongest feature of the nightglow.

To illustrate the profusion of analytical techniques that have been successfully applied to reactions of ground state atoms and radicals in discharge flow systems we refer briefly to a number of other studies. Del Greco and Kaufman[13] have used the absorption of 306.4 nm resonance radiation by OH radicals to follow the consumption of ground state OH in a flow system; a similar method was used by Iwai, Pratt and Broida[14] to monitor the concentration of CN radicals. Clyne and co-workers[15] have used the absorption of atomic resonance radiation in the vacuum ultraviolet to measure halogen atom concentrations in flow systems; they have also used absorption of near ultraviolet radiation to monitor concentrations of ClO and BrO radicals. Numerous workers have used the emission intensity of the N_2 first positive bands to follow changes in N atom concentrations in flow systems.[16]

A number of gas phase 'titration' methods, for determining concentrations of atoms in flow systems, have been described.[12,15,16] The classic example of such a titration is the NO titration for nitrogen atoms, which depends on the fast reaction

$$N + NO \rightarrow N_2 + O \tag{3.5}$$

($k_{3.5} = 2 \times 10^{-11}$ cm^3 molecule^{-1} s^{-1}). If the flow of NO is less than the N atom flow, the O atoms produced from the NO by reaction 3.5 react with N according to

$$N + O + M \rightarrow NO^* + M \tag{3.6}$$

$$NO^* \rightarrow NO + h\nu \tag{3.7}$$

and the result is a blue emission characteristic of excited NO. If NO is in excess, however, all of the N atoms are used up by reaction 3.5 and the O atoms react with excess NO according to

$$NO + O + M \rightarrow NO_2^* + M \tag{3.8}$$

$$NO_2^* \rightarrow NO_2 + h\nu \tag{3.9}$$

and the result is the green-white 'air afterglow' emission characteristic of excited NO_2. At the 'end-point', when the flows of NO and N are equal, the flowing gas contains only O atoms and N_2 and the tube appears dark. Reaction 3.5 has also been employed as a means of generating O atoms in the absence of O_2, and as a source of vibrationally excited N_2.

Atomic resonance fluorescence has been used to measure atom concentrations in static systems[17]; the application to flow systems should be fairly straightforward.* The work of Westenberg and co-workers[18] has led to a number of studies in which quantitative electron spin resonance spectroscopy has been used to determine atom concentrations during measurements of reaction rates for atoms and radicals in the gas phase. Jonathan and co-workers[19] have recently begun to use photoelectron spectroscopy for the same purpose.

A similar diversity of methods have been used to study reactions of long-lived excited species in flow systems. Vibrationally and electronically excited molecules have been estimated by measuring the heat released when they are deactivated at the catalytic surface of a calorimeter probe[20]; this method was first used to estimate atom concentrations from the heat given up during surface recombination.[21] If a vibrationally excited molecule emits radiation in an allowed transition in the infrared this provides a means of monitoring both its concentration and its energy level distribution, as Polanyi and co-workers[22] have demonstrated. Vibrationally excited molecules often play a significant part in reaction kinetics; thus the vibrationally excited OH produced by reaction 3.4 is able to destroy a further O_3 molecule by transfer of vibrational energy in a collision. Phillips and Schiff[23] used measurements of O_3 destruction to estimate concentrations of vibrationally excited N_2 in levels with $v \geqslant 4$. Ogryzlo and co-workers[24] have studied the important metastable species $O_2(^1\Delta_g)$ in flow systems, using a catalytic HgO surface to remove O atoms from the gas stream, thus leaving $O_2(^1\Delta_g)$ as the sole product of a microwave discharge through O_2. Changes in the concentration of $O_2(^1\Delta_g)$ were followed by measuring the intensity of the red emission band arising from the process:

$$2O_2(^1\Delta_g) \rightarrow 2O_2(^3\Sigma_g{}^-) + h\nu \qquad (3.10)$$

The intensity of the infrared atmospheric bands has also been used to monitor $O_2(^1\Delta_g)$.[25] McNeal and Cook[26] and Wayne[26a] have capitalized on the fact that the ionization potential of $O_2(^1\Delta_g)$ is less than that of $O_2(^3\Sigma_g{}^-)$, i.e. the photoionization threshold is at a longer wavelength, to devise an ingenious photoionization system for measuring its concentration in the presence of excess ground-state ($^3\Sigma_g{}^-$) oxygen molecules. Young and Black[27] showed that $O_2(^1\Sigma_g{}^+)$ is formed during the rapid quenching by O_2

---

* Since this was written Clyne and Cruse (*J. Chem. Soc. Faraday Trans. II*, **68**, 1281, 1377 (1972)) have reported kinetic studies of oxygen and halogen atom reactions, using atomic resonance fluorescence in a flow system.

of the $O(^1D)$ atoms which result from 147 nm photolysis of O_2; Izod and Wayne[28] used this as a means of generating $O_2(^1\Sigma_g^+)$ in a flow system so that its reactions could be studied under flow conditions. Setser and co-workers[29] have generated metastable rare gas atoms in a flow system by means of a low current d.c. discharge. Reactions of the metastable atoms themselves have been studied, and also the reactions of other excited species, notably $N_2(A^3\Sigma_u^+)$ and $CO(a^3\Pi)$, which are able to be produced in high yield by energy transfer from the excited rare gas atoms. Finally we should point out that a flow system need not operate at a very low pressure if other factors guarantee a supply of reactive species. Thus, for example, Sugden and co-workers[30] showed that a hydrogen–oxygen flame at atmospheric pressure could be used as a fast-flow system for measuring rates of reaction of atoms and radicals at temperatures up to 2500 K.

3.5.2 FLOW SYSTEM STUDIES OF ION-MOLECULE REACTIONS

It is only since the mid 1950's that ion–molecule processes such as charge transfer

$$A^\pm + BC \rightarrow BC^\pm + A \tag{3.11}$$

and ion–atom interchange

$$A^\pm + BC \rightarrow AB^\pm + C \tag{3.12}$$

have been considered as important processes in the upper atmosphere. Initially it was thought that ions such as O^+, produced by direct photo-ionization of atomic oxygen, would recombine rapidly with atmospheric electrons. Once the importance of ion–molecule reactions occurring within the ionosphere became appreciated a rapid upsurge of laboratory interest was evident, and during the 1960's the rates of many of these ion–molecule reactions were measured. Many different experimental methods for studying reactions between molecules and ions have been used, including the observation of secondary processes in mass spectrometers operated at high pressure, observation of collisions of ions with molecules using tandem mass spectrometers, observation of reactions in ion cyclotron resonance spectrometers, and studies of stationary afterglows, of flowing afterglows, and of reactions in drift tubes. It is mainly by the flowing afterglow (FA) technique that the majority of reaction rates relevant to atmospheric processes have been determined. The FA method (Fig. 3.6) of Ferguson, Fehsenfeld, Schmeltekopf and their collaborators,[31] is a development of the mass spectrometric discharge flow system that we have discussed in connection with the study of atom–molecule reactions. A fast (100 m s^{-1}) flow of carrier gas, usually helium, is established in a 1 metre long, 10 centimetre diameter reaction tube, by means of a Roots-type pump (backed by a large mechanical fore pump) at a pressure between 10^{-2} and 10 torr. The helium carries ions,

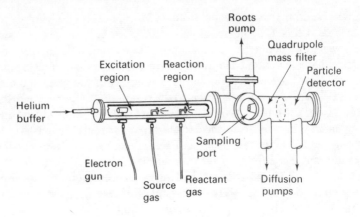

Fig. 3.6 Flowing afterglow system of Ferguson and co-workers.

produced by an electron gun at the upstream end, along the tube, and neutral reactant gases can be added at one or more of the inlet ports. Ion concentrations are determined by a differentially pumped quadrupole mass spectrometer which monitors samples withdrawn through a small aperture on the axis of the reaction tube. The extent of reaction is changed by varying the neutral reactant flow or, alternatively, by varying the sample to port distance, thereby changing the reaction time. Because the neutral reactants are normally present in large excess over the primary ion, the concentration of the primary ion decreases exponentially as either time or reactant pressure increases. This decrease, and the simultaneous appearance of product ions, can be observed quantitatively over a range of several decades (Fig. 3.7) and in favourable cases rate constants accurate to $\pm 10\%$ can be obtained. Normally an accuracy of 20–30% is claimed for rate constants obtained by the FA method. A useful account of other methods for measuring ion–molecule reaction rates is given in the book by McDaniel *et al.*[33]

Complications may arise in connection with numerical values listed for the rates of ion–molecule reactions. The FA system conveniently produces rate constants k in cm^3 particle^{-1} s^{-1} for a 'bimolecular' process. Mass spectrometer ion source measurements, where the ions have a non-Maxwellian translational energy distribution, yield cross sections q, which have until recently been expressed in Å^2 (the appropriate SI unit would be the nm$^2 = 100$ Å^2). For reactions which have no activation energy the relationship between k, the rate constant, and $\bar{q}$ the reaction or 'phenomenological' cross section is

$$k = \bar{q}\bar{v} \tag{3.13}$$

where $\bar{v}$ is the average velocity of the ion. In the FA system the ionic and neutral reactants come together with thermal translational energy and so

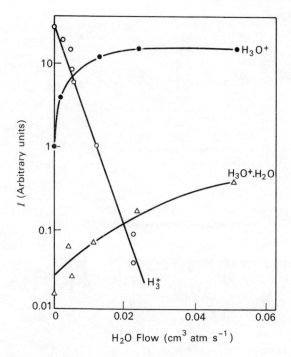

Fig. 3.7 The decay of H_3^+ ions in the presence of water vapour in a FA reaction tube, from Burt *et al.*[32]

have a Maxwellian velocity distribution. Substituting for $\bar{v}$ from the Maxwell–Boltzmann formula then gives

$$k = \bar{q}\left(\frac{8RT}{\pi\mu}\right)^{1/2} \tag{3.14}$$

where μ is the reduced mass, in atomic mass units, of the reacting species. However, to convert from the cross section obtained at a non-thermal velocity to a specific reaction rate (which will normally be a function of the average kinetic energy of the ions) is more difficult. In situations where the thermal energy is negligible compared with the total ion energy, for example in a mass spectrometer ion source (Fig. 3.8), it can be shown[34] that

$$\bar{v} = \left(\frac{eEd}{2M_p}\right)^{1/2}$$

and

$$k = \bar{q}\left(\frac{eEd}{2M_p}\right)^{1/2} \tag{3.15}$$

In eqn. 3.15 e is the charge on the ion, E the strength of the electric field by which ions are repelled towards the ion-source exit slit, d the distance from the ion exit slit of the plane in which ions are continuously formed,

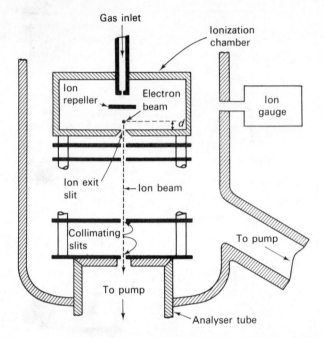

Fig. 3.8 Mass spectrometer ion source, as used in measurements of ion–molecule reaction rates. Reaction occurs while the ions drift the distance d under the influence of the weak repeller field.

and M_p the mass of the primary ion. Eqn. 3.15 can be applied when it has been shown experimentally that $\bar{q}$ varies as $(E)^{-1/2}$.

3.5.3 STATIC PHOTOLYSIS EXPERIMENTS

In this section we consider experiments in which a fixed sample of gas is illuminated with a steady radiation source. The information which may be obtained from such experiments includes absorption cross sections, photoionization cross sections and photoionization yields, identification of excited states of products of the primary process, photofragmentation yields of primary processes, fluorescence yields and rate constants for quenching of fluorescence, and quantum yields of the final photolysis products.

Absorption and photoionization cross sections for atmospheric gases are discussed in Chapter 2; further reference may be made to reviews by Hudson[35] and Samson[36] and to the book by Samson.[37] The states of excited products of the primary process can be determined from observations of fluorescence and, if the photon energy is high enough, by photoelectron spectroscopy.[38] Photoelectron spectroscopy has the advantage of being able to detect transitions to states that do not radiate. Information about the products of photofragmentation processes can be obtained for both neutral[39] and charged[40] species. For neutral species the work has been performed with molecular beams; for charged fragments photoionization mass spectrometry is used. The subject of fluorescence and fluorescence quenching has been discussed many times; basic references are the books by Pringsheim[41] and by Mitchell and Zemansky.[42] In recent years many workers have exploited the favourable characteristics of continuous wave lasers as excitation sources, especially the argon-ion laser, which has a number of convenient lines in the blue and green regions of the spectrum. The determination of product yields is the main preoccupation of classical photochemistry; reference may be made to the book by Calvert and Pitts[43] and the review by McNesby and Okabe.[44] For most of the experimental methods mentioned above we are content to refer the reader to previously published material. However, there is one topic which appears to merit a brief discussion, namely, the study of photolysis and fluorescence quenching in the extreme ultraviolet.

Recent advances in extreme ultraviolet photolysis have derived from studies by Ausloos and co-workers,[45] who showed how to use a thin film aluminium window to isolate the photolysis cell from a helium or neon microwave discharge lamp. Photolysis in the extreme ultraviolet is interesting in that the primary process usually involves photoionization, rather than simple fragmentation into neutral radicals. Studies of ionizing photolysis therefore constitute a bridge between ordinary photochemistry and high energy radiolysis. A typical photolysis set-up,[46] which has been used to measure absorption coefficients and ionization yields at 58.4 nm, as well as product yields, is shown in Fig. 3.9. Wauchop and Broida[47] employed a similar lamp and window combination in determining cross sections for emission of fluorescence from excited ions formed by He 58.4 nm irradiation of CO_2 and other gases. Rate constants for quenching reactions of excited ions produced in this way have been determined by Winkler et al.,[48] who observed the effect of added gases on the intensity of the ion fluorescence.

3.5.4 EXPERIMENTS WITH MODULATED PHOTOLYSIS LAMPS

A standard method[49] of determining the radiative lifetime of an excited state is to excite the ground state species with a modulated light source, i.e. one which is caused to undergo large, periodic fluctuations of intensity at a constant frequency f, and to measure the phase-shift δ between the fundamental components of the exciting light and the resulting modulated fluorescence. If τ is the mean lifetime of the excited state, then

$$\tan \delta = 2\pi f \tau \qquad (3.16)$$

Rate constants for parallel and consecutive reactions can be determined from measurements of the lifetime and intensity of fluorescence under appropriate conditions.[50]

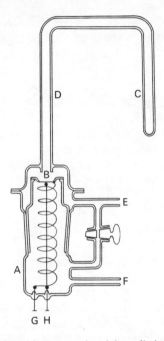

Fig. 3.9   Photolysis cell for use with extreme ultraviolet radiation. A microwave discharge in the quartz tube D produces helium resonance radiation which is transmitted by the thin film aluminium window at B. The cold finger C allows the trapping of impurities such as water liberated from the heated glass. The body of the cell, A, has a volume of about 150 cm^3. E and F are filling tubes; G and H are electrodes for ion current measurements.

The classical rotating sector technique,[43] for determining rate constants of intermediate reactions during photolysis, is closely related to the phase-shift method of determining lifetimes. The difference is that in the rotating sector technique one follows the amplitude of the output signal as a function of modulation frequency, rather than the phase. In most applications the 'amplitude' which is measured is given by the relative yield of some chemical product, but measurements of the intensity of emitted light as a function of modulation frequency have also been used.[51] Modulated emission was also studied in experiments in which a modulated (100 Hz) Lyman-α lamp was used to excite atomic hydrogen to the ^{2}P level in the presence of various gaseous reagents.[52] Then, for example, in the presence of oxygen the occurrence of the reaction

$$H(^2P) + O_2(^3\Sigma_g^-) \rightarrow OH(^2\Sigma) + O(^3P) \tag{3.17}$$

was demonstrated by the observation of modulated 306.4 nm emission, arising from the transition $A^2\Sigma \rightarrow X^2\Pi$ of OH. In this work the modulation provided a useful diagnostic tool.

Johnston and co-workers[53] developed a phase-shift system using low frequency (typically up to 32 Hz) modulation of a photolysis lamp, together with phase sensitive detection of absorption spectra of transient intermediates. By this means absorption spectra were determined for reactive intermediates such as HO_2 and $ClOO$, and values of rate constants for intermediate reactions were found. Phillips and co-workers used microwave powered lamps modulated at up to 150 kHz to determine reaction rates for excited mercury and xenon atoms, from phase-shift measurements of emission spectra.[54] A diagram of the phase-shift apparatus used in this work is given in Fig. 3.10. The main advantage of the method over the ordinary rotating sector technique, apart from the higher modulation frequency, is that it can easily be used with sources of vacuum ultraviolet radiation. Hunziker[55] has described an rf powered discharge lamp, modulated at frequencies up to 250 kHz, which he used to determine lifetimes and spectra of triplet states of organic compounds such as benzene and naphthalene, from phase-shift measurements of absorption spectra. Rate

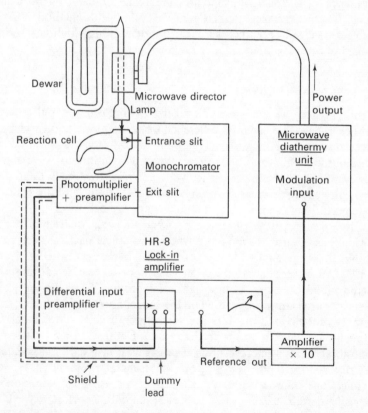

Fig. 3.10 Phase-shift system using a microwave powered photolysis lamp modulated at up to 150 kHz. The reference signal from the PAR lock-in amplifier drives a switching circuit which periodically disconnects the plate of the magnetron from ground, thereby cutting off the microwave output to the lamp.

constants for reactions of the excited species were also obtained from phase-shift measurements. Atkinson and Cvetanovic[56] subsequently employed Hunziker's system to study, for example, the reaction kinetics of oxygen atoms generated by mercury sensitized decomposition of nitrous oxide. The atoms were produced according to

$$Hg(^3P_1) + N_2O \rightarrow Hg(^1S_0) + N_2 + O(^3P) \tag{3.18}$$

and their relative concentration was determined from the intensity of the 'air afterglow' luminescence in the presence of added nitric oxide:

$$NO + O + M \rightarrow NO_2^* + M \tag{3.8}$$

$$NO_2^* \rightarrow NO_2 + h\nu \tag{3.9}$$

To determine rate constants for O atom reactions the phase-shift between the exciting light and the emission given by reaction 3.9 was measured in the presence of a known partial pressure of an added reagent such as ethylene. These examples suggest that the phase-shift method has considerable unrealized potential as a means of determining lifetimes, spectra, and reaction rates for intermediates in photochemical systems.

3.5.5   RELAXATION METHODS

In this category we include studies using flash photolysis, pulsed lasers, and shock tubes. The essential difference from the modulation techniques considered in Section 3.5.4 is that there the input to the system of interest, and the output which resulted, were treated as continuous wave signals, having a fundamental component with characteristic frequency f and phase angle δ, whereas here we are concerned with the response of the system to a single perturbing event such as a light flash or a shock wave.

The technique of flash photolysis,[57] after almost a quarter century of operation, has matured to the point where a complete packaged set-up can be obtained commercially.* A diagram of the basic apparatus is given in Fig. 3.11. An intense, short duration flash is produced by causing a few thousand joules of energy from a bank of capacitors to be dissipated in a tubular discharge lamp. Depending on the capacitance used and the residual inductance of the circuit the duration of the flash (usually stated in terms of the time for the peak intensity to decay by a factor of e) may range from about 5 to about 500 microseconds. The gas used in the discharge lamp is commonly xenon, which gives strong continuous emission in the near ultraviolet, but other gases have been tried; for example, Claesson and Lindqvist[58] used a mixture of 20 torr of argon with 2 torr of oxygen. The material to be photolysed is contained in a cell adjacent to the flash lamp,

---

* From Applied Photophysics Ltd., 20 Albemarle Street, London W1X 3HA, England.

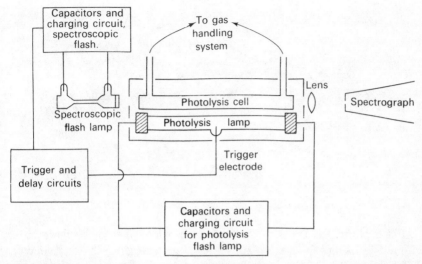

Fig. 3.11 Basic flash photolysis apparatus, with photographic recording of transient spectra.

and during the flash it may absorb about 10^{18} photons per cm^3 in the near ultraviolet and blue part of the spectrum. One result of the absorption of so much energy is a temperature rise, which can amount to 1000 K in a gas at low pressure. To avoid unwanted thermal effects it is usual to buffer the absorption cell with a few hundred torr of an inert gas such as nitrogen. The absorption spectrum of the photolysis products is determined as a function of time after the initial flash, either photoelectrically, with a continuously operating light source and a photomultiplier monitoring a chosen band of wavelengths, or photographically, with the aid of a 'spectroscopic flash' lamp which is fired a predetermined interval after the photolysis flash. The system in Fig. 3.11 uses a spectroscopic flash lamp; with this method the history of the photolysis products generated by the primary flash is assembled from a number of experiments with different delay times between the photolysis and spectroscopic flashes. This kind of experimental system has been used to make many important advances in kinetics and photochemistry, and has also permitted detailed spectroscopic studies to be made of important intermediates, such as the free radicals CH_3 and CH_2 and triplet states of aromatic molecules, which had not previously been available in sufficient concentration to be observed spectroscopically. Developments of the basic flash photolysis method include the use of high intensity spark discharges[59] and exploding wires[60] in place of the spectroscopic flash, and the use of lithium fluoride windows to extend the method into the vacuum ultraviolet.[61] Callear and co-workers[62] have described microwave powered flash lamps which contain the vapour of a metal such as mercury or cadmium, and emit the metallic resonance lines with high intensity in the near ultraviolet.

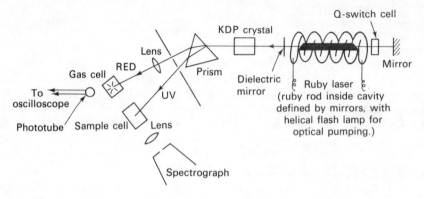

Fig. 3.12   Principle of nanosecond flash photolysis apparatus.

Flash photolysis experiments have been extended into the nanosecond domain by using a Q-switched ruby laser to generate a brief pulse of coherent red light, and then using a non-linear crystal such as potassium dihydrogen phosphate (KDP) as a frequency doubler, to obtain useful near ultraviolet radiation.[63] A diagram of a typical system is given in Fig. 3.12. In the experiments of Novak and Windsor, on which Fig. 3.12 is based, about 3% of the ruby light was converted into ultraviolet radiation. The remainder, after being separated from the frequency doubled beam, was focussed into a gas-filled cell. Here it produced an intense spark, whose duration could be varied by changing the nature of the filling gas. This spark served in place of the spectroscopic flash of Fig. 3.11. The main application of this technique so far has been to the determination of the absorption spectra and lifetimes of short-lived excited states, especially singlet states, of

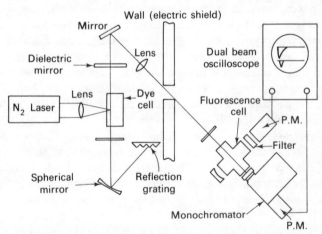

Fig. 3.13   Tuneable dye laser used by Broida and co-workers for fluorescence decay measurements.

aromatic molecules. A closely related experimental technique involves the use of a tuneable dye laser to excite selected energy levels of a molecule such as iodine or formaldehyde in the gas phase.[64] The laser operates in a pulsed mode, giving pulses that are typically of a few nanoseconds duration, and relaxation processes can be studied by observing the decay of fluorescence or phosphorescence. A typical set-up (Fig. 3.13) consists of an optical cavity, one end of which is defined by a diffraction grating and into which a cell containing a fluorescent dye such as Rhodamine 6G is inserted. The dye is optically pumped by another laser; for example, a pulsed nitrogen laser emitting at 337.1 nm was used to pump the dye in reference 64.

By the use of a mode-locked ruby or neodymium glass laser it is possible to obtain light pulses with durations of a few tens of picoseconds. As before, the laser output can be frequency doubled with a KDP crystal to give pulses of wavelength 347 nm (ruby) or 503 nm (Nd glass).[65] The duration of such a pulse can be determined by measuring its length in millimetres (1 picosecond = 0.3 mm in vacuo). The application of these ultra-short pulses to the study of fast gas phase processes is still in the future.

The basic principle of the shock tube is shown in Fig. 3.14. A high pressure of a light driver gas such as helium is suddenly released by rupture of a diaphragm which separates it from the shock tube proper. The resulting series of pressure waves heats the gas and, because the velocity of sound in a gas increases with temperature, the waves at the rear catch up to the ones at the front, where they unite to form a single intense shock wave. The shock front propagates through the low pressure gas with a velocity which is equal to the sound velocity plus the bulk flow rate of the released gas as it expands along the tube. After it has travelled along the tube for several metres the wave may either be reflected back from the end of the tube or 'dumped' into a large volume container. As the shock front passes through the low pressure gas the region just behind the front undergoes adiabatic, homogeneous heating to a temperature which can be as high as 20 000 K in a time of less than a microsecond. Chemical and physical processes occurring behind the shock front can be followed with the aid of a variety of diagnostic techniques, of which measurements of light emission or

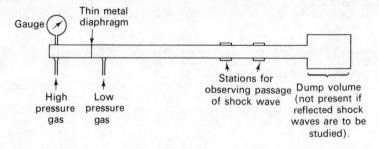

Fig. 3.14 Elements of a simple shock tube.

absorption are perhaps the most obvious. Shock tubes have proven extremely useful in studies of chemical kinetics and energy transfer in gases at temperatures above 1000 K; the shock tube has an advantage over flash photolysis in that it is not limited to systems which absorb light. For further details the reader is referred to books by Bradley[66] and by Gaydon and Hurle.[67]

References

1   Meinel, A. B., *Astrophys. J.*, **111**, 555 (1950); Petrie, W., *Astrophys. J.*, **116**, 433 (1952).
2   Noxon, J. F., *Space Sci. Rev.*, **8**, 92 (1968).
3   Hunten, D. M., *Space Sci. Rev.*, **6**, 493 (1967).
4   Noxon, J. F. and Vallance Jones, A., *Nature*, **196**, 157 (1962).
5   Narcisi, R. S. and Bailey, A. D., *J. geophys. Res.*, **70**, 3687 (1965).
6   Fehsenfeld, F. C. and Ferguson, E. E., *J. geophys. Res.*, **74**, 2217 (1969).
7   Evans, W. F. J., Hunten, D. M., Llewellyn, E. J. and Vallance Jones, A., *J. geophys. Res.*, **73**, 2885 (1968).
8   Hunten, D. M. and McElroy, M. B., *J. geophys. Res.*, **73**, 2421 (1968); Paulsen, D. E., Huffman, R. E. and Larrabee, J. C., *Radio Sci.*, **7**, 51 (1972).
9   Groves, G. V., *J. geophys. Res.*, **68**, 3033 (1963).
10 Spindler, G. B., *Planet. Space Sci.*, **14**, 53 (1966).
11 Bass, A. M. and Broida, H. P., '*The Formation and Trapping of Free Radicals*', Academic Press, New York, 1960.
12 Phillips, L. F. and Schiff, H. I., *J. chem. Phys.*, **37**, 1233 (1962).
13 Del Greco, F. P. and Kaufman, F., *Discuss. Faraday Soc.*, **33**, 128 (1962).
14 Iwai, T., Pratt, D. W. and Broida, H. P., *J. chem. Phys.*, **49**, 919 (1968).
15 Clyne, M. A. A. and Cruise, H. W., *Trans. Faraday Soc.*, **67**, 2869 (1971); Clyne, M. A. A., 'Reactions of atoms and free radicals studied in discharge-flow systems', in '*Physical Chemistry of Fast Reactions*', Ed. Levitt, B. P., Vol. 1, p. 245. Plenum Press, London, 1973; Clyne, M. A. A. and Coxon, J. A., *Proc. R. Soc. (London)*, **A303**, 207 (1968).
16 Wright, A. N. and Winkler, C. A., '*Active Nitrogen*', Academic Press, New York, 1968; Jennings, K. R., *Q. Rev. chem. Soc.*, **15**, 237 (1961).
17 Tellinghuisen, J. B., Lawrence Radiation Laboratory Report, UCRL-19112, University of California, Berkeley, 1969; Kurylo, M. J., Peterson, N. C. and Braun, W., *J. chem., Phys.*, **53**, 2776 (1970); Clyne, M. A. A. and Cruse, H. W., to be published.
18 Westenberg, A. A. and de Haas, N., *J. chem. Phys.*, **40**, 3081 (1964).
19 Jonathan, N., Ross, K. and Smith, D. J., *J. chem. Phys.*, **53**, 3758 (1970).
20 Elias, L., Ogryzlo, E. A. and Schiff, H. I., *Can. J. Chem.*, **37**, 1680 (1959); Morgan, J. E., Phillips, L. F. and Schiff, H. I., *Discuss. Faraday Soc.*, **33**, 118 (1962).
21 Tollefson, E. L. and LeRoy, D. J., *J. chem. Phys.*, **16**, 1057 (1948).
22 Anlauf, K. G., Kuntz, P. J., Maylotte, D. H., Pacey, P. D. and Polanyi, J. C., *Discuss. Faraday Soc.*, **44**, 183 (1967).
23 Phillips, L. F. and Schiff, H. I., *J. chem. Phys.*, **36**, 3283 (1962).
24 Bader, L. W. and Ogryzlo, E. A., *Discuss. Faraday Soc.*, **37**, 46 (1964).
25 Wayne, R. P. and Pitts, J. N., Jnr., *J. chem. Phys.*, **50**, 3644 (1969).
26 McNeal, R. J. and Cook, G. R., *J. chem. Phys.*, **47**, 5385 (1967).
26a Wayne, R. P., *Adv. Photochem.*, Volume 7, p. 311, Eds. Pitts, J. N., Jnr., Hammond, G. S. and Noyes, W. A., Jnr., Interscience, New York, 1969.
27 Young, R. A. and Black, G., *J. chem. Phys.*, **47**, 2311 (1967).
28 Izod, T. P. J. and Wayne, R. P., *Proc. R. Soc. (London)*, **A308**, 81 (1968).
29 Setser, D. W., Stedman, D. H. and Coxon, J. A., *J. chem. Phys.*, **53**, 1004 (1970).
30 Bulewicz, E. M. and Sugden, T. M., *Trans. Faraday Soc.*, **54**, 1855 (1958).
31 Ferguson, E. E., Fehsenfeld, F. C. and Schmeltekopf, A. L., *Adv. At. and Mol. Phys.*, Volume 5, p. 1, Ed., Bates, D. R., Academic Press, New York, 1969.

32   Burt, J. A., Dunn, J. L., McEwan, M. J., Sutton, M. M., Roche, A. E. and Schiff, H.
      I., *J. chem. Phys.*, **52,** 606 (1970).
33   McDaniel, E. W., Cermak, V., Dalgarno, A., Ferguson, E. E. and Friedman, L., *'Ion
      Molecule Reactions'*, Wiley-Interscience, New York, 1970.
34   Gioumousis, G. and Stevenson, D. P., *J. chem. Phys.*, **29,** 294 (1958); Lampe, F. W.,
      Franklin, J. L. and Field, F. H., 'Kinetics of Reactions of Ions with Molecules',
      Progress in Reaction Kinetics, Volume 1, p. 67, Ed. Porter, G. and Stevens, B., 1961.
35   Hudson, R. D., *Rev. Geophys. and Space Phys.*, **9,** 305 (1971).
36   Samson, J. A. R., *Adv. At. and Mol. Phys.*, Ed. Bates, D. R., Volume 2, p. 177,
      Academic Press, New York, 1966.
37   Samson, J. A. R., *'Techniques of Vacuum Ultraviolet Spectroscopy'*, John Wiley and
      Sons Inc., New York, 1967.
38   Turner, D. W., Baker, C., Baker, A. D. and Brundle, C. R., *'Molecular Photoelectron
      Spectroscopy'*, John Wiley and Sons Inc., New York, 1970.
39   Busch, G. E., Mahoney, R. T., Morse, R. I. and Wilson, K. R., *J. chem. Phys.*, **51,** 449
      (1969).
40   For example, Dibeler, V. H., Walker, J. A. and Rosenstock, H. M., *J. Res. nat. Bur.
      Stand.* **70A,** 459 (1966).
41   Pringsheim, P., *'Fluorescence and Phosphorescence'*, Interscience, New York, 1949.
42   Mitchell, A. C. G. and Zemansky, M. W., *'Resonance Radiation and Excited Atoms'*,
      Cambridge University Press, London, 1934.
43   Calvert, J. G. and Pitts, J. N., Jnr., *'Photochemistry'*, John Wiley and Sons Inc., New
      York, 1966.
44   McNesby, J. R. and Okabe, H., *Adv. Photochem.*, Eds., Noyes, W. A., Jnr., Hammond,
      G. S. and Pitts, J. N., Jnr., Volume 1, p. 157, John Wiley and Sons Inc., New York,
      1964.
45   Gordon, R., Jnr., Rebbert, R. E. and Ausloos, P., *'Rare Gas Resonance Lamps'*,
      National Bureau of Standards (U.S.A.) Technical Note 496, 1969.
46   Bennett, S. W., Tellinghuisen, J. B. and Phillips, L. F., *J. phys. Chem.*, **75,** 719 (1971);
      Tellinghuisen, J. B., Winkler, C. A. and Phillips, L. F., *J. phys. Chem.*, **76,** 298 (1972).
47   Wauchop, T. S. and Broida, H. P., *J. geophys. Res.*, **76,** 21 (1971).
48   Winkler, C. A., Tellinghuisen, J. B. and Phillips, L. F., *J. chem. Soc.*, *Faraday Trans-
      actions II*, **68,** 121 (1972).
49   Bailey, E. A. and Rollefson, G. K., *J. chem. Phys.*, **21,** 1315 (1953).
50   Metcalf, W. S., *J. chem. Soc.*, 3726 (1960).
51   For example, Newman, R. H., Freeman, C. G., McEwan, M. J., Claridge, R. F. C. and
      Phillips, L. F., *Trans. Faraday Soc.*, **66,** 2827 (1970).
52   Wauchop, T. S and Phillips, L. F., *J. chem. Phys.*, **47,** 4281 (1967); **51,** 1167 (1969).
53   Johnston, H. S., McGraw, G. E., Paukert, T. T., Richards, L. W. and Van den Bogaerde,
      J., *Proc. natn. Acad. Sci. (U.S.A.)*, **57,** 1146 (1967); Johnston, H. S., Morris, E. D., Jnr.
      and Van den Bogaerde, J., *J. Am. chem. Soc.*, **91,** 7712 (1969); Paukert, T. T., Ph.D.
      thesis, (Lawrence Radiation Laboratory Report UCRL-19109), University of California,
      Berkeley, 1969; Phillips, L. F., in *'Progress in Reaction Kinetics'*, Ed. Jennings, K. R.,
      and Cundall, R. B., Vol. 7, p. 83. Pergamon Press, London, 1973.
54   Phillips, L. F., *Rev. scient. Instrum.*, **42,** 1098 (1971); Freeman, C. G., McEwan, M. J.,
      Claridge, R. F. C. and Phillips, L. F., *Chem. Phys. Lett.*, **9,** 578 (1971); **10,** 530 (1971).
55   Hunziker, H. E., *Chem. Phys. Lett.*, **3,** 504 (1969); *IBM J. Res. Dev.*, **15,** 10 (1971).
56   Atkinson, R. and Cvetanovic, R. J., *J. chem. Phys.*, **55,** 659 (1971); **56,** 432 (1972).
57   Norrish, R. G. W. and Porter, G., *Nature*, **164,** 658 (1949); Porter, G., in *'Technique of
      Organic Chemistry'* Vol. VIII, Part 2, 2nd ed., p. 1055, Interscience, New York, 1963.
58   Claesson, S. and Lindqvist, L., *Ark. Kemi*, **11,** 535 (1957); **12,** 1 (1958).
59   Mains, G. J., Roebber, J. L. and Rollefson, G. K., *J. phys. Chem.*, **59,** 733 (1955).
60   Oster, G. K. and Marcus, R. A., *J. chem. Phys.*, **27,** 182, 472 (1957).
61   For example, Donovan, R. J., Husain, D. and Kirsch, L. J., *Trans. Faraday Soc.*, **66,**
      2551 (1970).
62   Callear, A. B. and McGurk, J., *Nature*, **226,** 844 (1970); Callear, A. B. and Wood, P. M.
      Trans. Faraday Soc., **67,** 598 (1971).
63   Novak, J. R. and Windsor, M. W., *J. chem. Phys.*, **47,** 3075 (1967); Porter, G. and
      Topp, M. R., *Proc. R. Soc. (London)*, **A315,** 163 (1970).

64   For example, Sakurai, K., Capelle, G. and Broida, H. P., *J. chem. Phys.*, **54,** 1220 (1971).
65   Rentzepis, P. M., *Chem. Phys. Lett.*, **2,** 117 (1968); Eisenthal, K. B., *Chem. Phys. Lett.*, **6,** 155 (1970); Rentzepis, P. M., *Science, N.Y.*, **169,** 139 (1970).
66   Bradley, J. N., *'Shock Waves in Chemistry and Physics'*, Methuen, London, 1962.
67   Gaydon, A. G. and Hurle, I. R., *'The Shock Tube in High Temperature Chemistry and Physics'*, Reinhold, New York, 1963.

4
Composition and Dynamics of the Chemosphere

4.1 Introduction

The region of the atmosphere where chemical reactions constitute the dominant type of activity is called the Chemosphere. The name is generally associated with the part of the atmosphere that lies between 20 and 110 kilometres altitude. For many purposes the chemosphere may be regarded as consisting of just the stable gases oxygen, nitrogen and argon, other constituents being present in only very minor amounts. However, it is not the bulk components which truly characterize the chemosphere, but the minor constituents O, O_2^* (electronically excited molecular oxygen), O_3, NO, NO_2, CO_2, H_2O, H and the various electrically charged species of the lower ionosphere. Table 1.2 gives the proportions of the major constituents of the atmosphere below about 90 km altitude; typical relative amounts of minor constituents such as water vapour, ozone and nitric oxide are given in Table 1.3. At the time of writing there have been relatively few measurements of composition within the chemosphere, and the concentrations of the minor constituents are not as well established as in other atmospheric regions. In addition, whereas at altitudes above 200 km numerous determinations of atmospheric parameters such as pressure and mean free path have been made by orbiting satellites, measurements within the chemosphere have been made only intermittently during short-lived rocket flights. Thus this region is to be regarded as relatively unexplored, and not properly understood despite considerable interest in its behaviour.

Values of important parameters in the chemosphere, such as air density, total particle concentration, mean free path, ambient temperature, pressure, and bimolecular collision frequency are given at a number of different altitudes in Table 4.1. In practice, the listed quantities vary in a predictable way with latitude, time of day, season and the level of solar activity. Above about 90–100 km the mean molecular mass varies with altitude because diffusion and dissociation become increasingly important.

Most of the solar radiation falling on the atmosphere passes into the chemosphere, where it acts as a broad photolyzing source. The resulting photolytically generated atoms, radicals, excited molecules and ions undergo a host of chemical reactions with themselves and with the stable atmospheric gases. These reactions are the major concern of this chapter. Some of the excited photolysis and/or reaction products undergo radiative transitions,

Table 4.1   Chemosphere parameters.

The data given apply to an idealized year-round mean, mid latitude (approx. 45°) location at mid solar cycle. The quantity n is the total particle density and T is the kinetic temperature (Data extracted from U.S. Standard Atmosphere tables, 1962).

Height (km)	Density (g cm^{-3})	n (particles cm^{-3})	Pressure (mm of Hg)[b]	Collision Frequency (collisions s^{-1})	Mean Free Path (cm)	T (Kelvin)
10	4.14 -4[a]	8.60 $+18$	1.99 $+2$	2.06 $+9$	1.96 -5	223
20	8.89 -5	1.85 $+18$	4.14 $+1$	4.35 $+8$	9.14 -5	217
30	1.84 -5	3.83 $+17$	8.98	9.22 $+7$	4.41 -4	227
40	4.00 -6	8.31 $+16$	2.15	2.10 $+7$	2.03 -3	250
50	1.03 -6	2.14 $+16$	5.98 -1	5.62 $+6$	7.91 -3	271
60	3.06 -7	6.36 $+15$	1.68 -1	1.63 $+6$	2.66 -2	256
70	8.75 -8	1.82 $+15$	4.14 -2	4.32 $+5$	9.28 -2	220
80	1.99 -8	4.16 $+14$	7.78 -3	8.94 $+4$	4.07 -1	181
90	3.17 -9	6.59 $+13$	1.23 -3	1.42 $+4$	2.56	181
100	4.97 -10	1.04 $+13$	2.26 -4	2.41 $+3$	1.63 $+1$	210
110	9.83 -11	2.07 $+12$	5.52 -5	5.36 $+2$	8.15 $+1$	257
120	2.44 -11	5.23 $+11$	1.89 -5	1.59 $+2$	3:23 $+2$	351

a $4.14 - 4 = 4.14 \times 10^{-4}$
b   1 mm of Hg = 1 torr = 1.3332 millibars = 133.32 Nm^{-2}

emitting light which can be detected at ground level. This is the phenomenon of the *airglow*, which forms the subject of Chapter 5. The types of chemical processes to be discussed in this and following chapters are summarized in Table 4.2. Chemical products of the type shown in Table 4.2 may be removed by ensuing chemical reactions, by physical transport of the species by winds and diffusive processes, or by a mixture of both. In the next section we consider the laws which govern these removal processes.

4.2   Physical transport processes and the continuity equation

Once the rate coefficients for the important production and loss processes involving a particular constituent have been determined it is possible to construct an equation showing the time variation of the constituent's concentration. An equation of this sort is called a *continuity* equation and usually includes terms of two types. One kind of term gives the molecular flux due to production and loss by chemical reactions, and the other gives the flux due to physical transport of the constituent by any of several kinds of diffusive process. Although we shall be mainly concerned with chemical production and loss in what follows, the importance of physical transport for species with long chemical lifetimes should not be over-looked. We therefore give a brief description of the three distinct types of physical transport that must be considered.

Molecular diffusion occurs when the concentration gradient of a particular gas in a mixture differs from its equilibrium value, which in the atmosphere

Table 4.2 Important types of chemical reaction in the atmosphere.

A representative reaction is given in the first column, the expression for the rate of the reaction is given in the second column, with conventional units for the rate coefficient (J or k) in parenthesis, and the third column contains the name of the process.

Reaction	Rate	Process
Neutral species		
$NO + h\nu \rightarrow NO^+ + e$	$J[NO] = [NO] \int I_\lambda \Phi_\lambda q_\lambda \, d\lambda^a \, (s^{-1})$	Photoionization
$O_2 + h\nu \rightarrow O + O$	$J[O_2] = [O_2] \int I_\lambda \Phi_\lambda q_\lambda \, d\lambda^b \, (s^{-1})$	Photodissociation
$H(^2S) + h\nu \rightarrow H(^2P)$	$J[H(^2S)] = [H(^2S)] \int I_\lambda \Phi_\lambda q_\lambda \, d\lambda^c \, (s^{-1})$	Photoexcitation
$Na(^2P) \rightarrow Na(^2S) + h\nu$	$k[Na(^2P)] \, (s^{-1})$	Fluorescence or Luminescence
$O(^1D) + O_2 \rightarrow$ products	$k[O(^1D)][O_2] \, cm^3 \, molecule^{-1} \, s^{-1}$	Quenching
$N + O_2 \rightarrow NO + O$	$k[N][O_2] \, cm^3 \, molecule^{-1} \, s^{-1}$	Bimolecular reaction
$O + O + M \rightarrow O_2 + M$	$k[O][O][M] \, cm^6 \, molecule^{-2} \, s^{-1}$	Termolecular reaction
$NO + O \rightarrow [NO_2^*]$ $\rightarrow NO_2 + h\nu$	$k[NO][O] \, cm^3 \, molecule^{-1} \, s^{-1}$	Radiative combination
$Na(^2S) + O_2^* \rightarrow Na(^2P) + O_2$	$k[Na(^2S)][O_2^*] \, cm^3 \, molecule^{-1} \, s^{-1}$	Energy transfer
Ions		
$N_2^+ + O_2 \rightarrow N_2 + O_2^+$	$k[N_2^+][O_2] \, (cm^3 \, molecule^{-1} \, s^{-1})^d$	Charge transfer
$N_2^+ + O \rightarrow NO^+ + N$	$k[N_2^+][O] \, (cm^3 \, molecule^{-1} \, s^{-1})$	Ion-atom interchange
$NO^+ + e \rightarrow N + O$	$k_{ei}[NO^+][e] \, (cm^3 \, molecule^{-1} \, s^{-1})$	Dissociative recombination
$O^+ + e \rightarrow O + h\nu$	$k_{ei}[O^+][e] \, (cm^3 \, molecule^{-1} \, s^{-1})$	Radiative recombination
$e + O_2 + O_2 \rightarrow O_2^- + O_2$	$k_{ea}[e][O_2][O_2] \, (cm^6 \, molecule^{-2} \, s^{-1})$	Three body attachment
$e + O_3 \rightarrow O^- + O_2$	$k_{ea}[e][O_3] \, (cm^3 \, molecule^{-1} \, s^{-1})$	Dissociative attachment
$e + O \rightarrow O^- + h\nu$	$k_{ea}[e][O] \, (cm^3 \, molecule^{-1} \, s^{-1})$	Radiative attachment
$O_2^- + O_2 \rightarrow O_2 + O_2 + e$	$k_{cd}[O_2^-][O_2] \, (cm^3 \, molecule^{-1} \, s^{-1})$	Collisional detachment
$O^- + O \rightarrow O_2 + e$	$k_{ad}[O^-][O] \, (cm^3 \, molecule^{-1} \, s^{-1})$	Associative detachment
$O^- + h\nu \rightarrow O + e$	$J[O^-] = [O^-] \int \Phi_\lambda I_\lambda q_\lambda \, d\lambda \, (s^{-1})$	Photodetachment

a I_λ is the radiation intensity in the range λ to $\lambda + d\lambda$, q_λ is the absorption cross section in this wavelength range, and Φ_λ the quantum efficiency for photoionization. The product $\Phi_\lambda q_\lambda$ is the photoionization cross section of the neutral molecule, in this case NO, at wavelength λ.
b   As above, except that the product $\Phi_\lambda q_\lambda$ is now the photodissociation cross section.
c   As above, but $\Phi_\lambda q_\lambda$ is the excitation cross section.
d   We use the term molecule in its most general sense, to include ions, atoms and excited atomic or molecular species.

is governed by the scale height H. (see Section 1.3). If the concentration gradient responsible for the diffusion is dc/dz, and f is the flux of molecules through unit area (measured at right angles to the flux) in unit time, then f and dc/dz are related by

$$f = -D\frac{dc}{dz} \tag{4.1}$$

which defines the diffusion coefficient D. Eqn. 4.1 is known as Fick's law of diffusion. The molecular diffusion coefficient characterizes the transport of a gas when the transport results from the average molecular velocity of the gas being different from that of other constituents. If the constituent under consideration is charged its diffusion in the partially ionized gas, or 'plasma', is strongly influenced by the electrostatic forces of the plasma as a whole. The diffusion resulting from a concentration gradient under these circumstances is called *ambipolar diffusion*. Thus, for example, the electrons in the ionosphere have a greater diffusion coefficient than the ions. However, electrostatic forces between the electrons and ions prevent any macroscopic separation of electrons and positive ions, and the plasma diffuses as a whole with a diffusion coefficient D_A, the ambipolar diffusion coefficient.

Eddy diffusion describes the transport of gases due to turbulent mixing in the presence of a composition gradient. Transport of constituents arising from eddy diffusion occurs whether or not there is any difference in average molecular velocity between the various species present; it can be treated similarly to molecular diffusion, by an equation analogous to 4.1. A vertical eddy diffusion coefficient, K, is defined by considering the nett flow of one constituent relative to the others.

Thermal diffusion arises from the presence of large temperature gradients in a gas mixture. The thermal diffusion coefficient (α_T) is proportional to the difference between the molecular weight of the diffusing species and the medium through which it is diffusing. As a consequence, thermal diffusion is likely to be important in the atmosphere only for the light gases hydrogen and helium. In addition, temperature gradients are small below 110 km, so that under most circumstances thermal diffusion can be neglected in the chemosphere.

When transport as well as chemical processes are included, the continuity equation for the ith constituent is[1]

$$\frac{\partial[i]}{\partial t} + \frac{\partial}{\partial z}([i]V_i) = \mathcal{P}_i - \mathcal{L}_i[i] \tag{4.2}$$

Here $[i]$ is the concentration of the constituent at time t and altitude z, $\mathcal{P}_i$ and $\mathcal{L}_i[i]$ are chemical production and loss rates respectively and V_i is the mean vertical velocity due to diffusion. The second term in eqn. 4.2, $(\partial/\partial z)([i]V_i)$, is essentially the divergence of $V_i[i]$, $(\partial/\partial x)V_i[i]$ and $(\partial/\partial y)V_i[i]$ being assumed negligible. For a neutral constituent whose concentration is small this term may be written

$$\frac{\partial}{\partial z}([i]V_i) = -(D_i + K)\left[\frac{\partial[i]}{\partial z} + \frac{[i]}{T}\frac{\partial T}{\partial z}\right] - \left(\frac{D_i}{H_i} + \frac{K}{H_{av}}\right)[i] \quad (4.3)$$

where D_i is the molecular diffusion coefficient, K the eddy diffusion coefficient, T the absolute temperature, H_i the scale height of the ith constituent and H_{av} the scale height for a constituent having the mean molecular mass.[2]

When the rate of production of constituent i from all sources equals its loss rate, the time derivative of $[i]$ vanishes, and eqn. 4.2 reduces to

$$\frac{\partial}{\partial z}([i]V_i) = \mathscr{P}_i - \mathscr{L}_i[i]. \quad (4.4)$$

With the aid of eqns. 4.2 and 4.4 models may be constructed giving the concentrations of minor constituents as a function of altitude. It must be remembered, however, that in a real atmosphere the situation resulting from the operation of these equations will be severely perturbed by winds and convection. The assumption of ignoring horizontal winds is usually valid at mid latitudes, but in certain areas such as the polar regions horizontal advection may be the principal source of a constituent. This is particularly true during the polar winter. In discussing diffusion processes we have implied a separation of the vertical velocity (V_i) of a minor neutral constituent into three components v_{av}, v_{diff} and v_{turb}, related to V_i by eqn. 4.5,

$$V_i = v_{av} + v_{diff} + v_{turb} \quad (4.5)$$

where v_{av} is the macroscopic velocity of the atmosphere, v_{diff} the molecular diffusion velocity of a minor constituent, and v_{turb} is the velocity gained by the constituent from turbulent processes. Of these three components of V_i, the two most difficult to evaluate are v_{turb} and v_{av}. Often, in discussions of turbulent mixing, the vertical eddy diffusion constant K is used as an adjustable parameter, the value of K being chosen in order to give the best fit between the model and experimental observations. An estimate of the variation of K with altitude is shown in Fig. 4.1. Little is known about planetary-scale circulation systems, and consequently contributions to V_i from v_{av} are usually assumed to be negligible. The turbulent velocity is of the order of K/H_{av} and in the upper chemosphere $H_{av} \sim 10^6$ cm and $K \sim 5 \times 10^6$ cm^2 s^{-1}.[4] Thus $v_{turb} \sim 5$ cm s^{-1}. In comparison, macroscopic velocities may reach and even exceed v_{turb}. It is therefore to be expected that increased attention will be devoted to large scale circulation systems in the future.[4]

At present an 'ab initio' solution of the continuity equations which will hold in general for all atmospheric constituents is out of the question. Simplifying approximations such as the neglect of convection, and in some cases diffusion, are commonly introduced. There are in the literature many models of varying degrees of sophistication to account for the behaviour

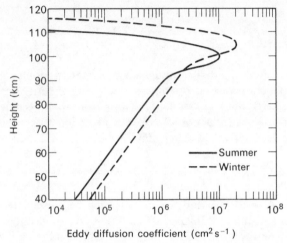

Fig. 4.1 Altitude profile of the eddy diffusion coefficient K.[3]

of species of importance in the chemosphere. A comparison of two such models with an observed concentration profile for nitric oxide is shown in Fig. 4.2. The model corresponding to the dashed line in Fig. 4.2 was based on a scheme utilizing only chemical production and loss processes, whereas the model corresponding to the dotted line includes the effect of eddy diffusion as well as emphasizing differently the chemical production processes

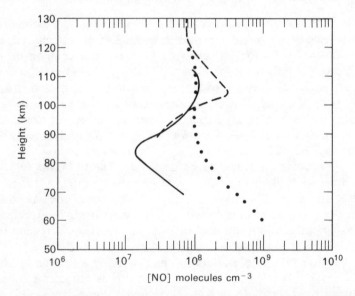

Fig. 4.2 Measured and calculated profiles of nitric oxide. The solid curve represents the observed results of ref. 5. The dashed and dotted curves are the calculated profiles of refs. 6 and 7, respectively.

for NO. The question of which process, physical or chemical, is the more important in determining the concentration profile of a constituent, is related to the constituent's lifetime with respect to both processes. If, for example, the chemical loss process of nitric oxide around 80 km occurs mainly by

$$O_3 + NO \rightarrow NO_2 + O_2, \tag{4.6}$$

$(k_{4.6} = 9.5 \times 10^{-13} \exp(-1241/T) cm^3 \text{ particle}^{-1} s^{-1})[8]$

an estimate of the lifetime of nitric oxide with respect to this process, without considering physical transport, can be obtained by dividing the concentration of NO by the rate of the loss process, i.e.

$$\tau_{NO}^{chemistry} = \frac{[NO]}{k[O_3][NO]} = \frac{1}{k[O_3]}$$

$\sim 7 \times 10^5$ s, if an ozone concentration of 1×10^8 molecule cm^{-3} is assumed.

The mean lifetime of NO with respect to transport is equal to the time required for the species to move vertically through a distance equal to its own scale height, i.e.

$$\tau_{NO}^{transport} = \frac{H}{V} = \frac{H^2}{D} \tag{4.7}$$

where H is the scale height, V the diffusion velocity and D the diffusion coefficient. In the mesosphere at 80 km, K the eddy diffusion coefficient is $\sim 10^6$ cm^2 s^{-1} [9] and H the scale height for NO is ~ 5.7 km. Thus,

$$\tau_{NO}^{transport} \approx 3.2 \times 10^5 \text{ s}.$$

For a photochemical steady state to be maintained, it is necessary that chemical processes to be much more rapid than vertical transport, i.e.

$$\tau^{chemistry} \ll \tau^{transport}.$$

In the mesosphere it appears that this condition is not satisfied for NO (see Section 4.3.3). We are concerned primarily with chemical rather than physical processes, and it is not proposed to pursue the subject of physical transport any further. The main point we wish to convey here is that physical transport processes must not be overlooked in the case of species having long photochemical lifetimes.

4.3 Production and loss processes of minor constituents

Studies of the positive and negative ion chemistry of the atmosphere have demonstrated the importance of many of the minor constituents O_3, O, $O_2(^1\Delta_g)$, H_2O, H_2O_2, HO_2, OH, H_2, H, NO, NO_2 and CO_2 in controlling the concentrations of charged species in the lower ionosphere, and studies of the airglow (Chapter 5) indicate that many of these species are involved in important light emitting processes. We have already noted that they dominate the neutral chemistry of this region. In this section we summarize the important production and loss processes of these minor constituents, and consider their distributions as a function of altitude up to and including the lower thermosphere.

4.3.1 SPECIES DERIVED FROM OXYGEN*

(i) *Ozone.* Ozone has long been recognized as an important minor constituent of the stratosphere. The region centred near 30 km has, in fact, been termed the ozonosphere because of the relatively high concentration of ozone (peak concentration at 30 km $\approx 2 \times 10^{-4}$ torr ≈ 15 ppm). In the troposphere ozone can be formed by photochemical processes involving nitrogen oxides in polluted air; we discuss this type of production in Chapter 7. In the upper atmosphere ozone is formed as a consequence of dissociation of molecular oxygen by solar ultraviolet radiation:

$$O_2 + h\nu(\lambda < 242.4 \text{ nm}) \rightarrow O(^3P) + O(^3P \text{ or } {}^1D) \tag{4.8}$$

$$O(^3P) + O_2 + M (M = N_2 \text{ or } O_2) \rightarrow O_3 + M \tag{4.9}$$

$$k_{4.9} = 1.1 \times 10^{-34} \exp(510/T) \text{ cm}^6 \text{ molecule}^{-2} \text{ s}^{-1}\dagger$$
(the value of $k_{4.9}$ given is for $M = N_2$)

Thus, the chemistry of ozone in the upper atmosphere cannot be separated from that of atomic oxygen.

The minimum energy necessary to dissociate O_2 is 5.11 eV (41 260 cm^{-1}), which corresponds to a quantum of wavelength 242.4 nm. Reference to the energy level diagram for O_2 (Fig. 2.7) indicates that this wavelength corresponds to the onset of the Herzberg continuum of molecular oxygen. As noted in Chapter 2, ozone is a strong absorber of ultraviolet radiation and a weak absorber in the visible region. The practical importance of the strong ultraviolet absorption of ozone is that it shields the earth's surface from potentially lethal solar radiation between 242 and 290 nm. Energetically, ozone is able to be photolyzed at all wavelengths below the infrared. Thus, in addition to eqns. 4.8 and 4.9, we must also consider the processes

$$O_3 + h\nu \rightarrow O_2 + O \tag{4.10}$$

where $J_{4.10} = 10^{-2} \text{ s}^{-1}$, integrated over radiation in the ultraviolet and visible regions of the solar spectrum,[10]

$$O + O_3 \rightarrow 2O_2 + 392 \text{ kJ} \tag{4.11}$$

---

† R. E. Huie, J. T. Herron and D. D. Davis, *J. Phys. Chem.*, **76**, 2653 (1972).
* More recent results for the rate coefficients of many of the reactions in this chapter are listed in the appendix.

where $k_{4.11} = 1.1 \times 10^{-11} \exp(-2150/T) \, cm^3 \, molecule^{-1} \, s^{-1}$,[11] and

$$O + O + M \rightarrow O_2 + M + 498 \, kJ \qquad (4.12)$$

where $k_{4.12} = 3.6 \times 10^{-31} \, T^{-1} \exp(-170/T) \, cm^6 \, molecule^{-2} \, s^{-1}$.[8]

With the scheme comprising reactions 4.8 to 4.12, which is applicable to an 'oxygen only' atmosphere, and assuming photochemical equilibrium, the time derivatives of the concentrations of ozone and of atomic and molecular oxygen are

$$\frac{d[O_3]}{dt} = k_{4.9}[O][O_2][M] - J_{4.10}[O_3] - k_{4.11}[O][O_3] \qquad (4.13)$$

$$\frac{d[O]}{dt} = 2J_{4.8}[O_2] + J_{4.10}[O_3] - 2k_{4.12}[O][O][M]$$

$$- k_{4.9}[O][O_2][M] - k_{4.11}[O][O_3] \qquad (4.14)$$

and $\quad \dfrac{d[O_2]}{dt} = J_{4.10}[O_3] + k_{4.12}[O][O][M] + 2k_{4.11}[O][O_3]$

$$- J_{4.8}[O_2] - k_{4.9}[O][O_2][M] \qquad (4.15)$$

The conditions for simultaneous variation of $[O]$ and $[O_3]$ can be written

$$\frac{d[O]}{dt} + \frac{d[O_3]}{dt} + 2k_{4.11}[O][O_3] + 2k_{4.12}[O]^2[M] = 2J_{4.8}[O_2] \quad (4.16)$$

For certain atmospheric regions, simplifying approximations can be made in eqn. 4.16. At high altitudes (above 90 km) $[O_3] \ll [O]$, and a steady state concentration may be assumed for O_3. Eqn. 4.16, therefore, reduces to

$$\frac{d[O]}{dt} + 2k_{4.12}[O]^2[M] = 2J_{4.8}[O_2] \qquad (4.17)$$

During the daytime in the stratosphere $[O] \ll [O_3]$, and a steady state concentration may be assumed for $[O]$. This gives

$$\frac{d[O_3]}{dt} + 2k_{4.11}[O][O_3] = 2J_{4.8}[O_2] \qquad (4.18)$$

If we assume in the stratosphere that $d[O]/dt = 0$, and that processes 4.11 and 4.12 which remove O atoms are slow compared with 4.9, then eqn. 4.14 becomes

$$k_{4.9}[O][O_2][M] = 2J_{4.8}[O_2] + J_{4.10}[O_3] \qquad (4.19)$$

At 30 km the photolysis rate coefficient $J_{4.10}$ is many orders of magnitude greater than $J_{4.8}$. Eqn. 4.19, therefore, reduces to

$$[O] = \frac{J_{4.10}[O_3]}{k_{4.9}[M][O_2]}$$

which upon substitution into eqn. 4.18 gives the conventional stratospheric equation for ozone,

$$\frac{d[O_3]}{dt} + \frac{2k_{4.11}J_{4.10}[O_3]^2}{k_{4.9}[O_2][M]} = 2J_{4.8}[O_2] \qquad (4.20)$$

Models based on the reaction scheme 4.8 to 4.12 for an 'oxygen only' atmosphere have not been particularly successful in reproducing the observed ozone profiles. (For an example, see Fig. 4.3.)

Until about 1964 it was generally believed that the reaction sequence 4.8–4.12 gave an adequate description of the photochemistry of ozone in the stratosphere. It is now apparent the theory outlined must be modified by including reactions of hydrogen species and nitrogen oxides. Since 1964 various attempts have been made to construct a more realistic model for atmospheric ozone by including the effect of small concentrations of the hydrogen-containing species (H, H_2O and CH_4) which are expected to have a significant influence on the ozone concentration.[12] Two recent models have demonstrated that this inclusion of hydrogen diminishes the ozone concentration appreciably. The ozone concentration is also likely to

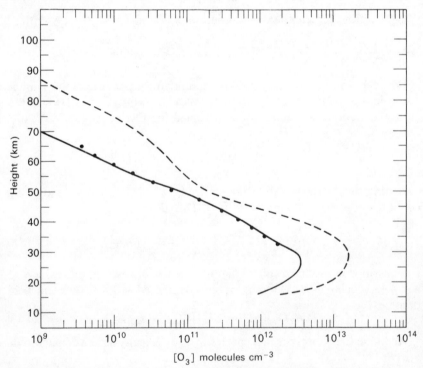

Fig. 4.3 A comparison of a photochemical profile of O_3 for an oxygen only atmosphere[12] (dashed line) with the experimental results of Johnson et al.[15] (solid curve) and Hilsenrath (dotted curve).[13]

be strongly affected by the presence of nitrogen oxides,[14] a subject which is considered further in Chapter 7 in connection with the possible effects of pollution by supersonic transport aircraft. Nicolet included with reactions 4.8–4.12 the reaction scheme shown in Table 4.3, and concluded that

Table 4.3   Processes (additional to reactions 4.8 to 4.12) considered in reference 11 for a hydrogen–oxygen atmosphere. The units of k are cm^3 molecule^{-1} s^{-1} for a bimolecular process and cm^6 molecule^{-2} s^{-1} for a three-body process.

$H + O_2 + M \rightarrow HO_2 + M + 192.5 \, kJ$	(4.21)
$\quad k_{4.21} = 6.7 \times 10^{-33} \exp(238/T)$	
$\quad$ (M = Ar, N_2 is 3.4 times as efficient)[a]	
$H + O_3 \rightarrow O_2 + OH^*_{v \leqslant 9} + 322 \, kJ$	(4.22)
$\quad k_{4.22} = 1.5 \times 10^{-12} \, T^{1/2}$	
$OH + O \rightarrow H + O_2 + 69.5 \, kJ$	(4.23)
$\quad k_{4.23} = 3 \times 10^{-12} \, T^{1/2}$	
$OH + O_3 \rightarrow HO_2 + O_2 + 160 \, kJ$	(4.24)
$\quad k_{4.24} = 1.3 \times 10^{-12} \exp(-956/T)^{21b,b}$	
$OH + H_2O_2 \rightarrow H_2O + HO_2 + 125 \, kJ$	(4.25)
$\quad k_{4.25} = 4 \times 10^{-13} \, T^{1/2} \exp(-600/T)$	
$OH + H_2 \rightarrow H_2O + H + 63 \, kJ$	(4.26)
$\quad k_{4.26} = 7 \times 10^{-12} \exp(-2000/T)$	
$OH + OH \rightarrow H_2O + O + 71.1 \, kJ^c$	(4.27)
$\quad k_{4.27} = 7.5 \times 10^{-13} \, T^{1/2} \exp(-500/T)$	
$OH + HO_2 \rightarrow H_2O + O_2 + 301 \, kJ$	(4.28)
$\quad k_{4.28} = 2 \times 10^{-10 \; 21a}$	
$HO_2 + HO_2 \rightarrow H_2O_2 + O_2 + 178 \, kJ$	(4.29)
$\quad k_{4.29} \approx 1.7 \times 10^{-11} \exp(-500/T)$	
$H_2O_2 + hv \rightarrow OH + OH$	(4.30)
$O + HO_2 \rightarrow O_2 + OH^*_{v \leqslant 6} + 230 \, kJ$	(4.31)
$\quad k_{4.31} > 10^{-11}$	

a   M. J. Kurylo, *J. Phys. Chem.*, **76**, 3518 (1972).
b   The rate coefficient for the reaction $HO_2 + O_3 \rightarrow OH + 2O_2$ is also in doubt. Anderson and Kaufman put an upper limit on the rate coefficient for this reaction of $k < 5 \times 10^{-15} \, cm^3$ molecule^{-1} s^{-1}.[21b] If the rate coefficient is close to this upper limit, then the reaction between HO_2 and O_3 may be important in the stratosphere.
c   At sufficiently high pressures (e.g. the lower stratosphere), reaction 4.27 proceeds as the termolecular reaction

$$OH + OH + M \rightarrow H_2O_2 + M \qquad (4.27a)$$
$$k_{4.27a} = 4 \times 10^{-30}$$

Table 4.4 Processes considered by Crutzen[19] as influencing the ozone concentration below 60 km, additional to reactions 4.8–4.12. (Rate coefficients are expressed in cm^3 molecule^{-1} s^{-1}).

$HNO_3 + O \rightarrow OH + NO_3 + 4\,kJ$	(4.32)
$\quad k_{4.32} < 2 \times 10^{-14}$	
$HNO_2 + O \rightarrow OH + NO_2 + 98\,kJ$	(4.33)
$\quad k_{4.33} \approx 1.7 \times 10^{-11}$	
$OH + O \rightarrow H + O_2 + 69.5\,kJ$	(4.23)
$\quad k_{4.23} = 5 \times 10^{-11}$	
$O + HO_2 \rightarrow OH + O_2$	(4.31)
$\quad k_{4.31} \sim 2 \times 10^{-11}$	
$H + O_3 \rightarrow OH + O_2$	(4.22)
$\quad k_{4.22} = 2.6 \times 10^{-11}$	
$HO_2 + h\nu(\lambda < 454\,nm) \rightarrow OH + O$	(4.34)
$NO_2 + O \rightarrow NO + O_2 + 192.5\,kJ$	(4.35)
$\quad k_{4.35} = 9.2 \times 10^{-12}$ between 235–350 K[11]	
$NO + O_3 \rightarrow NO_2 + O_2 + 200\,kJ$	(4.36)
$\quad k_{4.36} = 1.7 \times 10^{-12} \exp(-1310/T)$	
$NO_2 + h\nu(\lambda < 397.5\,nm) \rightarrow NO + O$	(4.37)

while there was no important difference between a pure oxygen atmosphere and a hydrogen–oxygen atmosphere in the region of the stratosphere, the difference in the mesosphere would be as large as a factor of 100 in the calculated ozone concentration.[11,16] Crutzen[19] also modified the reaction scheme 4.8–4.12 to include the processes 4.32–4.37 shown in Table 4.4 and considered to be important below 60 km.

If we include this reaction sequence (Table 4.4), eqn. 4.16 becomes

$$
\left[\frac{d[O]}{dt} + \frac{d[O_3]}{dt} \right] = 2J_{4.8}[O_2] - 2k_{4.11}[O][O_3] - k_{4.32}[O][HNO_3]
$$

$$
- k_{4.33}[O][HNO_2] - k_{4.23}[OH][O]
$$

$$
- k_{4.31}[O][HO_2] - k_{4.22}[H][O_3] + J_{4.34}[HO_2]
$$

$$
- k_{4.35}[NO_2][O] - k_{4.36}[NO][O_3] + J_{4.37}[NO_2]
$$

$$
(4.38)
$$

To further complicate the ozone distribution, mixing by wind occurs; however, between 30 and 80 km it is not expected that winds will alter the mean ozone concentration from its photoequilibrium value by more than about 20%.[19] In the lower stratosphere, there is a complete departure from photochemical equilibrium as the chemical equilibrium times are greater than one year, and therefore physical transport is the controlling factor. Shimazaki and Laird included the effect of eddy and molecular diffusion in calculating diurnal variations of minor neutral constituents including ozone.[20] The complexity of such an attempt at a complete treatment of atmospheric minority species is illustrated by Table 4.5 which lists the reactions considered important by Shimazaki and Laird and the rate coefficients they chose. At night the ozone concentration increases because of the absence of the photodissociative process 4.10.

A comparison of some experimental and calculated day and night ozone profiles is given in Fig. 4.4. These results indicate that quite large discrepancies exist between the (few) measured concentrations and the calculated profiles. Unfortunately, this situation is typical for all minor constituents of the chemosphere.

(ii) *Atomic oxygen.* Although oxygen atoms play a very important role in the chemosphere, their concentration profile has not been well established. Few experimental measurements have been made in the mesosphere, and the various models proposed lead to quite widely differing conclusions. A further complication is the need to consider both ground state (^{3}P) atoms and excited atoms in the ^{1}D and ^{1}S states*.

$\underline{O(^3P)}$.   The important processes for the production of atomic oxygen in the lower chemosphere are reactions 4.8a and 4.8b:

$$O_2 + hv(\lambda < 175 \text{ nm}) \rightarrow O(^1D) + O(^3P) \tag{4.8a}$$

Schumann–Runge Continuum

$$O_2 + hv(\lambda < 242 \text{ nm}) \rightarrow O(^3P) + O(^3P) \tag{4.8b}$$

Herzberg Continuum

The oxygen atoms once formed may recombine by the sequence, 4.9, 4.10, 4.11 and 4.12, as discussed earlier in connection with ozone. The absorption cross sections for processes 4.8a and 4.8b are very different (cf. Figs. 2.8, 2.9 and 2.10), and, as a consequence, there is a great difference in the depth of penetration of radiation into the atmosphere in these two wavelength ranges (Fig. 2.3). At night there must be a large decrease in the concentration of atomic oxygen because of the cessation of primary production and the continuation of recombination processes. Height profiles of $O(^3P)$ for both day and night, calculated on the basis of two different models, are shown in Fig. 4.5. Reliable atomic oxygen profiles have proven extremely difficult to obtain from rocket-borne mass spectrometers. If the data of Offerman and von Zahn[23] are accepted, the ratio of [O] to [O_2] at 120 km is about 3.5. Current model atmospheres assume this ratio to be 1. There is thus

---

* Atomic oxygen is in photochemical equilibrium up to the mesopause. In the thermosphere its distribution is controlled by eddy diffusion (M. Nicolet, *Can. J. Chem.*, **52**, 1381 (1974)).

Table 4.5 The reactions and rate coefficients used in reference 3 in constructing a model for calculating diurnal and seasonal variations of minor neutral constituents in the mesosphere and lower thermosphere. (Rate coefficients are expressed in cm^3 particle^{-1} s^{-1} for bimolecular processes and cm^6 particle^{-2} s^{-1} for three-body processes.)

Photodissociation Processes

$O_2 + hv$ (135 $< \lambda <$ 175 nm S–R cont) $\rightarrow O(^1D) + O(^3P)$
$O_2 + hv$ (175 $< \lambda <$ 200 nm S–R bands) $\rightarrow O(^3P) + O(^3P)$
$O_2 + hv$ (175 $< \lambda <$ 250 nm, Herzberg) $\rightarrow O(^3P) + O(^3P)$
$O_3 + hv$ (200 $< \lambda <$ 266 nm) $\rightarrow O_2(^1\Sigma_g) + O(^1D)$
$O_3 + hv$ (266 $< \lambda <$ 320 nm) $\rightarrow O_2(^1\Delta_g) + O(^1D)$
$H_2O + hv$ (135 $< \lambda <$ 190 nm plus Lyα) $\rightarrow OH + H$
$H_2O_2 + hv$ (190 $< \lambda <$ 285 nm) $\rightarrow OH + OH$
$NO + hv(Ly\alpha) \rightarrow NO^+ + e \rightarrow N + O$
$NO_2 + hv \rightarrow NO + O$
$N_2O + hv$ (135 $< \lambda <$ 210 nm) $\rightarrow N + NO$
$N_2 + hv \rightarrow N + N$

Chemical Reactions	*Rate Coefficients*
$O(^3P) + O(^3P) + M \rightarrow O_2 + M$	$3 \times 10^{-33} (T/300)^{-2.9}$
$O(^3P) + O_2 + M \rightarrow O_3 + M$	$5.5 \times 10^{-34} (T/300)^{-2.6}$
$O(^3P) + O_3 \rightarrow O_2(^1\Delta_g) + O_2$	$1.2 \times 10^{-11} \exp(-2000/T)$
$O(^1D) + O_3 \rightarrow O_2 + O_2$	3×10^{-10}
$O(^1D) + N_2 \rightarrow O(^3P) + N_2$	5×10^{-11}
$H + O_3 \rightarrow O_2 + OH$	2.6×10^{-11}
$OH + O(^3P) \rightarrow H + O_2$	5×10^{-11}
$OH + O_3 \rightarrow HO_2 + O_2$	5×10^{-13}
$H + O_2 + M \rightarrow HO_2 + M$	3×10^{-32}
$HO_2 + O(^3P) \rightarrow OH + O_2$	10^{-11}
$HO_2 + O_3 \rightarrow OH + 2O_2$	10^{-14}
$OH + OH \rightarrow H_2O + O(^3P)$	2.0×10^{-12}
$OH + HO_2 \rightarrow H_2O + O_2$	10^{-11}
$H + HO_2 \rightarrow H_2 + O_2$	2×10^{-13}
$H + HO_2 \rightarrow 2OH$	10^{-11}
$O(^3P) + H_2 \rightarrow OH + H$	$7 \times 10^{-11} \exp(-5150/T)$
$HO_2 + HO_2 \rightarrow H_2O_2 + O_2$	1.5×10^{-12}
$OH + H_2O_2 \rightarrow H_2O + HO_2$	4×10^{-13}
$O(^3P) + H_2O_2 \rightarrow OH + HO_2$	10^{-15}
$H + H_2O_2 \rightarrow H_2 + HO_2$	10^{-13}
$O(^1D) + H_2 \rightarrow OH + H$	10^{-11}
$O(^1D) + H_2O \rightarrow 2OH$	10^{-11}
$NO + O + M \rightarrow NO_2 + M$	1.1×10^{-31}
$N + O + M \rightarrow NO + M$	9×10^{-33}
$O + NO_2 \rightarrow NO + O_2$	$3.2 \times 10^{-11} \exp(-300/T)$
$O_3 + NO \rightarrow NO_2 + O_2$	$9.5 \times 10^{-13} \exp(-1240/T)$
$N + NO \rightarrow N_2 + O$	2.2×10^{-11}
$N + O_2 \rightarrow NO + O$	$1.4 \times 10^{-11} \exp(-3580/T)$
$N_2 + O \rightarrow N_2O + hv$	10^{-24}
$O_2(^1\Delta_g) \rightarrow O_2 + hv$ (1.27 μ)	2.8×10^{-4}
$O_2(^1\Delta_g) + O_3 \rightarrow 2O_2 + O$	$5 \times 10^{-13} \exp(-1500/T)$
$O_2(^1\Delta_g) + N_2 \rightarrow O_2 + N_2$	1.1×10^{-19}
$O_2(^1\Delta_g) + O_2 \rightarrow O_2 + O_2$	2.4×10^{-18}
$O(^1D) + O_2 \rightarrow O_2(^1\Delta_g) + O$	10^{-12}
$O_2(^1\Delta_g) + O \rightarrow O_2 + O$	1.3×10^{-16}

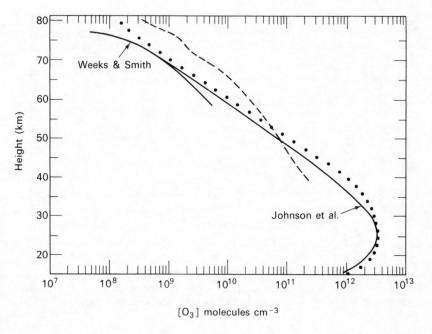

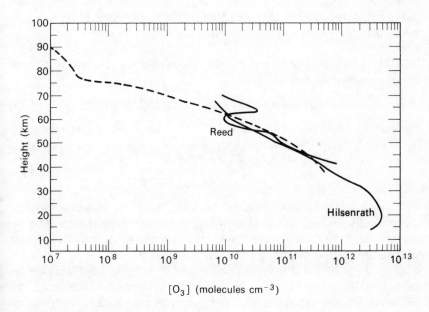

Fig. 4.4 (a) Daytime ozone profile. A comparison of the observed (Johnson *et al.*,[15] Weeks and Smith[17]) and calculated profiles (dotted line is from ref. 16 and dashed line is from ref. 3). (b) Night-time ozone profile. A comparison of the observed (Reed,[18] Hilsenrath[13]) and calculated[3] profiles.

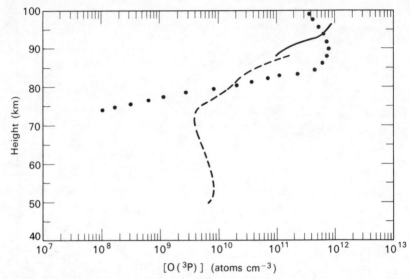

Fig. 4.5 O(^{3}P) altitude profile. The measured day profile (solid curve) is that of Henderson;[22] the calculated profiles are for day conditions[16] (dashed curve) and night conditions[3] (dotted curve).

considerable uncertainty as to atomic oxygen densities in the chemosphere and lower thermosphere. It is possible that the main source of error is that the model atmospheres have used eddy diffusion coefficients which are too large.

O(^{1}D). Two reactions are mainly responsible for the production of O(^{1}D), namely 4.8a and

$$O_3(^1A_1) + h\nu(\lambda < 310 \text{ nm}) \rightarrow O_2(^1\Delta_g) + O(^1D) \qquad (4.39)$$

The principal loss process is rapid collisional deactivation (quenching) by N_2 or O_2,

$$O(^1D) + N_2 \text{ (or } O_2) \rightarrow O(^3P) + N_2 \text{ (or } O_2)$$

$$k_{4.40} > 10^{-11} \text{ cm}^3 \text{ molecule}^{-1} \text{ s}^{-1 \; 26} \qquad (4.40)$$

As a result of undergoing process 4.40, the quencher molecule may be electronically excited, in the case of O_2, or vibrationally excited. Nicolet has calculated the O(^{1}D) concentration as a function of altitude for different solar zenith angles, assuming photochemical equilibrium for O(^{1}D) and a value for $k_{4.40}$ of 5×10^{-11} cm^3 molecule^{-1} s^{-1} (Table 4.6). The increase above the mesopause (85 km) is the result of photodissociation in the Schumann–Runge system of O_2. The relatively long radiative lifetime of O(^{1}D) ($\tau = 110$ s) means that the probability of radiative decay is low compared with that of chemical deactivation.

O(^{1}S). The most likely reaction for producing O(^{1}S) in the chemosphere is the Chapman reaction 4.41 which results when the third body M in

Table 4.6 Concentrations of $O(^1D)$ in a sunlit atmosphere, calculated in reference 16 for various solar zenith angles.

Altitude (km)	0° (overhead sun) (cm^{-3})	60° (cm^{-3})	90° (sun at horizon) (cm^{-3})
20	2.5	0.6	
25	1.1×10^1	3.2	
30	4.4×10^1	1.6×10^1	2.7×10^{-9}
35	1.5×10^2	6.6×10^1	4.0×10^{-4}
40	4.4×10^2	2.1×10^2	5.2×10^{-1}
45	7.8×10^2	4.3×10^2	8.0
50	7.9×10^2	6.0×10^2	2.7×10^1
55	5.6×10^2	5.1×10^2	5.7×10^1
60	3.3×10^2	3.2×10^2	1.0×10^2
65	2.0×10^2	1.9×10^2	1.3×10^2
70	1.2×10^2	1.2×10^2	1.0×10^2
95	4×10^2		
100	1×10^3		
105	2×10^3		
110	4×10^3		
115	5×10^3		
120	4×10^3		

reaction 4.12 is another oxygen atom,

$$O + O + O \rightarrow O_2 + O(^1S)$$
$$k_{4.41} = 4.8 \times 10^{-33} \text{ cm}^6 \text{ atom}^{-2} \text{ s}^{-1} \text{ [24]} \tag{4.41}$$

Additional daytime sources of $O(^1S)$ are, at higher altitudes, photodissociation of oxygen[25] and, perhaps in the stratosphere, photodissociation of ozone. In contrast to $O(^1D)$, $O(^1S)$ is relatively resistant to collisional deactivation. For

$$O(^1S) + N_2 \text{ or } O_2 \rightarrow \text{Products} \tag{4.42}$$
$$k_{4.42}^{N_2} < 5 \times 10^{-17} \text{ cm}^3 \text{ molecule}^{-1} \text{ s}^{-1}$$

and $k_{4.42}^{O_2} = 4.9 \times 10^{-12} \exp(-860/T) \text{ cm}^3 \text{ molecule}^{-1} \text{ s}^{-1}$ [26]

The quenching rate by atomic oxygen may be as high as 8×10^{-12} cm^3 molecule^{-1} s^{-1} (see Table 5.6) and, therefore, quenching by atomic oxygen may compete with quenching by molecular oxygen in the mesosphere.

(iii) *Excited molecular oxygen.* The most important electronically excited state of molecular oxygen in the chemosphere is $a^1\Delta_g$. Direct estimates of the concentration of $O_2(^1\Delta_g)$ have been obtained from rocket observations of the dayglow emission of the (0, 0) band of the transition

$$O_2(a^1\Delta_g) \rightarrow O_2(X^3\Sigma_g^-) + h\nu \tag{4.43}$$

at 1.27 μm[28] (see Fig. 4.6).

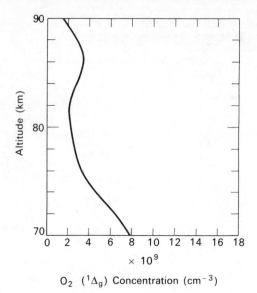

O_2 ($^1\Delta_g$) Concentration (cm^{-3})

Fig. 4.6   Daytime vertical distribution of $O_2(^1\Delta_g)$ at a solar zenith angle between 70° and 81°. The profile shown is the mean of four measured profiles of ref. 29.

$O_2(^1\Delta_g)$ is produced mainly by photolysis of ozone by sunlight:

$$O_3(^1A_1) + h\nu(\lambda < 310 \text{ nm}) \rightarrow O_2(^1\Delta_g) + O(^1D) \qquad (4.39)$$

Theoretical estimates of the night-time concentration of $O_2(^1\Delta_g)$ indicate that the concentration in the lower mesosphere should decrease during the course of the night,[30] which is in agreement with the few experimental observations (Fig. 5.15b). The main loss processes for $O_2(^1\Delta_g)$ are radiation (process 4.43, transition rate = $2.6 \times 10^{-4} \text{ s}^{-1}$) and deactivation by ground state O_2, reaction 4.44.[4]

$$O_2(^1\Delta_g) + O_2 \rightarrow O_2(^3\Sigma_g^-) + O_2 \qquad (4.44)$$

$$k_{4.44} = 2.4 \times 10^{-18} \text{ cm}^3 \text{ molecule}^{-1} \text{ s}^{-1}$$

4.3.2   SPECIES CONTAINING HYDROGEN

(i) H_2O, HO_2 *and* H_2O_2.   Continuous sources of water are available at ground level; consequently, water vapour is quite abundant at low altitudes (mole fraction $\approx 10^{-2}$). The proportion of water generally decreases in relation to the main atmospheric constituents as altitude increases, and in the stratosphere a figure of $3–10 \times 10^{-6}$ is commonly quoted as a reasonable value for the H_2O mixing ratio or mole fraction. Evidence for water vapour in the mesosphere has come from the detection by rocket-borne mass spectrometer probes of cluster ions such as $H_5O_2^+$, and from the visual

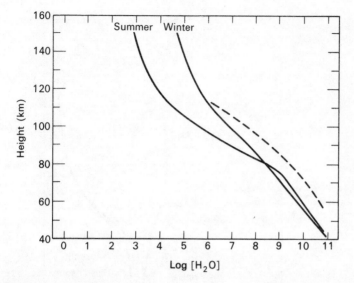

Fig. 4.7   Height profile for H_2O based on calculations in ref. 3 (solid line) and 31 (dashed line). Both models include the effect of eddy diffusion.

observation of noctilucent clouds. Noctilucent clouds are relatively rare high altitude clouds which may be visible at twilight in high latitudes during summer; the height of the clouds is typically about 80 km. The absence of direct measurements of water vapour concentration above the stratosphere has so far precluded any experimental tests of the various models that have been proposed. The results of calculations based on two of these models are shown in Fig. 4.7.

Water vapour is readily dissociated by sunlight (see Fig. 2.17) with a photodissociation coefficient $J_{4.45} = 6.5 \times 10^{-6}$ s^{-1} at 100 km,[16]

$$H_2O + h\nu(\lambda < 200 \text{ nm}) \rightarrow H + OH \tag{4.45}$$

At the mesopause level (85 km), the photodissociation coefficient decreases to 4.5×10^{-6} s^{-1}. Water vapour is dissociated by solar Lyman-α which penetrates down to the 70 km level and is responsible for the major part of the photodissociation of H_2O in the upper mesosphere. In the stratosphere and lower mesosphere the important process for dissociation of H_2O is reaction with $O(^1D)$ (produced by the photodissociation of ozone) according to

$$H_2O + O(^1D) \rightarrow OH + OH^\dagger(v \leqslant 2) + 120.5 \text{ kJ} \tag{4.46}$$

$$k_{4.46} = (3-5) \times 10^{-10} \text{ cm}^3 \text{ molecule}^{-1} \text{ s}^{-1} \text{ }[32]$$

The relative rates of removal of H_2O by both photodissociation and reaction with $O(^1D)$ are compared in Fig. 4.8 in the altitude range 50–100 km. The regeneration of H_2O is rapid in the stratosphere and lower mesosphere,

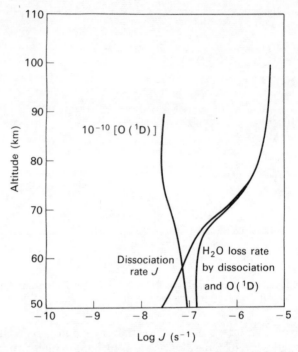

Fig. 4.8 The photodissociation rate of water vapour compared with its dissociation rate by $O(^1D)$ (process 4.46) calculated[33] for an overhead sun. (The rate coefficient chosen for reaction 4.46 was 1×10^{-10} cm^3 molecule^{-1} s^{-1}.)

so that its vertical distribution is not markedly affected by the dissociation process. In the lower thermosphere, water vapour which has travelled upwards by eddy diffusion is dissociated, and the H atoms so produced are carried up at a sufficient rate to replace H atoms being lost in the exosphere. H_2O is also formed in the stratosphere from the oxidation and dissociation products of the minority species H_2 and CH_4,* in addition to its regeneration from its own dissociation products. Important reactions leading eventually to the formation of water are

$$O(^1D) + CH_4 \rightarrow CH_3 + OH^\dagger(v \leqslant 4) + 182.0 \, kJ \qquad (4.47)$$

where $k_{4.47} \approx 3 \times 10^{-10}$ cm^3 molecule^{-1} s^{-1},

$$O(^1D) + H_2 \rightarrow H + OH^\dagger(v \leqslant 4) + 182.0 \, kJ \qquad (4.48)$$

where $k_{4.48} \approx 1 \times 10^{-10}$ cm^3 molecule^{-1} s^{-1} (Table 5.5), and the reaction between CH_4 and OH (reaction 4.55). Reaction 4.47 may account for the low mole fraction of CH_4 at the stratopause, less than 5×10^{-8}, compared

---

* The CH_4 is believed to have a biological origin, since the $C^{12}:C^{14}$ ratio in atmospheric methane is the same as in biological materials.

Table 4.7 Important secondary reactions of H_2/O_2 compounds controlling the concentration of water vapour in the chemosphere.

Reaction	Rate Coefficients*
$OH + OH \rightarrow H_2O + O + 69.9$ kJ	$k = 7.5 \times 10^{-13}\ T^{1/2} \exp(-500/T)^{11}$
$H + O_2 + M \rightarrow HO_2 + M + 192$ kJ	$6.7 \times 10^{-33} \exp(238/T)$ (Table 4.3)
$HO_2 + HO_2 \rightarrow H_2O_2 + O_2 + 175.7$ kJ	$1.7 \times 10^{-11} \exp(-500/T)^{11}$
$H + O_3 \rightarrow O_2 + OH^\dagger\ (v \leqslant 9) + 322.6$ kJ	2.6×10^{-11}
$H + O_3 \rightarrow O + HO_2 + 92$ kJ	unknown
$OH + O_3 \rightarrow HO_2 + O_2 + 160.2$ kJ	$1.3 \times 10^{-12} \exp(-956/T)^{21b}$
$OH + O \rightarrow H + O_2 + 69.5$ kJ	$(5 \pm 2) \times 10^{-11}$
$OH + HO_2 \rightarrow H_2O + O_2 + 301.7$ kJ	$2 \times 10^{-10\ 21a}$
$O + HO_2 \rightarrow O_2 + OH^\dagger\ (v \leqslant 6) + 230.1$ kJ	$> 10^{-11}$

* Rate coefficients are taken from reference 34 (unless referenced otherwise) and are expressed in cm^3 molecule^{-1} s^{-1} for bimolecular processes and cm^6 molecule^{-2} s^{-1} for termolecular processes at 300 K.
† A dagger is used to denote a vibrationally excited species.

with $\sim 1.5 \times 10^{-6}$ in the stratosphere.[4] Clearly methane must be removed by chemical reactions in the stratosphere, and oxidation by $O(^1D)$ and OH (reaction 4.55) are the most likely processes.

The primary photodissociation and oxidation reactions, 4.45–4.48, are followed by the sequence of reactions listed in Table 4.7. The presence of nitrogen and carbon oxides adds an additional complicating factor by modifying the rate of formation of water vapour through reactions such as:

$$H + NO_2 \rightarrow OH + NO + 122\ kJ \qquad (4.49)$$

where $k_{4.49} = 4.8 \times 10^{-11}$ cm^3 molecule^{-1} s^{-1},[34]

$$HO_2 + NO \rightarrow NO_2 + OH + 38\ kJ \qquad (4.50)$$

where $k_{4.50} \approx 5 \times 10^{-13}$ cm^3 molecule^{-1} s^{-1},[35]

$$OH + NO_2 + M \rightarrow HNO_3 + M \qquad (4.51)$$

where $k_{4.51} = 1.6 \times 10^{-30}$ cm^6 molecule^{-2} s^{-1},[36]

$$CO + OH \rightarrow H + CO_2 \qquad (4.52)$$

$$k_{4.52} = 2.1 \times 10^{-13} \exp(-115/T)^{37}$$

and possibly others. Two further reactions of NO and NO_2 with H_2O_2 are now known to have very small rate coefficients (less than 5×10^{-20} cm^3 molecule^{-1} s^{-1}) and are, therefore, unimportant in the stratosphere,[11]

$$NO + H_2O_2 \rightarrow HNO_2 + OH + 46\ kJ \qquad (4.53)$$

$$NO_2 + H_2O_2 \rightarrow HNO_3 + OH + 17\ kJ \qquad (4.54)$$

Nicolet[11] has proposed a reaction scheme which includes the main reactions likely to influence the H_2O concentration in the stratosphere; this scheme is shown in Fig. 4.9a. Any consideration of the photochemistry of atmospheric

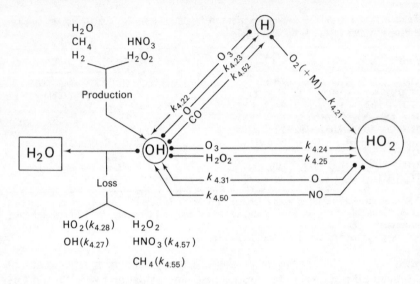

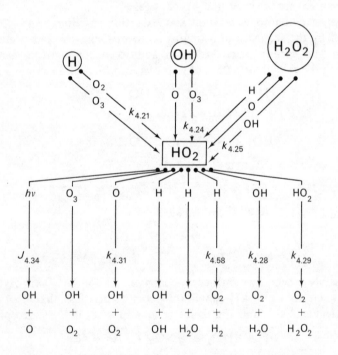

Fig. 4.9 (a) Basic reaction scheme of H_2O in the stratosphere, modified from Nicolet.[11] The arrow labelled production represents the production of OH from the species shown, by photodissociation and oxidation. (b) Reaction scheme of HO_2 in a hydrogen–oxygen atmosphere modified from Nicolet.[11]

water vapour in the lower stratosphere must therefore include the species H, H_2O, OH, HO_2, and H_2O_2 as well as the nitrogen and carbon oxides and methane.

Of the hydrogen-containing species only the hydroxyl radical has had its altitude distribution measured above the stratosphere. Indeed, of the odd hydrogen species H_2O_2, OH and HO_2, only OH has been positively identified above the stratosphere, although HO_2 and H_2O_2 clearly have important roles in atmospheric chemistry. It is possible, for example, that the reaction between OH and HO_2 (4.28) is the principal process leading to the reformation of water vapour in the mesosphere.[11] Also, in the stratosphere the rate of the reaction

$$H + O_2 + M \rightarrow HO_2 + M \qquad (4.21)$$

is sufficient to convert all hydrogen atoms into hydroperoxyl radicals. In the lower stratosphere the reaction scheme for HO_2 radicals shown in Figure 4.9b for a hydrogen–oxygen atmosphere must be modified to include reactions with oxides of carbon and nitrogen. Estimates of the concentration of some hydrogen oxides at different altitudes are given in Table 4.8. In preparing this table, Nicolet assumed photoequilibrium below 70 km and fixed the concentration of atomic oxygen above 85 km according to arbitrarily chosen eddy diffusion conditions.[16] The concentration of atomic hydrogen in the thermosphere was made to follow a mixing distribution calibrated to give a value of 3.5×10^7 atoms cm^{-3} for the hydrogen atom concentration at 100 km. For comparison with Table 4.8, the altitude profiles of OH, H, H_2O_2, HO_2, and H_2 obtained from a model based on the reaction sequence shown in Table 4.5, and incorporating eddy diffusion, are given in Fig. 4.10 for noon and midnight conditions. A feature of the model shown in Fig. 4.10 is the midnight concentrations of HO_2 and H_2O_2, which are about an order of magnitude larger than their daytime values in the 60–80 km region. This increase is most likely a consequence of the reduction in $O(^3P)$ densities

Table 4.8   Comparison of daytime concentrations of O_3, OH, and HO_2, for concentrations of O and H fixed by mixing conditions in the lower thermosphere, after Nicolet.[16]

Altitude km	[O] atoms cm^{-3}	[O_3] molecules cm^{-3}	[H] atoms cm^{-3}	[OH] molecules cm^{-3}	[HO_2] molecules cm^{-3}
100	3.2×10^{11}	1.6×10^6	3.5×10^7	1.7×10^2	6.5
95	3.3×10^{11}	1.3×10^7	8.7×10^7	3.2×10^3	1.4×10^2
90	3.0×10^{11}	1.1×10^8	2.3×10^8	7.0×10^4	4.5×10^3
85	3.0×10^{10}	1.0×10^8	6.9×10^8	2.0×10^6	5.3×10^5
80	1.4×10^{10}	1.4×10^8	8.6×10^8	2.2×10^7	1.7×10^7
75	3.8×10^9	3.2×10^8	5.1×10^7	1.8×10^7	1.3×10^7
70	4.1×10^9	1.0×10^9	1.0×10^7	9.5×10^6	7.0×10^6
65	5.1×10^9	3.2×10^9	6.0×10^6	1.3×10^7	8.7×10^6
60	6.6×10^9	1.0×10^{10}	4.4×10^6	1.9×10^7	1.2×10^7
55	8.5×10^9	3.2×10^{10}	3.3×10^6	2.8×10^7	1.6×10^7
50	6.5×10^9	1.0×10^{11}	1.2×10^6	4.4×10^7	2.8×10^7

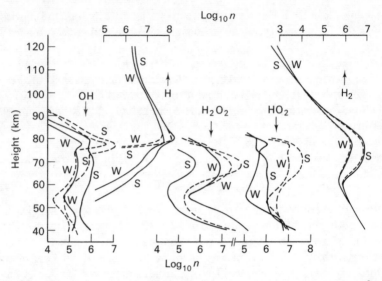

Fig. 4.10   Height profiles of OH, H, H_2O_2 and H_2 calculated by Shimazaki and Laird[3] for noon (solid curves) and midnight (dashed curve) conditions. The summer curve is indicated by S and the winter curve by W.

at night, and the resulting decrease in the rate of reaction of $O(^3P)$ with both HO_2 and H_2O_2.

(ii) **OH.** Anderson[38] used a rocket-borne ultraviolet spectrometer to measure the intensity of OH resonance fluorescence in the dayglow, in order to estimate the hydroxyl radical concentration as a function of altitude. His results for the altitude range between 45 and 70 km are presented in Fig. 4.11. Local OH densities were 4.4×10^6 cm^{-3} at 50 km, 5.5×10^6 cm^{-3}

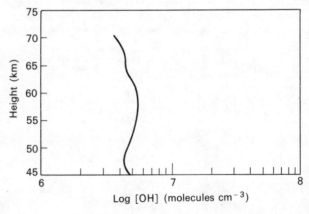

Fig. 4.11   Measured OH concentration in the mesosphere.[38]

at 60 km and 3.5×10^6 cm^{-3} at 70 km. Most of the important production and loss mechanisms of OH have been noted in the preceding section. but as hydroxyl radicals may react with carbon and nitrogen compounds, we should add the following processes,

$$OH + CH_4 \rightarrow H_2O + CH_3 + 63\,kJ \tag{4.55}$$

$$k_{4.55} = 5 \times 10^{-12} \exp(-1900/T) \text{ cm}^3 \text{ molecule}^{-1} \text{ s}^{-1}{}^{11}$$

$$HNO_3 + h\nu \rightarrow OH + NO_2 \tag{4.56}$$

and

$$HNO_3 + OH \rightarrow H_2O + NO_3 + 63\,kJ \tag{4.57}$$

$$k_{4.57} \approx 1.5 \times 10^{-13} \text{ cm}^3 \text{ molecule}^{-1} \text{ s}^{-1}{}^{11}$$

A reaction scheme of the hydroxyl radical in the lower stratosphere, which includes the more important of the possible reactions of OH, is given in Fig. 4.12. After sunset, OH is expected to convert rapidly into H_2O.

iii) *Molecular hydrogen.* Molecular hydrogen, like H_2O, has a continuous source at ground level. It is also produced in the stratosphere and mesosphere, where probably the most important process is 4.58:

$$H + HO_2 \rightarrow H_2 + O_2 + 239.7\,kJ \tag{4.58}$$

where $k_{4.58} > 3 \times 10^{-12}$ cm^3 molecule^{-1} s^{-1}.[34] A major loss process for H_2 is reaction 4.48:

$$H_2 + O(^1D) \rightarrow H + OH + 182\,kJ \tag{4.48}$$

$$k_{4.48} \approx 1 \times 10^{-10} \text{ cm}^3 \text{ molecule}^{-1} \text{ s}^{-1}$$

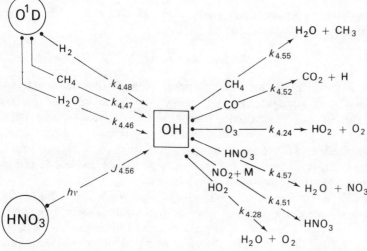

Fig. 4.12 Reaction scheme of OH in the lower stratosphere (modified from Nicolet[11]).

A comparison of the production and loss rates shows that below the meso-pause more molecular hydrogen is produced than is destroyed. The excess molecular hydrogen from this region flows downwards into the stratosphere (where it is converted into H and OH radicals by reaction 4.48) and upwards into the thermosphere.

(iv) *Atomic hydrogen.* We have noted that the three stable hydrogen-containing species, H_2O, CH_4 and H_2, have continuous production processes at ground level. The concentration of H_2O varies through the troposphere, the mixing ratio falling to $\sim 3 \times 10^{-6}$ in the stratosphere, but CH_4 and H_2 maintain their mole fractions, 1.5×10^{-6} and 0.5×10^{-6} respectively, up to the tropopause. The total mole fraction of hydrogen atoms present in any form above the tropopause can, therefore, be expected to be about 10^{-5}. In the stratosphere, the reaction of H_2O with $O(^1D)$ (reaction 4.46) does not constitute a significant source of H atoms, as regeneration of H_2O is rapid. Above the stratosphere, photodissociation of H_2O by light of wavelength less than 175 nm occurs, and to compensate for this loss, an upward flow of H_2O is required. At the mesopause most of the hydrogen is already in the form of either atomic or molecular hydrogen. In the lower thermosphere the remaining H_2O is rapidly dissociated by Lyman-α. In the middle thermosphere the temperature is high enough to make the endothermic reaction

$$H_2 + O \rightarrow OH + H - 7.9 \text{ kJ} \qquad (4.59)$$

$$k_{4.59} = 7 \times 10^{-11} \exp(-5130/T) \text{ cm}^3 \text{ molecule}^{-1} \text{ s}^{-1} \text{ [34]}$$

an important process for converting H_2 to H. The OH radicals produced are rapidly converted to H by the reaction

$$OH + O \rightarrow O_2 + H \qquad (4.23)$$

so that in the upper thermosphere atomic hydrogen is the dominant hydrogen species. The hydrogen atoms finally escape from the exosphere. Experimental estimates of atomic hydrogen densities from Lyman-α and Balmer-α scattering data[39] have not proven sufficiently reliable to provide a good test for models of a hydrogen–oxygen atmosphere. A calculated profile for atomic hydrogen based on the reaction sequence of Table 4.5, and incorporating eddy diffusion, is shown in Fig. 4.13.

Finally, in this section we summarize in Fig. 4.14 the important processes involving hydrogen-containing species in the atmosphere, from the production of H_2O, CH_4 and H_2 at ground level to the eventual loss of atomic hydrogen from the exosphere. It should be noted, however, that in practice this scheme will be modified by the presence of nitrogen oxides.

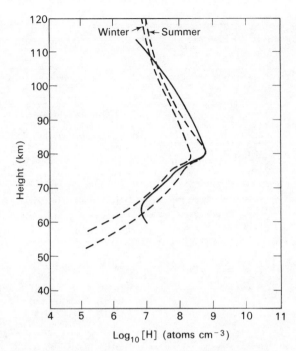

Fig. 4.13 Calculated profile of H after Bowman *et al.*[31] (solid line) and Shimazaki and Laird[3] (dashed line).

4.3.3 OXIDES OF NITROGEN

(i) *Nitric oxide.* Measurements of the nitric oxide concentration in the lower thermosphere have indicated higher NO concentrations than were predicted from photochemical theory. Whereas earlier photochemical calculations gave the concentration of NO at 80 km as being around 10^5 to 10^6 molecules cm^{-3},[40] values of 2×10^7 and 3×10^9 molecule cm^{-3} have been obtained experimentally.[5,41] The second value may be too large because of an underestimate of the background due to Rayleigh scattering in the vicinity of the (1, 0) γ–band of NO from whose intensity the concentration was deduced.[5]

Observed concentration profiles of nitric oxide in the mesosphere and lower themosphere are shown in Fig. 4.15. The principal reactions originally used to explain these profiles were

$$N + O_2 \rightarrow NO + O + 133.5 \text{ kJ} \qquad (4.60)$$

where $k_{4.60} = 1.4 \times 10^{-11} \exp(-3570/T) \text{ cm}^3 \text{ molecule}^{-1} \text{ s}^{-1}$, and

$$N + NO \rightarrow N_2 + O + 313.8 \text{ kJ} \qquad (4.61)$$

where $k_{4.61} = 2.2 \times 10^{-11} \text{ cm}^3 \text{ molecule}^{-1} \text{ s}^{-1}$,[8] which led to a steady

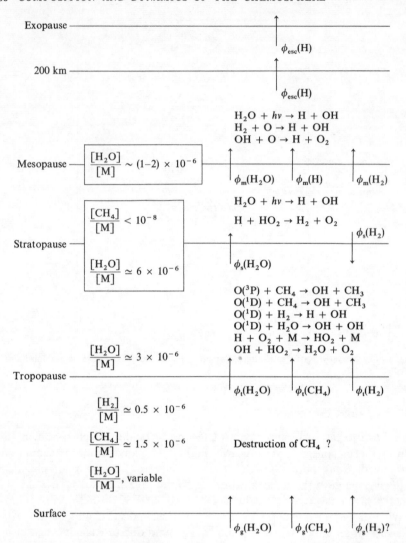

Fig. 4.14 Photochemistry and transport of hydrogen constituents in the atmosphere after Strobel.[4] Transport processes are shown by vertical arrows.

state nitric oxide concentration

$$[NO] = \frac{k_{4.60}}{k_{4.61}}[O_2] \approx 10^{-5}[O_2]$$

In view of the high measured NO concentrations in the mesosphere, these reactions (4.60 and 4.61) need to be supplemented by a more extensive scheme which provides an additional source of NO. The most likely process

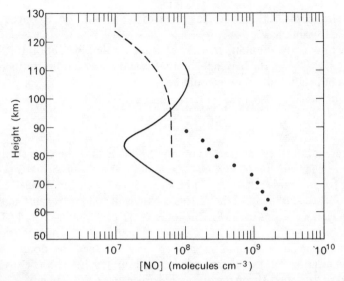

Fig. 4.15 Measured nitric oxide profiles of Meira (solid line),[5] Barth (dashed line)[42] and Pearce (dotted line).[41]

to provide the extra source of NO is the reaction of O_2 with the metastable species $N(^2D)$;

$$N(^2D) + O_2 \rightarrow NO + O \tag{4.62}$$

$$k_{4.62} = 1.4 \times 10^{-11} \text{ cm}^3 \text{ molecule}^{-1} \text{ s}^{-1} \text{ }^{43}$$

Reaction 4.62 offers an increase in the rate of a factor of 10^4 over the analogous reaction for ground state (^4S) nitrogen atoms. The major processes responsible for producing $N(^2D)$ are not clearly established and may be different at different altitudes. Likely processes are

$$NO^+ + e^- \rightarrow N(^2D) + O \tag{4.63a}$$

$$\rightarrow N(^4S) + O \tag{4.63b}$$

where $k_{4.63(a+b)} = 1 \times 10^{-7} (T/1000)^{-1.5} \text{ cm}^3 \text{ particle}^{-1} \text{ s}^{-1}$,

$$N_2^+ + O \rightarrow NO^+ + N(^4S) \tag{4.64a}$$

$$\rightarrow NO^+ + N(^2D) \tag{4.64b}$$

where $k_{4.64(a+b)} = 1.4 \times 10^{-10} \text{ cm}^3 \text{ particle}^{-1} \text{ s}^{-1}$,[9] and also the dissociation of N_2 by photoelectrons

$$N_2 + e^-(\text{fast}) \rightarrow N(^2D) + N + e^- \tag{4.65}$$

Another potentially important source of $N(^2D)$ is predissociation in the N_2 absorption bands between 80 and 100 nm:

$$N_2 + h\nu(\lambda 80\text{–}100 \text{ nm}) \rightarrow 2N(^4S \text{ or } ^2D) \tag{4.66}$$

The relative contributions of these processes to the total $N(^2D)$ concentration are presently unknown.

It was formerly thought, in view of the considerable concentrations of $O_2(^1\Delta_g)$ and O_2^+ in the mesosphere, that two other reactions for increasing the rate of production of NO would be

$$N(^4S) + O_2(^1\Delta_g) \rightarrow NO + O \qquad (4.67)$$

and
$$O_2^+ + N_2 \rightarrow NO + NO^+ \qquad (4.68)$$

However, laboratory measurements have shown that the rate coefficients are too low for either 4.67 or 4.68 to be an important source of NO ($k_{4.67} = 3 \times 10^{-15}$ cm^3 molecule^{-1} s^{-1} [9] and $k_{4.68} < 10^{-15}$ cm^3 molecule^{-1} s^{-1} [6]).

The overall statement of the nitrogen oxides or 'odd nitrogen'[*] budget in the atmosphere begins in the thermosphere with processes such as 4.63–4.65. The products of these reactions, $N(^4S$ or $^2D)$ and NO^+, undergo exchange reactions such as 4.62, and sooner or later produce the dominant odd nitrogen species of the thermosphere and mesosphere, nitric oxide. A downward flux of NO from the thermosphere into the mesosphere ($\sim 2 \times 10^8$ cm^{-2} s^{-1} [4]) then provides the principal source of odd-nitrogen compounds in the mesosphere. The downward flux from the thermosphere may be supplemented by a small production of N atoms due to predissociation of N_2 in Lyman–Birge–Hopfield bands having $v' > 6$.[9] Strobel[4] in his model estimates a downward flux of odd nitrogen from the lower mesosphere into the stratosphere, but G. Brasseur and M. Nicolet (*Planetary and Space Science*, **21**, 939 (1973) in a study of chemospheric processes of nitric oxide in the mesosphere and stratosphere conclude the opposite occurs. They show that in the mesosphere the loss of NO by predissociation exceeds the production rate so there is a small upward flux of NO in the lower mesosphere and upper stratosphere. In the stratosphere and the troposphere the odd nitrogen oxide flux is supplemented by the oxidation of N_2O by $O(^1D)$

$$N_2O + O(^1D) \rightarrow 2NO \qquad (4.69)$$

where $k_{4.69} = 9 \times 10^{-11}$ cm^3 molecule^{-1} s^{-1},[19] by dissociation of molecular nitrogen by cosmic radiation, and possibly by air pollution (see Chapter 7).[†] Like H_2O, H_2 and CH_4, N_2O is continuously produced at ground level. Finally, NO and NO_2 are converted into NO_3, N_2O_5, HNO_2 and HNO_3. The main loss of odd nitrogen species to the ground is probably the removal of HNO_2 and HNO_3 from the troposphere in rainfall. This reaction sequence, from the production of odd nitrogen in the thermosphere to its removal to the ground from the troposphere, is summarized in Fig. 4.16.

---

[*] The term 'odd nitrogen' is a convenient designation for nitrogen compounds derived from NO and NO_2, and including NO_3, N_2O_3, N_2O_4, N_2O_5, HNO_2 and HNO_3, but not N_2O.
[†] Another possible source of NO at low altitudes is the reaction of OH with NH_3. J. C. McConnell, *J. geophys. Res.*, **78**, 7812 (1973).

THERMOSPHERE

Exobase

$O^+(^4S) + N_2 \rightarrow NO^+ + N(^4S)$
$N_2^+ + O \rightarrow NO^+ + N(^4S \text{ or } ^2D)$
$N_2^+ + e \rightarrow 2N(^4S \text{ or } ^2D)$
$e(\text{fast}) + N_2 \rightarrow 2N(^4S \text{ or } ^2D)$
$N_2 + h\nu(\lambda 80\text{–}100 \text{ nm}) \rightarrow 2N(^4S \text{ or } ^2D)$

$NO^+ + e \rightarrow N(^4S \text{ or } ^2D) + O$
$N_2^+ + O_2 \rightarrow NO^+ + O$
$O_2^+ + N(^4S) \rightarrow NO^+ + O$
$N(^2D) + O_2 \rightarrow NO^+ + O$
$e + N(^2D) \rightarrow e + N(^4S)$
$O_2^+ + NO \rightarrow NO^+ + O_2$

$N(^4S) + NO \rightarrow N_2 + O$
$N(^2D) + NO \rightarrow N_2 + O$

Mesopause

$\downarrow \phi_1(NO) \sim 2.5 \times 10^8 \text{ cm}^{-2} \text{ sec}^{-1}$

$\downarrow \phi_1(NO) \sim 2.5 \times 10^8 \text{ cm}^{-2} \text{ sec}^{-1}$

Mesosphere and Upper Stratosphere

$N_2 + h\nu(\lambda 110\text{–}125 \text{ nm}) \rightarrow 2N(^4S)$

$NO + h\nu(\lambda 175\text{–}191 \text{ nm}) \rightarrow N(^4S) + O$
$NO + h\nu(\lambda 121.6 \text{ nm}) \rightarrow NO^+ + e$
$NO^+ + e \rightarrow N(^4S \text{ or } ^2D) + O$
$N(^4S) + O_3 \rightarrow NO + O_2$
$N(^4S) + OH \rightarrow NO + H$
$N(^4S) + O_2(^1\Delta) \rightarrow NO + O$
$N(^2D) + O_2 \rightarrow NO + O$
$O_3 + NO \rightarrow NO_2 + O_2$
$NO_2 + h\nu(\lambda < 397.5 \text{ nm}) \rightarrow NO + O$
$\quad\quad + h\nu(\lambda < 275 \text{ nm}) \rightarrow N(^4S) + O_2$
$O + NO_2 \rightarrow NO + O_2$
$N(^4S) + NO_2 \rightarrow 2NO$

$N(^4S) + NO \rightarrow N_2 + O$
$N(^4S) + NO_2 \rightarrow N_2 + O_2$
$\quad\quad\quad\quad \uparrow \rightarrow N_2 + 2O$
$\quad\quad\quad\quad\quad \rightarrow N_2O + O$

Middle Stratosphere

$\uparrow \phi_2(NO + NO_2)$

$\uparrow \phi_2(NO + NO_2)$

Lower Stratosphere and Troposphere

$O(^1D) + N_2O \rightarrow 2NO$
$N_2 + \text{cosmic rays} \rightarrow N + N$
$N_2O + h\nu(\lambda < 250 \text{ nm}) \rightarrow N(^4S) + NO$

$O_3 + NO \rightarrow NO_2 + O_2$
$NO, NO_2 \rightarrow HNO_2, HNO_3, NO_3, N_2O_5$

$N(^4S) + NO \rightarrow N_2 + O$
$N(^4S) + NO_2 \rightarrow \text{see above}$
removal of HNO_2 and HNO_3 by precipitation processes

Air pollution

Ground

$\uparrow \phi N_2O$

$\downarrow \phi(HNO_2 + HNO_3) = ??$

Fig. 4.16   Sources, sinks and exchange reactions of odd nitrogen in the atmosphere. Transport processes are shown by vertical arrows. (Modified from Strobel[4]).

Strobel has attempted a complete analysis of odd nitrogen in the mesosphere[9,7] and the reactions he includes in calculating the nitric oxide profile are given in Table 4.9. In addition to these reactions, it is also necessary to consider the rates of production of the ions O^+, O_2^+ and N_2^+ by photoionization and by photoelectron ionization, the effect of molecular and eddy diffusion, and an assessment of the rate of production of $N(^2D)$ by photoelectrons (4.65). The production rates of odd nitrogen compounds from different sources are shown in Fig. 4.17, and the production of $N(^2D)$ is shown for two different models in Fig. 4.18. An important feature of Strobel's

Table 4.9 **Chemical processes and rate coefficients used by Strobel[7,9] for an analysis of odd nitrogen in the mesosphere.**

Reaction		Rate Coefficient
$O^+(^4S) + N_2$	$\rightarrow NO^+ + N(^4S)$	$6 \times 10^{-13} (600/T)$
$O^+(^4S) + O_2$	$\rightarrow O_2^+ + O$	$1 \times 10^{-9} (T)^{-0.7}$
$O^+(^4S) + NO$	$\rightarrow NO^+ + O$	2×10^{-11}
$O^+(^2D) + N_2$	$\rightarrow N_2^+ + O$	1×10^{-9}
$O^+(^2D) + e$	$\rightarrow O^+(^4S) + e$	1×10^{-7}
$N_2^+ + O$	$\rightarrow NO^+ + N(^4S)$ ⎫ $\rightarrow NO^+ + N(^2D)$ ⎬	1.4×10^{-10}
$N_2^+ + e$	$\rightarrow N(^4S) + N(^4S)$	2.8×10^{-7}
$N_2^+ + O_2$	$\rightarrow O_2^+ + N_2$	4.7×10^{-11}
$N_2^+ + NO$	$\rightarrow NO^+ + N_2$	3.3×10^{-10}
$O_2^+ + N(^4S)$	$\rightarrow NO^+ + O$	1.8×10^{-10}
$O_2^+ + NO$	$\rightarrow NO^+ + O_2$	6.3×10^{-10}
$O_2^+ + e$	$\rightarrow O + O$	$6.6 \times 10^{-5} (T)^{-1}$
$NO^+ + e$	$\rightarrow N(^4S) + O$	$2.5 \times 10^{-8} (T/1000)^{-1.5}$
	$\rightarrow N(^2D) + O$	$7.5 \times 10^{-8} (T/1000)^{-1.5}$
$N^+ + O_2$	$\rightarrow NO^+ + O$	3×10^{-10}
	$\rightarrow O_2^+ + N$	3×10^{-10}
$N(^2D) + O_2$	$\rightarrow NO + O$	6×10^{-12}
$N(^2D) + O$	$\rightarrow N(^4S) + O$	2×10^{-13}
$N(^4S) + NO$	$\rightarrow N_2 + O$	2.2×10^{-11}
$N(^4S) + O_2$	$\rightarrow NO + O$	$2.4 \times 10^{-11} \exp(-3975/T)$
$N_2 + hv$	$\rightarrow N(^4S) + N(^4S)$	(Figure 2 of ref. 7)
$N(^4S) + O_2(^1\Delta)$	$\rightarrow NO + O$	3×10^{-15}
$N(^4S) + O_3$	$\rightarrow NO + O_2$	$3 \times 10^{-11} \exp(-1200/T)$
$N(^4S) + O + M$	$\rightarrow NO + M$	1.0×10^{-32}
$NO + hv (175 < \lambda < 190 \text{ nm})$	$\rightarrow N(^4S) + O$	(Figure 2 of ref. 7)
$NO + hv (\lambda = 121.6 \text{ nm})$	$\rightarrow NO^+ + e$	(Figure 2 of ref. 7)
$O_3 + NO$	$\rightarrow NO_2 + O_2$	$9.5 \times 10^{-13} \exp(-1240/T)$
$O + NO + M$	$\rightarrow NO_2 + M$	6.8×10^{-32}
$O + NO$	$\rightarrow NO_2 + hv$	6.4×10^{-17}
$NO_2 + hv (\lambda < 397.5 \text{ nm})$	$\rightarrow NO + O$	$8 \times 10^{-13} \text{ s}^{-1}$
$N(^4S) + NO_2$	$\rightarrow N_2O + O \quad 43\%$ ⎫ $\rightarrow N_2 + 2O \quad 13\%$ ⎪ $\rightarrow N_2 + O_2 \quad 10\%$ ⎬ $\rightarrow NO + NO \quad 33\%$ ⎭	1.8×10^{-11}
$O + NO_2$	$\rightarrow NO + O_2$	$3.2 \times 10^{-11} \exp(-300/T)$
$N(^4S) + OH$	$\rightarrow NO + H$	6.8×10^{-11}

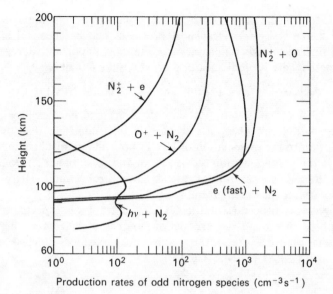

Fig. 4.17 Calculated production rates of odd nitrogen species at a solar zenith angle of 61.2° and 1400 LT (= 1400 hours local time).[9]

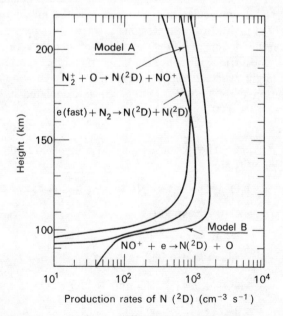

Fig. 4.18 Calculated production rates of $N(^2D)$ for two different models at 1400 LT and solar zenith angle of 61.2°.[7] The curves labelled model A assume the sources of $N(^2D)$ are photoelectron dissociation of N_2 and ion–atom interchange; the curve labelled model B assumes all NO^+ recombination results in $N(^2D)$.

model is that in the mesosphere the NO concentration is found to be controlled primarily by eddy transport processes i.e. the chemical lifetime is long compared to eddy diffusion. Also, the most important source of $N(^4S)$ in the mesosphere is predissociation of NO in the $\delta(0, 0)$ band

$$NO + h\nu \,(\lambda \sim 191 \text{ nm, and } 182.8 \text{ nm}) \rightarrow N(^4S) + O \qquad (4.70)$$

$J_{4.70} \sim 10^{-5} \text{ s}^{-1}$ at zero optical depth. Cieslik and Nicolet* in a study of the dissociation processes of nitric oxide also conclude that the major contribution to the dissociation of NO in both the mesosphere and stratosphere is the predissociation of the $\delta(1, 0)$ and $(0, 0)$ bands. A comparison between the NO concentrations calculated by Strobel and the experimental results of Meira[5] has been given in Fig. 4.2.

At night very little variation in NO concentration is expected *above* the stratosphere. Although the production sources of NO which depend on solar radiation are removed, the two main chemical loss processes for NO, reaction with $N(^4S)$ (eqn. 4.61) and with O_2^+;

$$NO + O_2^+ \rightarrow NO^+ + O_2 \qquad (4.71)$$

$$k_{4.71} = 8 \times 10^{-10} \text{ cm}^3 \text{ molecule}^{-1} \text{ s}^{-1} \text{ [7]}$$

are slowed down because after sunset the concentrations of both O_2^+ and $N(^4S)$ decrease. Also, as the concentration of NO is much greater than those of O_2^+ and $N(^4S)$, molecular diffusion becomes the most important loss process of NO at night. In the stratosphere, where the ozone concentration is appreciable, the lifetime of NO is short because of its reaction with ozone to form NO_2. The nitric oxide concentration is therefore expected to decrease markedly at night where ozone is present in sufficient abundance. The diurnal variation of NO, $N(^2D)$ and $N(^4S)$ as calculated by Strobel are shown in Figs. 4.19, 4.20 and 4.21 respectively.

(ii) *Nitrogen dioxide.*   The important reactions involved in the production and removal of NO_2 are:

$$NO + O + M \rightarrow NO_2 + M + 306 \text{ kJ} \qquad (4.72)$$

where $k_{4.72} = 2.9 \times 10^{-33} \exp(941/T)$,

$$NO_2 + O \rightarrow NO + O_2 + 192 \text{ kJ} \qquad (4.73)$$

where $k_{4.73} = 9 \times 10^{-12} \text{ cm}^3 \text{ molecule}^{-1} \text{ s}^{-1}$ and

$$NO_2 + h\nu(\lambda < 399 \text{ nm}) \rightarrow NO + O \qquad (4.74)$$

where $J_{4.74} = 7 \times 10^{-3} \text{ s}^{-1}$.[44]

---

* S. Cieslik and M. Nicolet, *Planetary and Space Science*, **21**, 925 (1973).

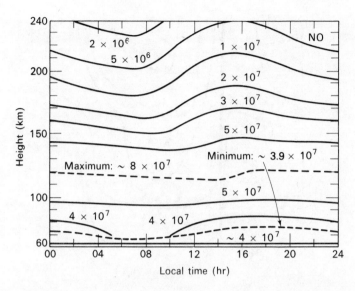

Fig. 4.19 Calculated diurnal variation of [NO].[4] The dashed lines indicate the height of maximum and minimum nitric oxide concentrations.

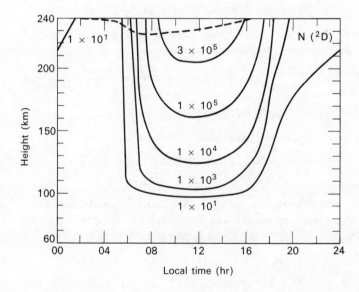

Fig. 4.20 Calculated diurnal variation of $N(^2D)$ for a model combining both the model A and model B $N(^2D)$ production rates of Fig. 4.18.[4] The dashed line indicates the height of the maximum value of $[N(^2D)]$.

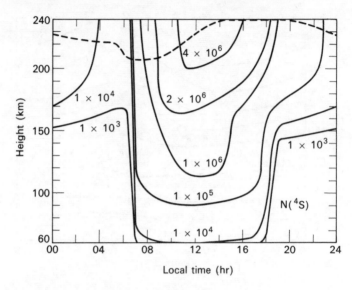

Fig. 4.21 Calculated diurnal variation of N(^{4}S) for the model of Fig. 4.20.[4] The dashed line indicates the height of maximum value of [N(^{4}S)].

When the ozone concentration begins to be important (below about 80 km) reaction 4.75 must also be considered:

$$NO + O_3 \rightarrow NO_2 + O_2 + 200\,kJ \tag{4.75}$$

where $k_{4.75} = 0.95 \times 10^{-12} \exp(-1240/T)$ cm^3 molecule^{-1} s^{-1}.[8]
For daytime photoequilibrium conditions,

$$\frac{d[NO_2]}{dt} = k_{4.72}[NO][O][M] + k_{4.75}[NO][O_3] - J_{4.74}[NO_2]$$
$$- k_{4.73}[NO_2][O] \tag{4.76}$$

Under steady state conditions, eqn. 4.76 reduces to

$$\frac{[NO_2]}{[NO]} = \frac{k_{4.72}[O][M] + k_{4.75}[O_3]}{J_{4.74} + k_{4.73}[O]} \tag{4.77}$$

which, upon substitution for [O], [M] and [O$_3$], indicates that *above* the stratopause nitrogen dioxide can be neglected in the analysis of nitric oxide reactions (Fig. 4.22).[40] The termolecular oxidation of NO,

$$NO + NO + O_2 \rightarrow 2NO_2 + 114.4\,kJ \tag{4.78}$$

where $k_{4.78} = 9 \times 10^{-39} \exp(413/T)$,[45] is also sufficiently slow to be neglected.

After twilight, when atomic oxygen is removed from the stratosphere by association with O$_2$ (reaction 4.9) the importance of reaction 4.75 increases.

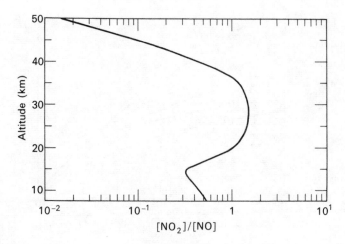

Fig. 4.22 Relative NO and NO_2 densities in the stratosphere at a solar zenith angle of 60°. (After G. Brasseur and M. Nicolet, *Planetary and Space Science*, **21**, 939 (1973).

Consequently it is believed that at night the NO_2 concentration increases markedly at the expense of NO.

(iii) *Other oxides of nitrogen.*   In the chemosphere in addition to NO and NO_2 already considered, smaller amounts of N_2O, NO_3, N_2O_5, etc. will be formed. Nitrous oxide, because it is very inert chemically, can be treated separately. The main process leading to production of N_2O in the atmosphere is thought to be microbiological: nitrous oxide and molecular nitrogen are released from nitrogenous matter when soil bacteria are actively growing under anaerobic conditions. It is unlikely that significant amounts of N_2O can be formed from chemical sources as its formation from N_2 and O in their normal states is not possible.[46] Also, the production of N_2O from $O(^1D)$ and N_2 is too small to provide a significant source.[47]

$$O(^1D) + N_2 + M \rightarrow N_2O + M \tag{4.79}$$

It therefore appears that no important chemical process produces N_2O in the stratosphere. Instead N_2O is transported up into the stratosphere from the troposphere. The main importance of N_2O as a minor species of the chemosphere thus lies in its loss mechanism of photodissociation (reactions 4.80*a* and 4.80*b*) and reaction with $O(^1D)$ (reaction 4.69a and 4.69b):

$$N_2O(X^1\Sigma) + h\nu(\lambda < 337 \text{ nm}) \rightarrow N_2(X^1\Sigma) + O(^1D) \tag{4.80a}$$

$$N_2O(X^1\Sigma) + h\nu(\lambda < 210 \text{ nm}) \rightarrow N_2(X^1\Sigma) + O(^1S) \tag{4.80b}$$

$$O(^1D) + N_2O \rightarrow N_2 + O_2 \tag{4.69a}$$

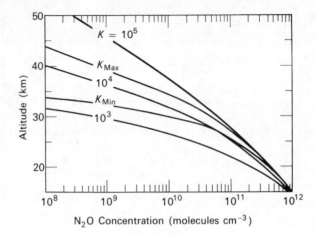

Fig. 4.23 Vertical distribution of N_2O in the stratosphere for various values of the eddy diffusion coefficient $K(cm^2 s^{-1})$ and for an overhead sun. (After M. Nicolet and W. Peetermans, *Annales de Geophysique*, **28**, 751 (1972).)

with $k_{4.69a} = 9 \times 10^{-11}$ cm³ molecule⁻¹ s⁻¹,[19] and

$$O(^1D) + N_2O \rightarrow 2NO \qquad (4.69b)$$

with $k_{4.69b} = 9 \times 10^{-11}$ cm³ molecule⁻¹ s⁻¹.[19]

Although the photodissociation rate is high it is limited by the rate at which N_2O is transported up through the stratosphere. The estimated dissociation rate is 2–3×10^9 N_2O molecules s⁻¹, which means there must be a source providing 2–3×10^9 N_2O molecules cm⁻² s⁻¹ over the surface of the earth[48] and presumably a microbiological source is capable of this rate of N_2O production.[49] Calculated altitude profiles of Nicolet and Peetermans for N_2O in the stratosphere are given in Fig. 4.23. Calculated mole fractions of nitrous oxide as a function of height are shown in Fig. 4.24. The behaviour of N_2O in the stratosphere is discussed in relation to NO production in section 7.4.

Where appreciable quantities of ozone are present the nitrogen trioxide radical NO_3 may be formed by the reactions

$$NO_2 + O_3 \rightarrow NO_3 + O_2 + 105\,kJ \qquad (4.81)$$

where $k_{4.81} = 9.8 \times 10^{-12} \exp(-3520/T)$ cm³ molecule⁻¹ s⁻¹,[44]

$$NO_2 + O(^3P) \rightarrow NO_3^* + 211\,kJ \qquad (4.82)$$

where $k_{4.82} = 1 \times 10^{-11}$ cm³ molecule⁻¹ s⁻¹,[50] and

$$NO_3^* + M \rightarrow NO_3 + M \qquad (4.83)$$

where $k_{4.83} = 1.3 \times 10^{-12}$ cm³ molecule⁻¹ s⁻¹.[50]

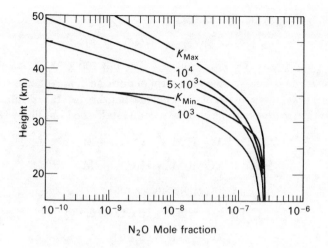

Fig. 4.24 Calculated mole fraction of N_2O as a function of altitude for various values of the eddy diffusion coefficient $K(cm^2\ s^{-1})$ and for a solar zenith angle of 60°. (After M. Nicolet and W. Peetermans, *Annales de Geophysique*, **28**, 751 (1972).)

NO_3 is rapidly removed, in a sunlit atmosphere, by the reactions

$$NO_3 + h\nu(\lambda 500–700\ nm) \rightarrow NO + O_2 \qquad (4.84)$$

where $J_{4.84} > 10^{-2}\ sec^{-1}$,[44] and

$$NO_3 + NO \rightarrow 2NO_2 + 92\ kJ \qquad (4.85)$$

where $k_{4.85} \sim 1 \times 10^{-11}\ cm^3\ molecule^{-1}\ s^{-1}$.[50]

At night in the stratosphere, because of the large reduction in NO concentration by reaction with ozone, reaction 4.85 is not as important as reaction 4.86 in removing NO_3. Other processes which can occur are:

$$NO_3 + NO_2 \rightarrow N_2O_5 + 88.0\ kJ \qquad (4.86)$$

$$k_{4.86} \sim 5 \times 10^{-12}\ T^{1/2} \exp(-1000/T)\ cm^3\ molecule^{-1}\ s^{-1}$$

$$NO_3 + NO_2 \rightarrow NO + NO_2 + O_2 - 16.7\ kJ \qquad (4.87)$$

$$k_{4.87} \sim 3 \times 10^{-14}\ T^{1/2} \exp(-2000/T)\ cm^3\ molecule^{-1}\ s^{-1}$$

and $\qquad NO_3 + NO_3 \rightarrow 2NO_2 + O_2 + 75.3\ kJ \qquad (4.88)$

$$k_{4.88} \sim 1.5 \times 10^{-13}\ T^{1/2} \exp(-3600/T)\ cm^3\ molecule^{-1}\ s^{-1}\ [40]$$

Nicolet estimates that the rate of production of NO_3 by reaction 4.81 at night decreases from 1.7×10^{-17} at the stratopause to $1.7 \times 10^{-18}\ cm^3\ molecule^{-1}\ s^{-1}$ near 30 km, so that the total production of nitrogen trioxide is very small in the stratosphere during a night of twelve hours.[40] N_2O_5 is converted into nitric acid by the reaction

$$H_2O + N_2O_5 \rightarrow 2HNO_3 + 40\,kJ \qquad (4.89)$$

where $k_{4.89} < 2 \times 10^{-18}\,cm^3\,molecule^{-1}\,s^{-1}$ at 300 K.[19]

Odd nitrogen species, once in the form of nitric acid, may then be removed from the atmosphere in rainfall. The mechanism for the production of HNO_3 in the stratosphere has not been well established but likely reactions are

$$OH + NO_2 + M \rightarrow HNO_3 + M \qquad (4.51)$$

$$HO_2 + NO + M \rightarrow HNO_3 + M \qquad (4.90)$$

$$(k_{4.90}\text{ not measured})$$

and

$$CH_3O_2 + NO_2 \rightarrow H_2CO + HNO_3 \qquad (4.91)$$

$$(k_{4.91}\text{ not measured})$$

Experimental evidence of HNO_3 in the stratosphere has been provided by observations of IR emission spectra obtained from a balloon-borne Czerny–Turner spectrometer.[51] The concentration profile of HNO_3 derived from this experiment is characterized by a negligible concentration below 14 km, a maximum concentration $\sim 1.5 \times 10^{10}$ molecules cm^{-3} at 19 ± 5 km and a steadily diminishing concentration above these altitudes.[51]

4.3.4 OXIDES OF CARBON AND METHANE

Carbon dioxide is important in that, although only a minor constituent, it plays a major role in controlling the transfer of infrared radiation through the earth's atmosphere. It is continuously produced at the earth's surface by biological action, by combustion processes, and from the oceans. In the mesosphere and thermosphere CO_2 can be dissociated to some extent by solar radiation, but it is often assumed that it maintains its ground level mole fraction of 3.15×10^{-4} [52] up to about the mesopause, and is in diffusive equilibrium above this.

Carbon monoxide has been measured at concentrations between 0.1 and 0.2 ppm (mole fraction $1-2 \times 10^{-7}$) in the troposphere and its chief source, which was once believed to be the internal combustion engine, is now thought to be the daytime oxidation of tropospheric methane.[53] The main recombination process of CO is [52,53,54]

$$CO + OH \rightarrow CO_2 + H \qquad (4.52)$$

$$k_{4.52} = 5.1 \times 10^{-13} \exp(-300/T)\,cm^3\,molecule^{-1}\,s^{-1}\ [54]$$

which may be supplemented to a small extent by

$$CO + O + M \rightarrow CO_2 + M \qquad (4.92)$$

$$k_{4.92} = 2.2 \times 10^{-36}\,cm^6\,molecule^{-2}\,s^{-1}\ [53].$$

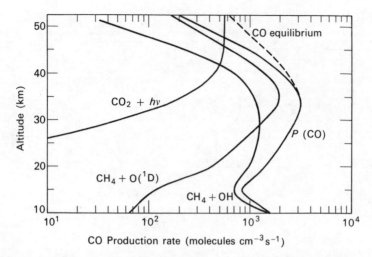

Fig. 4.25 A comparison of rates of CO production from photodissociation of CO_2, and reaction of $O(^1D)$ and OH with CH_4, for a maximum value of the eddy diffusion coefficient and a solar zenith angle of 60°. The curve labelled P(CO) represents the total production rate of CO. (After Nicolet and Peetermans[54].)

Transport processes also play an important part in determining the vertical distribution of carbon monoxide.[53,54]

Methane is produced at ground level by anaerobic decomposition of organic matter at an estimated production rate of 3×10^{11} molecules $cm^{-2} s^{-1}$ [53] over the earth's surface. This rate of methane production is sufficient to account for the major fraction of CO up to about 35–40 km. Photolysis of CO_2 becomes more important at higher altitudes (Fig. 4.25). It is generally assumed that methane maintains its ground level mole fraction in the troposphere but is photodissociated in the mesosphere and dissociated in the stratosphere by oxidation processes such as:

$$CH_4 + OH \rightarrow H_2O + CH_3 \tag{4.55}$$

$$k_{4.55} = 5.5 \times 10^{-12} \exp(-1900/T) \text{ cm}^3 \text{ molecule}^{-1} \text{ s}^{-1} \text{ [54]}$$

$$CH_4 + O(^1D) \rightarrow OH + CH_3 \tag{4.47}$$

$$k_{4.47} \sim 3 \times 10^{-10} \text{ cm}^3 \text{ molecule}^{-1} \text{ s}^{-1} \text{ [54]}$$

$$CH_3 + O_2 + M \rightarrow CH_3O_2 + M \tag{4.93}$$

$$k_{4.93} \sim 10^{-31} \text{ cm}^6 \text{ molecule}^{-2} \text{ s}^{-1} \text{ [54]}$$

$$CH_3O_2 + NO \rightarrow CH_3O_2NO \tag{4.94a}$$

$$CH_3O_2 + NO \rightarrow H_2CO + HNO_2 \tag{4.94b}$$

$$CH_3O_2 + NO_2 \rightarrow CH_3O_2NO_2 \tag{4.91a}$$

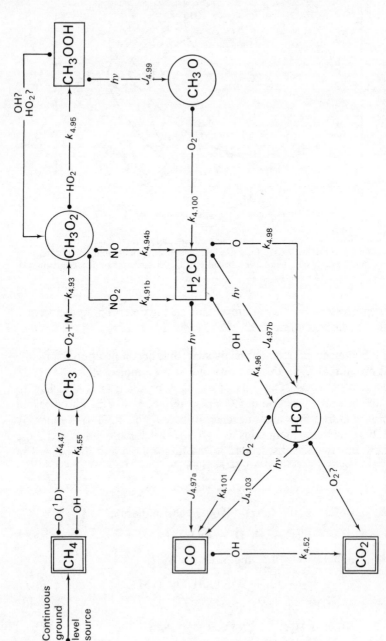

Fig. 4.26 Possible reaction scheme for the oxidation of CH_4 in the atmosphere, modified from Wofsy et al.[53] The squares bordered with double lines represent species with lifetimes > 1 month (observable gas). The rectangles bordered with single lines represent species with lifetimes ~ 1 day (observable in rain). The circles represent species with lifetimes < 10 min (reactive intermediates).

$$CH_3O_2 + NO_2 \rightarrow H_2CO + HNO_3 \qquad (4.91b)$$

$$CH_3O_2 + HO_2 \rightarrow CH_3OOH + O_2 \qquad (4.95)$$

$$H_2CO + OH \rightarrow HCO + H_2O \qquad (4.96)$$

$$k_{4.96} = 1.5 \times 10^{-11} \text{ cm}^3 \text{ molecule}^{-1} \text{ s}^{-1} \text{ at } 300 \text{ K}^{54}$$

$$H_2CO + hv \rightarrow H_2 + CO \qquad (4.97a)$$

$$H_2CO + hv \rightarrow H + HCO \qquad (4.97b)$$

$$H_2CO + O \rightarrow OH + HCO \qquad (4.98)$$

$$k_{4.98} = 1.5 \times 10^{-13} \text{ cm}^3 \text{ molecule}^{-1} \text{ s}^{-1} \text{ at } 300 \text{ K}^{54}$$

$$CH_3OOH + hv \rightarrow CH_3O + OH \qquad (4.99)$$

$$CH_3O + O_2 \rightarrow HO_2 + H_2CO \qquad (4.100)$$

$$HCO + O_2 \rightarrow CO + HO_2 \qquad (4.101)$$

$$HCO + OH \rightarrow H_2O + CO \qquad (4.102)$$

$$k_{4.102} > 1.5 \times 10^{-11} \text{ cm}^3 \text{ molecule}^{-1} \text{ s}^{-1} \text{ at } 300 \text{ K}^{54}$$

$$HCO + hv \rightarrow CO + H \qquad (4.103)$$

and probably others. Undoubtedly modifications to the oxidation scheme for methane, shown in Fig. 4.26, will be necessary when laboratory rate coefficients for the many possible reactions become better established.

References

1 Ratcliffe, J. A., *Physics of the Upper Atmosphere*, Chapter 9, Academic Press, New York, 1960.
2 Colegrove, F. D., Johnson, F. S. and Hanson, W. B., *J. geophys. Res.*, **71**, 2227 (1966).
3 Shimazaki, T. and Laird, A. R., *Radio Sci.*, **7**, 23 (1972).
4 Strobel, D. F., *Radio Sci.*, **7**, 1 (1972).
5 Meira, L. G., *J. geophys. Res.*, **76**, 202 (1971).
6 Norton, R. B. and Barth, C. A., *J. geophys. Res.*, **75**, 3903 (1970).
7 Strobel, D. F., *J. geophys. Res.*, **76**, 2441 (1971).
8 Schiff, H. I., *Can. J. Chem.*, **47**, 1903 (1969).
9 Strobel, D. F., *J. geophys. Res.*, **76**, 8384 (1971).
10 Nicolet, M., *Discuss. Faraday Soc.*, **37**, 7 (1964).
11 Nicolet, M., *Planet. Space Sci.*, **20**, 1671 (1972).
12 Hunt, B. G., *J. geophys. Res.*, **71**, 1386 (1966).
13 Hilsenrath, E., *J. Atmos. Sci.*, **28**, 295 (1971).
14 Crutzen, P. J., *Q. Jl. R. Met. Soc.*, **96**, 320 (1970).
15 Johnson, F. S., Purcell, J. D., Tousey, R. and Watanabe, K. *J. geophys. Res.*, **57**, 157 (1952).
16 Nicolet, M., *Annls Géophys.*, **26**, 531 (1970).
17 Weeks, L. H. and Smith, L. G., *Planet. Space Sci.*, **16**, 1189 (1968).
18 Reed, E. I., *J. geophys. Res.*, **73**, 2951 (1968).
19 Crutzen, P. J., *J. geophys. Res.*, **76**, 7311 (1971), *Can. J. Chem.*, **52**, 1569 (1974).
20 Shimazaki, T. and Laird, A. R., *J. geophys. Res.*, **75**, 3221 (1970).
21a Hochanadel, C. J., Ghormley, J. A. and Ogren, P. J., *J. chem. Phys.*, **56**, 4426 (1972).

21b   Anderson, J. and Kaufman, F., *Chem. Phys. Lett.*, **19**, 483 (1973).
22   Henderson, W. R., *J. geophys. Res.*, **76**, 3166 (1971).
23   Offerman, D. and von Zahn, U., *J. geophys. Res.*, **76**, 2520 (1971).
24   Felder, W. and Young, R. A., *J. chem. Phys.*, **56**, 6028 (1972).
25   Lawrence, G. M. and McEwan, M. J., *J. geophys. Res.*, **78**, 8314 (1973).
26   Noxon, J. F., *J. chem. Phys.*, **52**, 1852 (1970).
27   Slanger, T. G., Wood, B. J. and Black, G., *Chem. Phys. Lett.*, **17**, 401 (1972).
28   Evans, W. F. J., Hunten, D. M., Llewellyn, E. J. and Vallance Jones, A., *J. geophys. Res.*, **73**, 2885 (1968).
29   Paulsen, D. E., Huffman, R. E. and Larrabee, J. C., *Radio Sci.*, **7**, 51 (1972).
30   Gattinger, R. L., *Annls Géophys.*, **25**, 829 (1969).
31   Bowman, M. R., Thomas, L. and Geissler, J. E., *J. atmos. terr. Phys.*, **32**, 1661 (1970).
32   Schiff, H. I., *Annls Géophys.*, **28**, 67 (1972).
33   Anderson, J. G., *J. geophys. Res.*, **76**, 4634 (1971).
34   Kaufman, F., *Can. J. Chem.*, **47**, 1917 (1969).
35   Levy, H., *Science*, **173**, 141 (1971); *Planet. Space Sci.*, **20**, 919 (1972).
36   Westenberg, A. A. and de Haas, N., *J. chem. Phys.*, **57**, 5375 (1972).
37   Greiner, N. R., *J. chem. Phys.*, **51**, 5049 (1969).
38   Anderson, J. G., *J. geophys. Res.*, **76**, 7820 (1971).
39   Tinsley, B. A. and Meier, R. R., *J. geophys. Res.*, **76**, 1006 (1971).
40   Nicolet, M., *J. geophys. Res.*, **70**, 679 (1965).
41   Pearce, J. B., *J. geophys. Res.*, **74**, 853 (1969).
42   Barth, C. A., *Annls Géophys.*, **22**, 198 (1966).
43   Slanger, T. G., Wood, B. J. and Black, G., *J. geophys. Res.*, **76**, 8430 (1971).
44   Johnston, H. S., Lawrence Radiation Laboratory Report UCRL-20568, University of California, Berkeley, 1971.
45   Schofield, K., *Planet. Space Sci.*, **15**, 643 (1967).
46   Nicolet, M., in '*Mesospheric Models and Related Experiments*', Ed. Fiocco, G., D. Reidel Publishing Co., Dordrecht, Holland, 1971.
47   Simonaitis, R., Lissi, E. and Heicklen, J., *J. geophys. Res.*, **77**, 4248 (1972).
48   Bates, D. R. and Hays, P. B., *Planet. Space Sci.*, **15**, 189 (1967).
49   Schutz, K., Junge, C., Beck, R. and Albrecht, B., *J. geophys. Res.*, **75**, 2246 (1970).
50   Berces, T. and Forgeteg, S., *Trans. Faraday Soc.*, **66**, 640 (1970).
51   Williams, W. J., Brooks, J. N., Murcray, D. G., Murcray, F. H., Fried, P. M. and Weinman, J. A., *J. Atmos. Sci.*, **29**, 1375 (1972).
52   Hays, P. B. and Olivero, J. J., *Planet. Space Sci.*, **18**, 1729 (1970).
53   Wofsy, S. C., McConnell, J. C. and McElroy, M. B., *J. geophys. Res.*, **77**, 4477 (1972).
54   Nicolet, M. and Peetermans, W., *Pure and Applied Geophysics*, in press 1974.

5
The Airglow

5.1 Introduction

This chapter could well be sub-titled: 'Excitation processes, emission spectra, and reactions of excited states of atmospheric species'. Excited species in the atmosphere make their presence felt most directly by the radiative transitions they undergo, which lead to the emission of characteristic spectral features in the airglow. The light emitted is usually, though not invariably, measurable at ground level, and in consequence airglow observations have long provided one of the most valuable sources of information about the composition and dynamics of the upper atmosphere. Current research in this field is most active in two main areas, the first being the extension and refinement of field measurements of airglow spectra, with the double aims of improving spectral resolution and the precision of intensity measurements, and of determining diurnal, seasonal, altitude and latitude variations of intensity. The second major area of interest is concerned with elucidating the nature of the physico-chemical processes by which excitation and deactivation occur, and with laboratory studies aimed at determining rates and mechanisms for particular excitation and deactivation processes.

Two main types of optical emission phenomena occur in the upper atmosphere, one being the *airglow* and the other the *aurora*. Auroral phenomena can be distinguished from the airglow on the basis of the criteria that auroras are irregular in form and occurrence, are much more intense than the airglow (usually being visible to the eye), and are mainly restricted to the vicinity of the magnetic poles. In contrast, the airglow occurs continuously, is extremely weak, and is not confined to regions of high latitude. Both phenomena result from the excitation of atmospheric species such as N_2, N_2^+, O_2, O, N, H and OH, but auroral emissions are excited mainly by the impact of energetic electrons and protons on atmospheric gases, whereas the airglow results from the interaction of normal atmospheric constituents with one another and with sunlight. Energy level diagrams of atmospheric species from whose excited states detectable light emission occurs are given in Figs. 5.1, 5.2 and 5.3.

The study of airglow emissions can be considered under three main headings, according to the time of day at which the observations are made. Emission that is detectable from the ground at night-time is termed the *nightglow*, and that observed during the day is termed the *dayglow*. When the sun is below the horizon at ground level but is in view from an altitude of

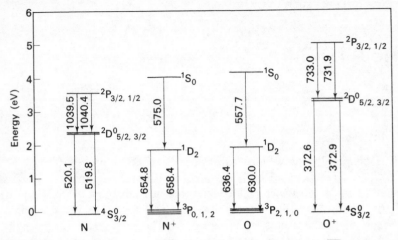

Fig. 5.1   Low lying energy levels of O, O$^+$, N and N$^+$. (1 eV $= 8065.7$ cm$^{-1} = 96.5$ kJ mol^{-1}.)

50–150 km the emission from the sunlit region of the atmosphere, as seen by an observer at ground level, is termed the *twilight glow* (see Fig. 5.13). The twilight glow can be regarded as being just the dayglow viewed from the earth's night side. Clearly it will often be an advantage to observe dayglow features at twilight, when it is less difficult to discriminate against direct solar radiation. The different types of excitation processes which can give rise to airglow emission are discussed in the next section. This is followed by a discussion of the observed dayglow, twilight glow and nightglow emission features, and of likely mechanisms for their production. In the last section of the chapter we consider some results of laboratory investigations of the chemistry of excited species derived from atmospheric gases.

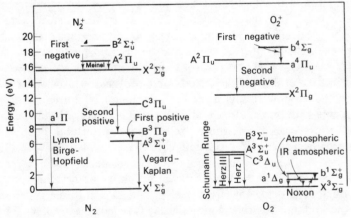

Fig. 5.2   Low lying energy states of O$_2$, O$_2{}^+$, N$_2$ and N$_2{}^+$.

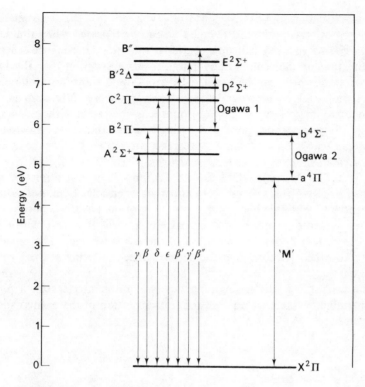

Fig. 5.3   Energy level diagram for NO[4].

5.2   Types of excitation process

5.2.1   FLUORESCENCE AND RESONANT SCATTERING

As discussed in Chapter 2, fluorescence is the emission process which results when an atom, molecule or ion is excited by light absorption. If the excited state is stable the result is simply a transition to the ground state, or to an intermediate excited state (thus possibly giving rise to a 'cascade' process), with emission of radiation characteristic of the absorbing species. Alternatively, the excited species may dissociate directly, may predissociate, or may ionize; if, in this case, an excited ion or radical is formed it may emit its own characteristic radiation. Interesting examples of both kinds of fluorescent process are provided by the airglow of Mars, which contains emission lines of atomic oxygen at 130.4 and 135.6 nm. The 135.6 line is produced by electron and photon impact on CO_2, whereas the 130.4 nm resonance line is the result of light absorption and re-emission by free oxygen atoms (see Chapter 8). In the special case where the excitation and emission wavelengths are identical the process is called resonance fluorescence, or resonance scattering. In a large absorbing volume, as in the atmosphere, the processes of absorption and re-emission can be repeated many times, thus giving rise to virtually isotropic scattered radiation. Resonance scattering by an atom or molecule strictly refers to an absorption into the lowest excited state which can be reached by an allowed transition, followed by re-emission back to the ground state. However, the use of the term has been extended to include any scattering from the ground state without change in energy. Important dayglow and twilight glow transitions involved in resonance scattering processes in the earth's atmosphere include the Lyman-α and Lyman-β lines of atomic hydrogen, $He(2^3P \rightarrow 2^3S)$, $O(^3S \rightarrow {}^3P)$, $N_2{}^+(B^2\Sigma_u{}^+ \rightarrow X^2\Sigma_g{}^+)$, $NO(A^2\Sigma \rightarrow X^2\Pi)$, and some transitions of metal atoms and ions.* It is interesting to note that when solar Lyman-β radiation is absorbed by atomic hydrogen about 12% of the absorbed photons take part in a cascade process, with initial emission of Balmer-α fluorescence and subsequent fluorescent emission of the Lyman-α line.[1] The remainder of the absorbed photons give rise to resonance scattering at the Lyman-β wavelength.

The probability that a single particle will resonantly scatter a photon from sunlight, assuming no collisional deactivation of the excited species, is given by[2,3]

$$g = \pi F \int q \, dv \tag{5.1}$$

$$= \pi F \frac{\pi e^2}{mc^2} \lambda f \tag{5.2}$$

where g is the probability of scattering in s^{-1}, q is the absorption cross section of the particle, πF is the solar photon flux at the wavelength of the transition, f is the oscillator strength of the transition, e and m are the electronic charge and mass, and c is the velocity of light. According to the classical picture, the oscillator strength is equal to the number of elastically bound electrons per particle which are responsible for the resonant scattering process. For a particle undergoing an electronic transition from an upper state 2 to a lower state 1, f is related to the quantum-mechanical transition probability by the equation[4]

$$f_{12} = 1.5 \times 10^{-14} \lambda_{12}{}^2 A_{21}$$

where λ_{12} is the wavelength of the absorbed light in nm and A_{21} the Einstein coefficient for a spontaneous radiative transition from the upper to the lower state, in reciprocal seconds. The natural radiative lifetime τ_2 of the upper

---

* The convention in spectroscopy is that the *upper* state of a transition is always written first, as $O(^3S \rightarrow {}^3P)$ in emission, or $H(^2P \leftarrow {}^2S)$ in absorption. Unfortunately, we must warn the reader that most books on aeronomy use the opposite convention and write the *lower* state first for atomic (but not molecular!) transitions.

state is related to the Einstein A coefficient by

$$1/\tau_2 = \sum_i A_{2i} \tag{5.3}$$

where the summation is over all lower states which are accessible from state 2. If non-radiative processes contribute to the deactivation of state 2 their rate coefficients must also be included in the sum in eqn. 5.3 (cf. eqn. 2.9).

A diffuse airglow source produced by resonance scattering has a brightness I (in photons $\text{cm}^{-2}\,\text{s}^{-1}\,\text{column}^{-1}$) given by

$$4\pi I = g \cdot \frac{N}{\mu} \tag{5.4}$$

where N is the column density of scattering species in cm^{-2} and μ is a factor which takes account of the slant angle of the incident beam relative to the scattering layer.[2]

5.2.2 EXCITATION BY CHARGED PARTICLES

There are two main processes involving charged particles which can give rise to excited species, namely inelastic collisions of photoelectrons, and electron–ion recombination reactions. Photoelectrons are produced in the daytime with a mean energy of around 10 eV at high altitudes. The kinetic temperature of ionospheric electrons remains appreciably higher than the ambient gas temperature, despite the tendency for energy to be lost rapidly by inelastic collisions with the molecular species present. An inelastic collision between a molecule and an electron commonly results in vibrational or rotational excitation of the colliding molecule; alternatively the 'hot' electrons may have sufficient energy to excite low lying electronic states, such as the ^1D state of atomic oxygen. Potentially important types of inelastic collision process are:

$$A + e \rightarrow A^* + e \tag{5.5}$$

$$AB + e \rightarrow AB^* + e \tag{5.6}$$

$$AB + e \rightarrow AB^{+*} + 2e \tag{5.7}$$

$$AB + e \rightarrow A + B^{+*} + e \tag{5.8}$$

$$A + B^+ \rightarrow A^+ + B^* \tag{5.9}$$

Photoelectron excitation is believed to be partly responsible for several dayglow features, including the emission lines arising from $O(^1S)$ at 557.7 nm, and $O(^1D)$ at 630 nm.

Recombination reactions can give rise to excited products by one of the steps:

$$AB^+ + e \rightarrow AB^* \tag{5.10}$$

$$\rightarrow A + B^* \tag{5.11}$$

$$\rightarrow A^* + B \tag{5.12}$$

When AB^+ in reactions 5.10–5.12 is O_2^+ or NO^+ the following processes are all energetically feasible

$$O_2^+ + e \rightarrow O + O + 6.95 \text{ eV} \tag{5.13}$$

$$\rightarrow O + O(^1D) + 4.99 \text{ eV} \tag{5.14}$$

$$\rightarrow O(^1D) + O(^1D) + 3.03 \text{ eV} \tag{5.15}$$

$$\rightarrow O + O(^1S) + 2.78 \text{ eV} \tag{5.16}$$

$$\rightarrow O(^1D) + O(^1S) + 0.82 \text{ eV} \tag{5.17}$$

$$NO^+ + e \rightarrow N + O + 2.76 \text{ eV} \tag{5.18}$$

$$\rightarrow N + O(^1D) + 0.80 \text{ eV} \tag{5.19}$$

$$\rightarrow N(^2D) + O + 0.39 \text{ eV} \tag{5.20}$$

Clearly there are likely to be problems involved in the assignment of an airglow feature to any one recombination process. Thus the 630.0, 636.4 nm doublet of atomic oxygen ($^1D_2 \rightarrow {}^3P_2, {}^3P_1$) which is observed in the nightglow could arise from any one of four of the above dissociative recombination processes. In this situation it is necessary to call on both theoretical arguments and the results of laboratory studies in order to assess the relative importance of the various alternative processes. Here, for example, reaction 5.19 could safely be ignored on the grounds that it is spin forbidden.*

5.2.3 CHEMICAL EXCITATION PROCESSES

In an exothermic chemical reaction, some of the excess energy of the reaction may go into rotational, vibrational or electronic excitation of one or more of the products. Reactions that result in the emission of photons are said to be chemiluminescent. Such reactions occur in the atmosphere and contribute to both the dayglow and the nightglow. The basic difference between dayglow and nightglow is that the main processes responsible for the dayglow, resonance scattering and photoelectron excitation, are almost entirely absent at night, so that the night airglow is essentially the result of

---

* The Wigner spin conservation rule requires that the total spin of a system be unaltered after a reaction when S is a good quantum number. (When S is not a good quantum number the total electronic angular momentum, denoted by J or Ω, is still conserved.) If two species have spins S_1 and S_2 before collision, the rule states that the resultant spin of the products formed is confined to the range $(S_1 + S_2), (S_1 + S_2 - 1), \ldots (|S_1 - S_2|)$. If the spins of the products are S_3 and S_4, the resultant spin is one of the numbers $(S_3 + S_4), (S_3 + S_4 - 1), \ldots (|S_3 - S_4|)$. For a simple interpretation see reference 5, p. 88.

chemiluminescent processes. Some important general types of chemi-luminescent reactions are represented in the processes 5.21–5.25.

$$A + B \quad \rightarrow AB^* \quad \rightarrow AB + hv \tag{5.21}$$

$$A + B + C \rightarrow AB + C^* \rightarrow AB + C + hv \tag{5.22}$$

$$A + B + C \rightarrow AB^* + C \rightarrow AB + C + hv \tag{5.23}$$

$$A + BC \quad \rightarrow AB^\dagger + C \rightarrow AB + C + hv \tag{5.24}$$

$$A + BC \quad \rightarrow AB + C^* \rightarrow AB + C + hv \tag{5.25}$$

An example of the type 5.21 in the atmosphere is provided by the reaction of NO with O;

$$NO + O \rightarrow NO_2 + hv \tag{5.26}$$

which is believed to be responsible for the weak visible continuum which underlies the discrete emission features of the nightglow. The low pressure radiative recombination process, which occurs without the mediation of a third body M, gives a slightly different emission spectrum from the termolec-ular recombination reaction which is normally observed under laboratory conditions. The green line of atomic oxygen at 557.7 nm is thought to arise mainly from the process

$$O + O + O \rightarrow O_2 + O(^1S) \tag{5.27}$$

which is of the type 5.22. The atmospheric bands of molecular oxygen, emitted by $O_2(b^1\Sigma_g{}^+)$ in the nightglow, are thought to be excited mainly by

$$O + O + M \rightarrow O_2(^1\Sigma_g{}^+) + M \tag{5.28}$$

which is of the type 5.23. In reactions 5.27 and 5.28 part of the energy released by the atom recombination process appears as electronic excitation of one of the products. Reaction 5.24, on the other hand, is representative of a large class of displacement reactions, of which reaction 5.29 is an example, in which a major part of the energy of reaction appears as vibrational excitation of the newly formed bond. Reaction 5.29 is believed to be the source of the Meinel OH bands which are the strongest feature of the night-glow. The reaction of NO with ozone is a rare example of such a displacement reaction in which the

$$H + O_3 \rightarrow OH^\dagger(v \leqslant 9) + O_2 \tag{5.29}$$

energy of reaction largely appears as electronic rather than vibrational excitation of the product NO_2; this reaction does not appear to make a significant contribution to the airglow. Reactions of the type 5.25, in which the energy of reaction appears as excitation of the molecular fragment left by a displacement reaction, are uncommon. The observation of weak emission of the atmospheric bands of O_2 from the H/O_3 flame in the laboratory suggests that reaction 5.29 is also an example of this type, with production of O_2 in the b-state occurring simultaneously with production of OH in the level $v = 4$.[6a] Other examples of excitation reactions involving neutral species are discussed by Rabinovitch and Flowers.[6b]

Some ion-molecule reactions are also potential sources of electronically excited species, for example

$$N_2^+ + O \rightarrow NO^+ + N + 3.08 \text{ eV} \tag{5.30}$$

where the reaction is sufficiently exothermic to result in electronic excitation of either NO^+ or N. However, at the time of writing no ion–atom interchange or charge–transfer process has been definitely established as a significant contributor to the airglow.

5.2.4   ENERGY TRANSFER IN COLLISIONS

Collisional deactivation (quenching), of an excited species can occur in a variety of ways, one of which involves excitation of the quencher by energy transfer. If the quenching species then radiates the result is sensitized fluorescence, as observed, for example, in the classic experiments of Cario and Franck.* Sensitized fluorescence normally involves the transfer of electronic energy, but examples are known in which the quencher molecule emerges from the collision process with vibrational rather than electronic excitation, and subsequently emits fluorescence in the infrared. One example, reported by J. Polanyi and coworkers,[7] is the quenching of Hg(6^3P) by CO, in which sufficient energy is transferred from the electronically excited mercury atom to populate levels up to $v = 9$ in the ground electronic state of CO:

$$Hg(^3P_1, {}^3P_0) + CO \rightarrow Hg + CO^\dagger(v \leqslant 9) \tag{5.31}$$

Much earlier, M. Polanyi et al.[8] had demonstrated that the reverse process occurred in the atomic flames produced by reacting sodium vapour with halogen compounds. In this case energy from vibrationally excited NaCl was transferred to atomic sodium, causing excitation to the ^{2}P state with subsequent emission of the sodium D-lines.

$$NaCl^\dagger + Na(^2S) \rightarrow NaCl + Na(^2P) \tag{5.32}$$

In the atmosphere energy transfer from $O(^1D)$ to ground state O_2 is believed to provide an important excitation mechanism for the O_2 atmospheric bands:

$$O(^1D) + O_2(X^3\Sigma_g^-) \rightarrow O(^3P) + O_2(b^1\Sigma_g^+) \tag{5.33}$$

$$O_2(b^1\Sigma_g^+) \rightarrow O_2(X^3\Sigma_g^-) + h\nu \text{ (Atmospheric bands.)} \tag{5.34}$$

Metastable species such as $O(^1D)$ and $O_2(^1\Delta_g)$, because of their long radiative lifetimes, are able to undergo a large number of collisions before radiating, so that there are many opportunities for energy transfer to occur.

* *Z. Physik*, **17**, 202 (1923).

5.3 Observed airglow features

All of the airglow emission features observed at night can be expected to be present, probably with considerably enhanced intensity, in the dayglow. Whether or not a particular feature is observed in the dayglow depends, of course, on how well it can be isolated from the strong background of scattered sunlight. The twilight glow has traditionally been treated separately from both dayglow and nightglow, largely because of experimental differences. This division of the airglow into three areas is somewhat artificial, but is convenient for purposes of interpretation, since different excitation mechanisms predominate at different times.

5.3.1 THE DAYGLOW

The mechanisms mainly responsible for dayglow are resonance scattering, fluorescence, and photoelectron excitation. Table 5.1 lists the observed emissions, together with their intensities (where data are available), important excitation processes, altitudes of emission and g values for photon scattering.

The ultraviolet dayglow ($\lambda < 395$ nm). Prominent ultraviolet dayglow emissions include the Lyman-α line of atomic hydrogen, the atomic oxygen lines near 130 nm, the NO γ-bands, the atomic nitrogen lines at 120 nm, and the second positive bands of N_2. Two spectra of the ultraviolet dayglow, covering the regions 110 to 200 nm and 190 to 430 nm, as obtained near 105 km during a rocket flight, are shown in Figs. 5.4 and 5.5 respectively.

The Lyman-α line of atomic hydrogen, at 121.6 nm, was first observed in the dayglow during a rocket flight in 1962.[10] This line is a very strong component of the solar spectrum, and the emission appears to originate from altitudes where the mole fraction of atomic hydrogen is largest. The atmosphere is optically thick to Lyman-α, so that its emission below 1000 km must be due to multiple scattering; scattered Lyman-α has been observed out to 10^5 km by Mariner 5 on its way to Venus.[1]

Atomic oxygen transitions which should contribute dayglow emissions are shown in Fig. 5.6. The observed ultraviolet dayglow features are the resonance triplet at 130.2, 130.4 and 130.6 nm, corresponding to the transitions ($^3S \rightarrow {}^3P_2, {}^3P_1, {}^3P_0$), the forbidden 135.6 nm line ($^5S \rightarrow {}^3P$), and the 297.2 nm line ($^1S \rightarrow {}^3P$). At moderate altitudes (150–200 km) the resonance triplet is produced mainly by photoelectron impact on atomic oxygen, and to a lesser extent by resonant scattering. Above 400 km resonant scattering becomes the dominant mechanism.[11] Probable minor daytime sources are absorption of the Lyman-β line by atomic oxygen,* radiative recombination of O^+,[12]

$$O^+ + e \rightarrow O^* \rightarrow O + h\nu \ (\lambda = 130 \text{ nm}) \tag{5.35}$$

* The energy of the Lyman-β line coincides almost exactly with that of an O atom transition from the ground state. The excited atom which results emits an 844.6 nm photon, and so populates the ^3S level. This process has become known as Bowen Fluorescence.

Table 5.1   Dayglow emission features.[a]

λ (nm)	Emitting Species	Intensity (Rayleighs)	Altitude (km)	Excitation[b] Mechanism	g (photons s^{-1} particle^{-1})	Notes
30.4	He$^+(^2$P)			R	1.1×10^{-4}	
58.4	He(^{1}P)			R	1.7×10^{-5}	
83.4	O$^+(^4$P)					
102.5	H(^{2}P)		200–10^4	R	2.6×10^{-6}	Lyβ
120.0	N(^{4}P)	400 R	180	R?e?		
121.6	H(^{2}P)	27 kR	100–10^5	R	2.1×10^{-3}	Lyα
130.2, .4, .6	O(^{3}S)	7.5 kR	190	eFR	1.0×10^{-4}	
135.6	O(^{5}S)	350 R	140	e		
130–150	N$_2$(a$^1\Pi_g$)			e		Lyman-Birge-Hopfield
149.3	N(^{2}P)			e		
174.4	N(^{2}P)			e		
200–400	N$_2$(A$^3\Sigma_u{}^+$)			e		Vegard-Kaplan
215 etc	NO(A$^2\Sigma_g{}^+$)	1 kR	70–150	R	4.0×10^{-6}	γ bands, g for (1, 0) band
279.5, 280.2	Mg$^+(^2$P$_{1/2, 3/2}$)	360 R	108	R	0.065	
297.2	O(^{1}S)		60–200	eDC		
306.4	OH(A$^2\Sigma^+$)	1 kR	50–100	R	4.3×10^{-4}	
337.1 etc	N$_2$(C$^3\Pi_u$)	900 R		e		2nd positive (0, 0) band
346.6	N(^{2}P)			e		
391.4	N$_2{}^+$(B$^2\Sigma_u{}^+$)	2.0 kR	150	RF	0.050	1st negative (0, 0) band
520.0	N(^{2}D)	90 R	~200	De	6×10^{-11}	
557.7	O(^{1}S)	3.0 kR	90, 175	FCDe	1×10^{-11}	
589.3	Na(^{2}P)	30 kR	92	R	0.8	
630.0, 636.4	O(^{1}D)	2–20 kR	250	FDe	4.5×10^{-10}	
761.9 etc	O$_2$(b$^1\Sigma_g{}^+$)	300 kR	40–120	RFE	6.3×10^{-9}	Atmospheric bands
777.4	O(^{5}P)	1.6 kR	~150	e		
844.6	O(^{3}P)	1.1 kR	~150	e		
1051 etc	N$_2$(B$^3\Pi_g$)	900 R	150	e		
1103.6 etc	N$_2{}^+$(A$^2\Pi_u$)	4 kR	150	RF	0.042	Meinel, g for (1, 0) band (120 μm)
1270 etc	O$_2$($^1\Delta_g$)	20 MR	50	F	9.4×10^{-11}	1 R Atm; (0, 1) band 1.58 μm
2800 etc	OH($v \leqslant 9$)	4.5 MR		C		Meinel; 4.5 μm to 381.6 nm

a   Modified from reference 1.
b   Excitation mechanisms are R = resonance scattering; e = photoelectron excitation; F = fluorescence and dissociative fluorescence; D = dissociative recombination; C = chemical association and E = energy transfer.

and dissociative fluorescence of O_2. The 135.6 nm line behaves very differently from the 130 nm triplet, in that it shows a sharp maximum at around 150 km with a decrease in intensity above 150 km that correlates with the estimated decrease in the ground state atomic oxygen concentration.[13] Thus the state giving rise to the 135.6 nm line, O(^{5}S), is thought to be produced by photoelectron excitation of O(^{3}P).[9] The forbidden line at 297.2 nm is also believed to arise from photoelectron impact on ground state atomic oxygen. Other transitions at 98.9 and 115.2 nm should be present, but so far have not been observed.[14]

Emission from NO during the day was first detected by Barth in 1963.[15] The emission is sufficiently intense to dominate the dayglow between 200

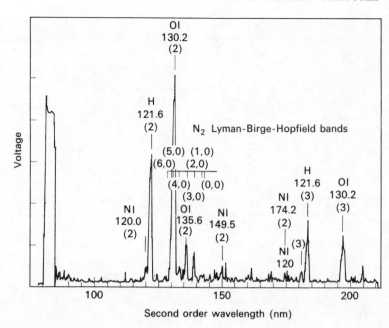

Fig. 5.4 Spectrum of the ultraviolet dayglow from 110 to 200 nm, after Pearce.[9] (The number in brackets is the diffraction order of the spectrometer.)

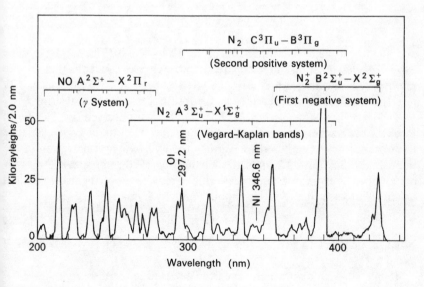

Fig. 5.5 Spectrum of the ultraviolet dayglow from 190 to 430 nm. (Kindly provided from unpublished data by Dr C. A. Barth.)

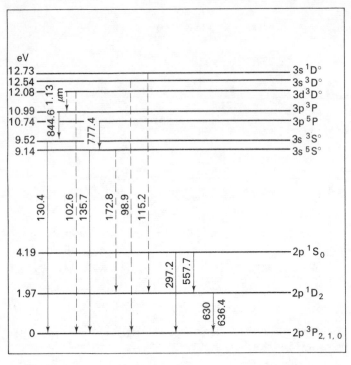

Fig. 5.6 Energy level diagram of atomic oxygen, showing potential and observed transitions.[11] The solid arrows are observed dayglow transitions and the dashed arrows signify transitions which could occur but which have not been observed. Wavelengths are in nanometres, except for the transition at 1.13 micrometres.

and 300 nm. Rocket measurements of the NO γ-bands (Fig. 5.3) have since provided the most useful estimate of nitric oxide concentration in the mesosphere and lower thermosphere (cf. Section 4.3.3). Excitation of NO into the $A^2\Sigma^+$ state is believed to be the result of resonance scattering.[13]

Hydroxyl radical emission $(A^2\Sigma^+ \rightarrow X^2\Pi_i)$ is observed at 306.4 nm. Rocket-borne scanning spectrophotometer observations of OH resonance fluorescence have been used to estimate the hydroxyl concentration in the mesosphere (Section 4.3.2(ii)). Although the bulk of OH emission at 306.4 nm certainly arises from resonance scattering, it may be supplemented to a small extent by dissociation of water vapour (5.36) by Lyman-α

$$H_2O + h\nu \rightarrow OH(A^2\Sigma^+) + H \tag{5.36}$$

which is believed to produce $OH(A^2\Sigma^+)$ with an efficiency of between 1 and 5%.[16]

Atomic nitrogen lines which have been detected in the ultraviolet dayglow are the 120 nm resonance triplet ($^4P \rightarrow {}^4S$), the triplet at 149.3 nm ($^2P \rightarrow {}^2D$), the quartet at 174.4 nm ($^2P \rightarrow {}^2D$) and the forbidden transition ($^2P \rightarrow {}^4S$)

at 346.6 nm. The triplet at 120 nm is somewhat analogous to the O atom resonance triplet at 130 nm, with the ^4P state being produced by both electron impact and resonance scattering. Photoelectron excitation of N and N_2 is believed to be responsible for the lines at 149.3, 174.4 and 346.6 nm.[17]

Bands of molecular nitrogen observed in the ultraviolet dayglow include the (0, 6) band of the Lyman–Birge–Hopfield system $(a^1\Pi_g \rightarrow X^1\Sigma_g^+)$ at 180.5 nm, the (1, 10) band of the Vegard–Kaplan system $(A^3\Sigma_u^+ \rightarrow X^1\Sigma_g^+)$ at 342.5 nm, and the (0, 0) band of the second positive system $(C^3\Pi_u \rightarrow B^3\Pi_g)$ at 337.1 nm. As the upper states of both the Vegard–Kaplan bands and the second positive bands cannot be excited by allowed transitions from the ground state, the excitation process is almost certainly photoelectron impact (eqn. 5.37).

$$N_2(X^1\Sigma_g^+) + e \rightarrow N_2(C^3\Pi_u, B^3\Pi_g, A^3\Sigma_u^+) + e \tag{5.37}$$

Interpretation of the observed emission intensities is complicated by the fact that the $B^3\Pi_g$ and $A^3\Sigma_u^+$ states may also be populated by cascading from higher energy levels (see Fig. 5.2), and by radiationless transitions which result from the crossing of the $A^3\Sigma_u^+$ and $B^3\Pi_g$ potential energy curves. It is even possible for vibrationally excited molecules in the $A^3\Sigma_u^+$ state to cross over to the $B^3\Pi_g$ state, then radiate in the first positive system $(B^3\Pi_g \rightarrow A^3\Sigma_u^+)$, and so arrive in lower vibrational levels of the A-state.[18]

The visible dayglow $(395 < \lambda < 700$ nm). Emissions from $N_2^+(B^2\Sigma_u^+)$, atomic oxygen $(^1D$ and $^1S)$ and sodium (^2P) dominate the visible dayglow.

N_2^+ *ions* in various states are produced from N_2 by extreme ultraviolet solar radiation:

$$N_2(X^1\Sigma_g^+) + h\nu(\lambda < 79.6 \text{ nm}) \rightarrow N_2^+(A^2\Pi_u, B^2\Sigma_u^+, X^2\Sigma_g^+) + e \tag{5.38}$$

by electron bombardment:

$$N_2(X^1\Sigma_g^+) + e \rightarrow N_2^+(A^2\Pi_u, B^2\Sigma_u^+, X^2\Sigma_g^+) + 2e \tag{5.39}$$

and possibly by the ion–molecule reaction[19]

$$N_2(X^1\Sigma_g^+) + O^+(^2D) \rightarrow N_2^+(A^2\Pi_u, v' = 1) + O(^3P) \tag{5.40}$$

Laboratory investigations of the process 5.38 using He 58.4 nm radiation have indicated quantum yields of approximately 30% for $N_2^+(X^2\Sigma_g^+)$, 50% for $N_2^+(A^2\Pi_u)$, and 10% for $N_2^+(B^2\Sigma_u^+)$.[20] Radiative transitions from the $A^2\Pi_u$ and $B^2\Sigma_u^+$ states to the ground $(X^2\Sigma_g^+)$ state give rise to the Meinel (785 nm) and first negative (391.4 nm) systems, respectively. Estimated contributions to the rate of production of the upper state of the first negative system, $N_2^+(B^2\Sigma_u^+)$, from various sources are shown in Fig. 5.7. Below about 150 km, the N_2^+ is rapidly destroyed by O_2 and O (Section 6.7); the chemical lifetime decreases from about 85 s at 500 km to ~1 s at 150 km.

Atomic nitrogen emits a closely spaced doublet at 520 nm as a result of

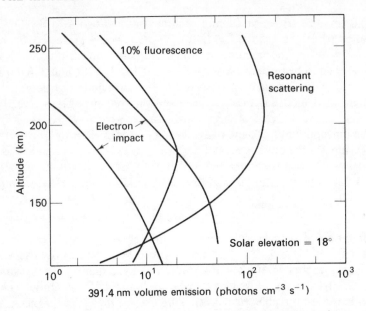

Fig. 5.7   Relative contributions of different mechanisms of excitation of $N_2^+(B^2\Sigma_u^+)$, which is the upper state of the first negative system of N_2^+ (0, 0 band at 391.4 nm).[11] The two curves marked 'electron impact' are two independent estimates of production of the $B^2\Sigma_u^+$ state of N_2^+ by electron impact.

the forbidden transition $(^2D_{5/2,3/2} \rightarrow {}^4S_{3/2})$. The metastable ^2D state has a radiative lifetime of around 26 hours,[22] so that quenching reactions largely govern its concentration. The bulk of 520 nm emission occurs between 150 and 250 km, but the signal is too weak (intensity ~ 100 R) to enable an accurate height profile to be determined. Possible production mechanisms of $N(^2D)$ were discussed in Section 4.3.3. Processes mentioned were dissociative recombination of N_2^+ and NO^+, the ion–molecule reaction

$$N_2^+ + O \rightarrow NO^+ + N(^2D) \tag{5.41}$$

and photoelectron excitation of N_2. Nitrogen atoms are not directly involved in any of these mechanisms, so that N atom concentrations cannot be deduced from the 520 nm emission intensity. Removal of $N(^2D)$ is most likely to occur by the processes[11,23]

$$N(^2D) + O_2(X^3\Sigma_g^-) \rightarrow NO + O \tag{5.42}$$

$$\rightarrow N(^4S) + O_2(^1\Delta_g, {}^1\Sigma_g^+) \tag{5.43}$$

$$N(^2D) + O \qquad \rightarrow N + O(^1D) \tag{5.44}$$

and $\qquad N(^2D) + e \qquad \rightarrow N(^4S) + e \tag{5.45}$

Once the production and loss mechanisms for $N(^2D)$ have been better characterized, the intensity of emission at 520 nm may yield valuable information regarding the nitric oxide concentration in the lower thermosphere, since NO is thought to be formed from $N(^2D)$ via reaction 5.42 (cf. Section 4.3.3).

Atomic oxygen transitions to the ground ^3P state from the 2p ^1S and 2p $^1D_2$ states produce the well known green (557.7 nm) and red (630 nm) lines of the aurora and airglow. For both of these species there are several alternative methods of excitation in the dayglow. The $O(^1S)$ state is populated by the Chapman reaction

$$O + O + O \rightarrow O_2 + O(^1S) \tag{5.27}$$

by photoelectron excitation of atomic oxygen, by dissociative recombination of $O_2{}^+$ [11] and by dissociative fluorescence.[24] The $O(^1D)$ state is produced by dissociation of molecular oxygen in the Schumann–Runge continuum, by photoelectron excitation, and possibly by energy transfer from $N(^2D)$ (reaction 5.44).

Observations of the green line emission at 557.7 nm as a function of altitude reveal the presence of two peaks in the intensity profile (Fig. 5.8). Therefore at least two different processes must be responsible for excitation of atomic oxygen to the ^1S state. Excitation in the lower layer at around 95 km has

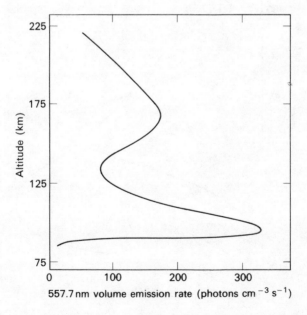

Fig. 5.8   Height variation of the volume emission rate of the 557.7 nm dayglow line of atomic oxygen, derived from rocket measurements of Wallace and McElroy.[21]

usually been attributed to reaction 5.27 or, alternatively,[25] to

$$O + O + M \rightarrow O_2^*(c^1\Sigma \text{ or } C\,^3\Delta) + M \qquad (5.46)$$

$$O_2^* + O \rightarrow O_2 + O(^1S) \qquad (5.47)$$

Both mechanisms predict that the green line intensity should be proportional to the third power of the oxygen atom concentration. However, new estimates of O atom densities and the measured rate of 5.27 show that the Chapman reaction is sufficient to account for observed green line intensities around 95 km. The intensity of emission in the upper layer, at around 170 km, is too great to arise from the same type of mechanism, in view of the low O atom concentrations. Photoelectron excitation of atomic oxygen and dissociative recombination of O_2^+ are considered to be likely sources.[24,26]

The red line of atomic oxygen at 630 nm is one of the brightest dayglow features. As shown in the energy level diagram of Fig. 5.6, the transition actually gives rise to a triplet (639.2, 636.4 and 630 nm) but only the two higher energy lines are observed in practice because the third transition violates the $\Delta J = 0, \pm 1$ selection rule. Measured height profiles of the 630 nm dayglow line are shown in Fig. 5.9. The discrepancy between the two sets of measurements may be due to differing degrees of interference from scattered sunlight.[26] A comparative estimate of three sources of $O(^1D)$, namely,

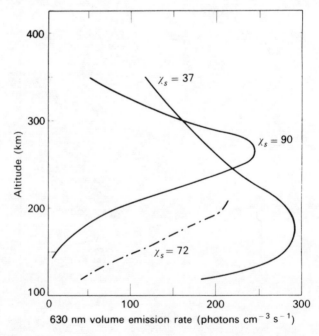

Fig. 5.9 Rocket measured height profiles of the 630 nm red line of atomic oxygen in the dayglow. The dashed curve is from ref. 21 and the solid curves are from ref. 27.

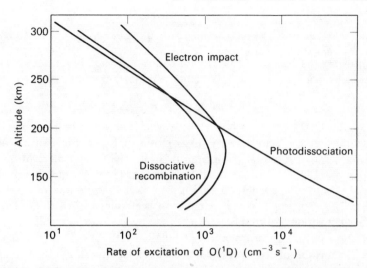

Fig. 5.10 Estimated relative contributions of three excitation mechanisms for the 630 nm dayglow.[21]

photodissociation of O_2, dissociative recombination of O_2^+, and photo-electron excitation of atomic oxygen, is given in Fig. 5.10. On the basis of this estimate it appears that photoelectron excitation predominates above 200 km, and photodissociation in the Schumann–Runge continuum predominates below 200 km.

The sodium D-lines at 589.0 and 589.6 nm are the strongest feature of the visible dayglow, and in fact were the first dayglow feature to be recognized. The excitation of ground state $^2S_{1/2}$ sodium atoms into the $^2P_{3/2}$ and $^2P_{1/2}$ levels is due to resonance absorption by a layer of free sodium near 90 km.[13] A surprising but reproducible observation is that of a marked daytime enhancement of the D lines, relative to the intensity expected from twilight measurements. There is also a large intensity maximum near noon. Another source of excitation may prove to be needed to supplement resonance scattering.

The infrared dayglow ($\lambda > 700$ nm). The emission features observed in this spectral region include the atmospheric and infrared atmospheric bands of molecular oxygen, and the OH vibration–rotation bands.

The atmospheric system of $O_2(b^1\Sigma_g^+ \rightarrow X^3\Sigma_g^-)$ has been observed both from rockets and from the ground. Rocket observations have shown the presence of the (0, 0) and (0, 1) bands at 761.9 and 864.5 nm, respectively; ground-based measurements have detected only the (0, 1) band because the (0, 0) band is reabsorbed by ground state O_2. Most of the observed emission originates below 120 km. The main processes responsible for exciting O_2 to the $b^1\Sigma_g^+$ state are thought to be resonant scattering, energy transfer from O^{1}D:

$$O(^1D) + O_2 \rightarrow O + O_2(b^1\Sigma_g^+) \qquad (5.48$$

and possibly the reaction of O^{1}D with ozone, although laboratory evidence is weighted against reaction 5.49 being an important source of $O_2(b^1\Sigma^+)$.[72]

$$O(^1D) + O_3(^1A_1) \rightarrow O_2(b^1\Sigma_g^+) + O_2 \qquad (5.49$$

The species O(^{1}D) can be produced by dissociation of O_2 in the Schumann–Runge continuum, and also by dissociation of O_3 in the Hartley bands (Section 2.3.3). Estimated contributions from each of the mechanisms are compared in Fig. 5.11, where it has been assumed that N_2 and O_2 are about equally effective in quenching O(^{1}D). The radiative lifetime of the $b^1\Sigma_g^+$ state is 12 s (Section 5.4) and is equal to the reciprocal of the quenching rate at an altitude of about 85 km.

The infrared atmospheric bands of $O_2(a^1\Delta_g \rightarrow X^3\Sigma_g^-)$ were first recognized in 1962 as a result of observations made from a jet aircraft flying at 13 km.[13] They are the strongest feature of the dayglow. The volume emission rate of $O_2(^1\Delta_g)$, as found from rocket measurements, is shown in Fig. 5.12. Photolysis of O_3 in the Hartley bands between 200 and 300 nm provides the main daytime production mechanism for $O_2(^1\Delta_g)$; above about 80 km

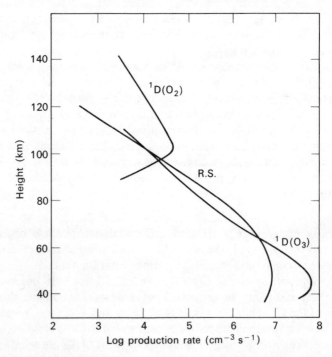

Fig. 5.11 Calculated production rates of $O_2(b^1\Sigma_g^+)$.[11] RS is production by resonance scattering, ^{1}D(O$_3$) is production by reaction 5.49, and ^{1}D(O$_2$) is production by reaction 5.48.

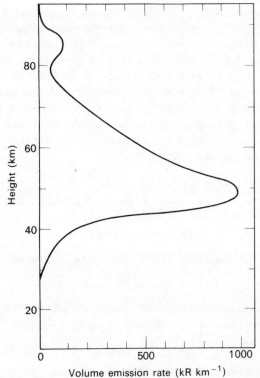

Fig. 5.12   Height profile of the volume emission rate of $O_2(^1\Delta_g)$, as measured by Evans *et al.*[28]

ozone photolysis may be supplemented by collisional deactivation of $O_2(b^1\Sigma_g^+)$, recombination of atomic oxygen, and possibly* by the energy transfer reaction[11]

$$OH^\dagger(v') + O_2 \rightarrow O_2(^1\Delta_g) + OH(v'') \tag{5.50}$$

The radiative lifetime of $O_2(^1\Delta_g)$ is 2.7×10^3 s, and the rate constant for quenching by ground state molecular oxygen is 2.4×10^{-18} cm^3 molecule^{-1} s^{-1} (Section 4.3.1). Thus the radiative and quenching rates are equal at about 70 km.

The Meinel bands of N_2^+ originate from the $A^2\Pi_u$ state, which is excited mainly by resonant scattering and electron impact.[11] Other possible excitation sources are photoionization (5.38) and the charge transfer reaction

$$O^+(^2D) + N_2 \rightarrow N_2^+(A^2\Pi_{v'=1}) + O \tag{5.51}$$

As the radiative lifetime of $N_2^+(A^2\Pi_u)$ is short (3×10^{-6} s), quenching is unimportant in comparison with radiative loss for $N_2^+(A^2\Pi)$ at 150 km.

---

* A recent laboratory investigation of reaction 5.50 has shown no evidence for energy transfer (H. I. Schiff, private communication).

The Meinel bands of OH (550–3000 nm) are always one of the strongest airglow features, during either day or night. They are emitted as fundamental ($\Delta v = 1$) and overtone ($\Delta v > 1$) transitions from vibrationally excited OH in its electronic ground state. The first dayglow measurements (of the $\Delta v = 1$ sequence near 3 μm) were reported in 1964. It was found that the day/night intensity ratio was near unity.[26] Subsequent observations indicated that the intensity of the morning twilight emission first decreased rapidly, and then recovered to approximately the night-time value.[3] Three mechanisms of excitation have been suggested: the chemiluminescent reaction of atomic hydrogen and ozone

$$H + O_3 \rightarrow O_2 + OH^{\dagger}(^2\Pi, v \leqslant 9) \tag{5.29}$$

the chemiluminescent reaction of atomic hydrogen with vibrationally excited O_2

$$H + O_2(v > 4) \rightarrow OH^{\dagger}(^2\Pi) + O \tag{5.52a}$$

and the chemiluminescent reaction between HO_2 and atomic oxygen*

$$HO_2 + O \rightarrow OH(v \leqslant 6) + O_2 \tag{5.52b}$$

Results from both laboratory and field experiments favour reaction 5.29 over 5.52a. There is no laboratory evidence to support reaction 5.52a, whereas vibrationally excited OH is known to be produced in reaction 5.29 with about 80% of the OH radicals in levels with $v > 4$.[29] The decrease in brightness of the OH emission at dawn has also been shown to favour the ozone mechanism.[30] Reaction 5.52b may provide an additional dayglow source of vibrationally excited OH at higher altitudes.

5.3.2 THE TWILIGHT GLOW

Whereas observations of dayglow features only began about 1961, on account of the experimental difficulties, twilight airglow emissions have a long observational history. The sodium twilight glow, for example, was first detected in 1938.[31] The study of twilight emissions provides valuable information on the transition period between the daytime, when resonant scattering and photoelectron excitation are dominant, and night-time, when chemiluminescence is most important. During this short twilight period a considerable change in observational conditions occurs. The sky brightness diminishes and weaker radiations become easier to distinguish from the background. A unique feature of twilight glow observations is that, as the earth's shadow scans through the emitting layer (Fig. 5.13), the variation in emission intensity with time can provide a sensitive indication of the height distribution of a particular emitting species. Because of this height effect,

---

* E. L. Breige, *Planetary and Space Science* **18**, 1271 (1970).

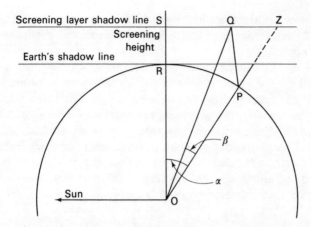

Fig. 5.13   Geometry of twilight illumination. An observer of emission from Q is at P on the earth's surface directly below Z. The tangent at R is the earth's shadow line. If an emission is totally absorbed by the atmosphere, the opaque planetary disc is effectively increased in radius and the shadow line is the screening layer shadow line SZ at a screening height SR. OR is the earth's radius; $(90 + \alpha)$ = solar zenith angle and α = solar depression angle.

twilight observations have been used in favourable cases as an inexpensive substitute for rocket investigations.

In Table 5.2 observed twilight emissions are listed together with the heights and intensities of emission where these have been established. Only those features which can be recognized from within the shadowed region of the atmosphere are included in Table 5.2; emissions from the sunlit area are included in the dayglow. Features observed in the twilight glow but not

Table 5.2   Twilight glow emission features.[a]

λ (nm)	Emitting Species	Intensity (Rayleighs)	Height (km)	Excitation[b] Mechanism	g (photons particle^{-1} s^{-1})	Remarks
388.9	He(^{3}P)	1 R	>400	R	0.1	
391.4 etc	N$_2^+$(B$^2\Sigma_u^+$)	200–500 R	300	RF	0.05	1st negative
393.3, 396.8	Ca$^+$(^{2}P)	≤100 R	80–200	R	0.3, 0.15	
436.8	O(4^3P)	1 R				
520.0	N(^{2}D)	10 R		D	6 × 10^{-11}	
557.7	O(^{1}S)	400 R	200?	e	1 × 10^{-11}	
589.3	Na(^{2}P)	1–4 kR	92	R	0.80	
630, 636.4	O(^{1}D)	1 kR	300	FDe	4.5 × 10^{-10}	
670.8	Li(^{2}P)	10–1000 R	∼90	R	16	May be of artificial origin
769.9	K(^{2}P)	40 R	∼90	R	1.67	
1083.0	He(^{3}P)	3 kR	500	R	16.8	Scatterer is He(^{3}S)
1270 etc	O$_2$(a$^1\Delta$)	5 MR	80	F	9.4 × 10^{-11}	IR atm. (0, 1) band at 1.58 μm
2800 etc	OH($^2\Pi_{v \leqslant 9}$)			C		Meinel

a   Modified from reference 1.
b   Excitation mechanisms are R = resonance scattering; e = photoelectron excitation; F = fluorescence; D = dissociative recombination and C = chemical reaction.

in the dayglow are the metallic resonance lines arising from $Li(^2P)$, $K(^2P)$ and $Ca^+(^2P)$, the helium lines at 388.9 and 1083 nm, and the atomic oxygen line at 436.8 nm.

Li, Na, K *and* Ca^+ *emission* has been detected in spite of the low concentrations of the metal atoms (1 part in 10^{10} or less) because of the very high efficiency of the resonance scattering mechanism. The three alkali metals form a homogeneous group, with similar excitation mechanisms and height distributions. Sodium emission has been extensively studied in both the northern and the southern hemisphere for a number of years. Some of the well-established results that have emerged are the existence of a definite layer of sodium atoms near 90 km (Fig. 5.14), seasonal variations in the sodium atom concentration with a maximum in winter, and diurnal

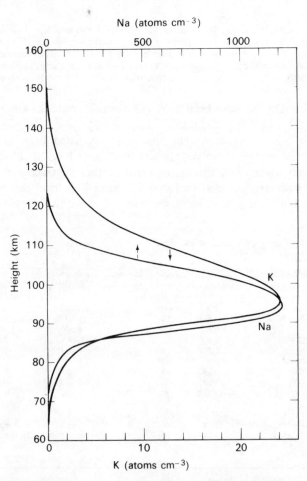

Fig. 5.14 Vertical distribution of atomic potassium and sodium over Saskatoon, from twilight emission measurements.[3]

variations which suggest enhanced daytime concentrations. The chemistry of atmospheric sodium is not well understood, but what evidence there is indicates that sodium atoms will react almost exclusively with the various forms of oxygen.[3] Important reactions are

$$Na + O + M \rightarrow NaO + M \tag{5.53}$$

$$Na + O_2 + M \rightarrow NaO_2 + M \tag{5.54}$$

and

$$Na + O_3 \rightarrow NaO + O_2 \tag{5.55}$$

Free sodium atoms can be regenerated by reaction of the oxides with atomic oxygen, thus

$$NaO_2 + O \rightarrow NaO + O_2 \tag{5.56}$$

$$NaO + O \rightarrow Na + O_2 \tag{5.57}$$

The rates of these reactions are not known under atmospheric conditions. If they are sufficiently large, the ratio of $[Na]$ to $[NaO]$ will be controlled by the $[O]/[O_3]$ ratio. Since the $[O]/[O_3]$ ratio is very large above 90 km, this implies that most of the neutral sodium should be in the form of free atoms. The ratio of $[Na^+]$ to $[Na]$ has not been accurately determined at the time of writing, but the Na^+ densities measured by mass spectrometers are small compared with Na densities derived from airglow observations. The loss rate of atomic sodium through ionization is thought to be small compared with its loss rate by eddy diffusion.[32]

Lithium and potassium emissions have not been as well studied as that of sodium. It appears that the ratio of atom densities is $Na:K:Li \sim 1000:5:1$, with the $[Na]/[K]$ ratio decreasing in the summer to below 100.[11] Possibly the most puzzling question concerning the alkali metals in the atmosphere is that of their origin. The cosmic $[Na]$ to $[K]$ ratio is between 7 and 10 to 1, with a similar ratio being typical of the earth's crust, meteoritic silicates, and the sun's photosphere. In sea water the ratio is 47.2 to 1, and it was once argued that the high atmospheric ratio implied a marine origin. However, if sodium and potassium are present in the atmosphere in dust particles around 10 nm diameter,[3] a lower rate of release of K atoms from the dust particles could also account for the observed $[Na]/[K]$ ratios. Ablation of meteors between 90 and 110 km is a possible source of free alkali metal atoms (Section 6.7.2). The observed meteor fluxes show no relationship to measured sodium atom concentrations,[11] but if eddy diffusion were the dominant mechanism controlling the concentrations of alkali metal atoms and ions, then the absence of a direct correlation between meteor influx and observed twilight glow would not be significant.

Lithium was first observed in August 1957, with an estimated intensity of about 120 R[3]. In August of the following year a much greater intensity was measured, and the increase was attributed to a high altitude thermonuclear explosion above Johnston Island, which took place on August 1st. (Lithium hydride is a constituent of fusion bombs.) Subsequent observations confirmed that lithium intensities increase markedly two to three days after a thermonuclear explosion. The occurrence of this artificial mode of injection has made it very difficult to assess the natural atmospheric lithium content.

Twilight emission has been detected from Ca^+ but not from Ca, presumably because the ratio $[Ca^+]/[Ca]$ is much greater than unity. The same applies to Mg^+ (Section 6.7.2). The intensity of Ca^+ emission behaves in an erratic fashion, which is inconsistent with the presence of a uniform layer of Ca^+, but which can be explained by the presence of discrete clouds of ions, most likely of meteoric origin.

Helium lines at 1083 and 388.9 nm, corresponding to transitions from the 2^3P and 3^3P levels, respectively, into the metastable 2^3S level, have been observed in the twilight glow. Excitation to the 2^3P and 3^3P states is by resonance scattering from the He 2^3S state, which is thought to be produced from ground state $He(1^1S)$ by photoelectrons of energy greater than 19.7 eV.[11] At 400 km $He(2^3S)$ has a lifetime of about 50 s with respect to Penning reactions such as

$$He(2^3S) + O \rightarrow He(^1S) + O^+ + e \qquad (5.58)$$

whereas its radiative lifetime is 4.5×10^4 s.[33]

Atomic oxygen lines which have been observed in the twilight glow are the red and green forbidden lines originating from $O(^1D)$ and $O(^1S)$ at F region altitudes, and a weak line at 436.8 nm arising from the allowed transition (4^3P-3^3S). As in the dayglow, the high altitude emission intensity of the red line at 630 nm is greater than that of the green line at 557.7 nm. For both lines the excitation mechanisms are the same as in the dayglow, namely photodissociation of molecular oxygen in the Schumann–Runge continuum for $O(^1D)$, and photoelectron excitation of O for $O(^1S)$, the latter possibly supplemented by dissociative fluorescence of O_2. The termolecular association of oxygen atoms (reaction 5.27), important to green line excitation in day- and nightglow, probably contributes little to the twilight glow because of the height of the emitting layer. Processes that produce enhancements of the red line are an evening twilight effect due to dissociative recombination of O_2^+

$$O_2^+ + e \rightarrow O + O(^1D) \qquad (5.14-5.17)$$

and a predawn enhancement caused by photoelectrons arriving from the magnetic conjugate point on the opposite side of the earth.[3]

The atmospheric system of O_2 $(^1\Sigma_g^+ - {}^3\Sigma_g^-$, (0, 0) band at 761.9 nm) cannot be seen from the ground because of absorption by O_2 in the lower atmosphere, and is not included in Table 5.2.

The infrared atmospheric system of O_2 $(^1\Delta_g - {}^3\Sigma_g^-)$ is present in the twilight glow, but again the (0, 0) band at 1.27 μm is heavily absorbed in the lower atmosphere, and only a very small fraction reaches the ground. The (0, 1) band at 1.58 μm was first observed in the twilight glow in 1958.[3] Although less intense than the (0, 0) band by about a factor of 70,[26] it is absorbed much less strongly. Measurements of the decay of this emission at evening twilight revealed that it was slower than the decay estimated on the assumption that only the daytime excitation process of photodissociation of ozone provided a source for $O_2(^1\Delta_g)$.[26] The subsequent discovery of an upper layer of $O_2(^1\Delta_g)$ in the daytime profile at 85 km has satisfactorily accounted for this observed excess emission. Ground based twilight observations of the 1.27 μm band from $O_2(^1\Delta_g)$ have also been used to provide a convenient measure of the ozone distribution above 80 km, on the basis of the known rate of photochemical production of $O_2(^1\Delta_g)$ from O_3.[34]

The $N_2(B^2\Sigma_u^+ \rightarrow X^2\Sigma_g^+)$ *first negative system* at 391.4 nm was the first twilight emission to be discovered; observations date back to 1933. As the bulk of the emission originates above the 300 km level, resonance scattering from existing nitrogen ions is considered to be the main excitation mechanism (Fig. 5.7).[3] In the twilight glow, as distinct from the dayglow, the height of the emitting N_2^+ ion layer rises as the solar depression angle increases. Associated with this rise is an increase in the chemical lifetime of N_2^+, up to a maximum of about 60 s (Section 5.3.1).

The Meinel bands of OH are believed to result from the chemiluminescent reaction of atomic hydrogen with ozone (reaction 5.29) during the twilight, just as during both day and night. As the steady state OH concentration is relatively small (Section 4.3.2), little twilight enhancement can be expected from resonance scattering. Balloon observations at morning twilight show a decrease in intensity as sunlight arrives in the 70 km region, followed by a slow build-up over a half-hour period.[3,35] This change in twilight emission is presumably related to changes in the concentrations of H and O_3 at sunrise.

5.3.3   THE NIGHTGLOW

During the day solar radiation is stored in photodissociation products of the atmosphere. This energy is released at night by various recombination and reaction processes, many of which result in light emission. The nightglow is thus largely the result of chemiluminescent reactions; resonance scattering plays only an extremely minor role. The few nightglow emissions which are produced by resonance scattering, notably Lyman-α, are in reality twilight emissions which are observable in the night sky through being emitted at a great altitude. The main problem associated with the study of low altitude nightglow emission is that of finding the correct channel through which excitation energy can be transferred from the main energy store, atomic oxygen, in order to produce the features observed. Spectra observed in the nightglow are listed in Table 5.3. We now consider the most important features in detail.

Band systems of O_2, are a strong feature of the nightglow. Three band systems, all originating below 100 km, have been observed. They are the

Table 5.3   Nightglow emission features.[a]

λ (nm)	Emitting Species	Intensity (Rayleighs)	Height (km)	Excitation[b] Mechanism	Remarks
30.4	$He^+(^2P)$	(4.8)			
58.4	$He(^1P)$	(12)			
102.5	$H(^2P)$	10 R	200	R	Ly-β
121.6	$H(^2P)$	2 kR	$100{-}10^5$	R	Ly-α
130.4	$O(^3S)$	~ 1.7 kR	<500	Ie?⎰	Observed in
135.6	$O(^5S)$	~ 1.4 kR	<500	Ie?⎱	equatorial nightglow
260–380	$O_2(A^3\Sigma_u{}^+)$	600 R	90	C	Herzberg bands
297.2	$O(^1S)$				
391.4 etc.	$N_2{}^+(B^2\Sigma_u{}^+)$	<1 R			1st negative bands
520	$N(^2D)$	1 R	~ 250	D	
500–600	NO_2?	10 R/nm	~ 90	C	continuum
557.7	$O(^1S)$	250 R	90, 300	C, D	
589.3	$Na(^2P)$	20–150 R	~ 92	C	
630, 636.4	$O(^1D)$	10–500 R	300	D	
656.3	$H(3^2P)$	3 R	200	F	
761.9	$O_2(b^1\Sigma_g{}^+)$	6 kR	~ 80	C	Atmospheric bands
1270 etc.	$O_2(a^1\Delta_g)$	80 kR	90?	C?	IR atmospheric bands
2800 etc.	$OH(^2\Pi_i\ v \leqslant 9)$	1 MR	90	C	Meinel OH bands

a   Modified from reference 1.
b   Excitation mechanisms are R = resonance scattering; F = fluorescence; I = radiative recombination; D = dissociative recombination; C = chemical reaction and e = electron impact.

Herzberg system arising from $O_2(A^3\Sigma_u{}^+)$, the atmospheric system from $O_2(b^1\Sigma_g{}^+)$, and the infrared atmospheric system from $O_2(a^1\Delta_g)$. The altitude profiles of the first two systems are shown in Fig. 5.15(a) and that of the infrared atmospheric system in Fig. 5.15(b). It is possible that the same mechanism of oxygen atom association (reaction 5.59) is responsible for all three systems.

$$O + O + M \rightarrow O_2(A^3\Sigma_u{}^+, b^1\Sigma_g{}^+, a^1\Delta_g) + M \qquad (5.59)$$

Laboratory evidence has indicated, however, that if current estimates of the concentration of atomic oxygen below 120 km are valid, process 5.59 is too slow, and must be supplemented by another source.[37] Wallace and Hunten maintain, despite some laboratory evidence to the contrary, that reaction 5.59 is sufficient to produce the observed intensities of the nightglow atmospheric bands.[38] Their calculated zenith intensity agrees well with measured values (Fig. 5.16). Another potential producer of the atmospheric bands is the reaction

$$O + O_3 \rightarrow O_2(^3\Sigma_g{}^-) + O_2(b^1\Sigma_g{}^+) \qquad (5.60)$$

although laboratory evidence indicates electronically excited O_2 is not produced in the reaction. In addition, the vertical distribution of emission based on process 5.60 does not give a good fit with the observed height profile.[38] Among the mechanisms suggested to account for the O_2 infrared

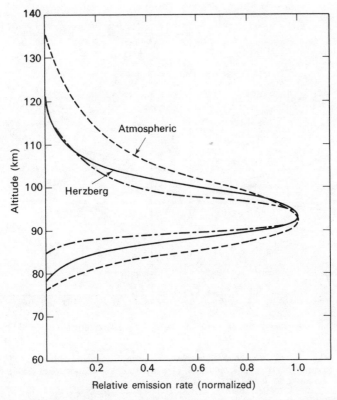

Fig. 5.15a Altitude profiles of the Herzberg and atmospheric bands in the nightglow, from ref. 36. The two curves marked Herzberg refer to two different rocket measurements.

atmospheric bands, in addition to reaction 5.59, are

$$H + O_3 \rightarrow O_2^* + OH^\dagger \tag{5.61}$$

or, alternatively, from the same reactants but in a later step by energy transfer from vibrationally excited OH:

$$OH^\dagger + O_2(^3\Sigma_g^-) \rightarrow O_2(^1\Delta_g) + OH \tag{5.62}$$

At the time of writing no process can be stated unequivocally to be directly responsible for a particular O_2 band system.

Lines of atomic oxygen at 130.4 and 135.6 nm (from the transitions $(^3S-^3P)$ and $(^5S-^3P)$ respectively) have been reported from satellite observations of the equatorial nightglow.[39] These emissions, which originate from altitudes below 500 km, are thought to arise either from radiative recombination of atomic oxygen ions (reaction 5.63) or from electron impact excitation of oxygen atoms.

$$O^+ + e \rightarrow O^* + h\nu \tag{5.63}$$

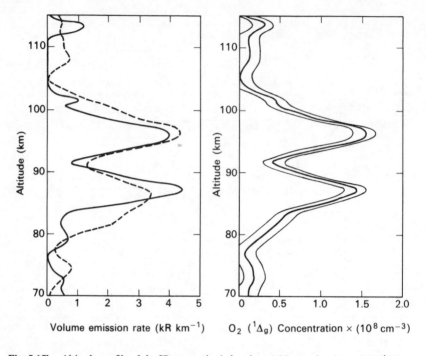

Fig. 5.15b Altitude profile of the IR atmospheric bands at 1.27 μ, and estimated $O_2(^1\Delta_g)$ concentrations at night (W. F. J. Evans, E. J. Llewellyn and A. Vallance Jones, *Journal of Geophysical Research*, **77**, 4899 (1972)).

The prominent atomic oxygen lines of the visible region of the nightglow are, as one would expect, the red and green lines at 630 and 557.7 nm, respectively. Both lines have emission sources at F region altitudes, but the main emitting layer for the green line is located near 90 km. The altitude profiles of emission rates from these lines, as measured by rocket-borne spectrometers, are shown in Fig. 5.17. Atomic oxygen association (reaction 5.27) is almost certainly responsible for the low altitude green line. Excitation mechanisms proposed for the high altitude emissions are the dissociative recombinations of NO^+ and O_2^+ (reactions 5.13–5.20)[40] but the $O(^1D)$ may also be produced by energy transfer from $N(^2D)$ to $O(^3P)$:

$$N(^2D) + O(^3P) \rightarrow N(^4S) + O(^1D) \tag{5.44}$$

With the assumptions that the production mechanism of the F region 630 and 557.7 nm lines is dissociative recombination of O_2^+, that O_2^+ is formed in the F region by a charge-transfer reaction of O^+ with O_2, that quenching of $O(^1D)$ by N_2 is important below 300 km, and that quenching of $O(^1S)$ by N_2 is important below 100 km, then the volume emission rates, R, of the

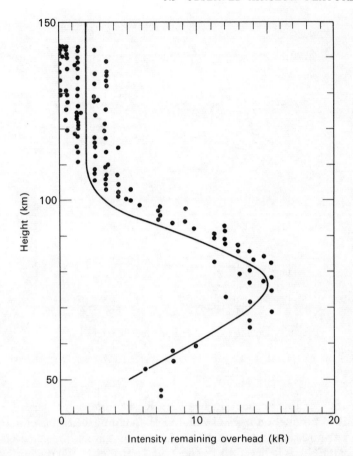

Fig. 5.16 Measured and estimated zenith intensities of the atmospheric bands of $O_2(b^1\Sigma_g^+)$ in the nightglow.[38] The circles refer to observed intensities and the solid curve to calculated intensities.

two lines can be shown to be[40]

$$R_{630} \propto \frac{[e][O_2]}{(1 + k[N_2])};$$ (5.64)

$$R_{630} \propto [e] \text{ below 300 km;}$$

$$R_{630} \propto [e][O_2] \text{ above 300 km}$$

$$R_{557.7} \propto [e][O_2] \text{ above 200 km}$$ (5.65)

Thus the ratio of volume emission rates, $R_{557.7}/R_{630}$, should be independent of altitude above 100 km, and should vary with $[N_2]$ below this altitude.

Atomic nitrogen emission at 520 nm, originating from $N(^2D)$ at F region altitudes, closely correlates with the atomic oxygen line at 630 nm.[23] An

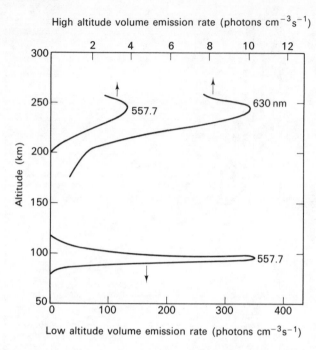

High altitude volume emission rate (photons cm^{-3}s^{-1})

Low altitude volume emission rate (photons cm^{-3}s^{-1})

Fig. 5.17 Altitude profiles of the red (630 nm) and green (557.7 nm) lines of atomic oxygen in the nightglow.[41]

explanation of this correlation has been found from a comparison of the methods of production of N(^{2}D) and O(^{1}D), and of their quenching rates. Dissociative recombination of NO$^+$ (formed by the ion molecule reaction of O$^+$ with N$_2$) is believed to be the main source of N(^{2}D) (reaction 5.20), with quenching by O$_2$ and inelastic collision with electrons the main loss mechanisms. Using these assumptions it can be shown that the volume emission rate is

$$R_{520} \propto \frac{[e][N_2]}{1 + k[O_2] + k'[e]};\qquad (5.66)$$

or

$$R_{520} \propto [e] \text{ below 250 km};$$

$$R_{520} \propto [N_2] \text{ above 250 km}$$

where k and k' are quenching rate constants.[40] From eqns. 5.64 and 5.66 it is evident that the ratio R_{520}/R_{630} should vary as $([N_2]/[O_2])^2$ at low altitudes, and is therefore relatively insensitive to height below about 250 km. At higher altitudes it is predicted that $R_{557.7}/R_{630}$ varies as $[N_2]/[e]$, and should therefore decrease markedly with increasing altitude.

The NO$_2$ emission continuum is a very well known phenomenon in the laboratory. The presence of a green continuum in the nightglow was first reported in 1951.[2] It was attributed to the chemiluminescent reaction between NO and O (reaction 5.26), which is responsible for the familiar 'air afterglow' observed in discharge tubes containing N_2 and O_2, and when excess reagent has been added during the 'titration' reaction of N atoms with NO (see Chapter 3). Laboratory measurements of the overall rate of the process

$$NO + O \xrightarrow{M} NO_2 + hv \qquad (5.26)$$

give $k_{5.26} \sim 6.4 \times 10^{-17}$ cm^3 molecule^{-1} s^{-1} (the actual value depends on the nature of the third-body molecule M), and indicate that the rate is sufficient to account for the observed nightglow intensities.[1] A possible alternative source of excitation of the NO_2 continuum is energy transfer to NO_2 from vibrationally excited O_2:[42]

$$O_2^\dagger + NO_2 \rightarrow NO_2^* + O_2 \qquad (5.67)$$

Atomic sodium emission has long been known to be present in the nightglow, but the mechanism for exciting Na atoms to the ^2P state is still uncertain, mainly because the relevant reaction rates are not known. The sodium nightglow originates from a layer at around 90 km, where the concentration of atomic oxygen is appreciable, and it is very likely that oxygen atoms are involved in the excitation process. One mechanism which has gained wide acceptance involves the formation of vibrationally excited NaO:

$$Na + O + M \rightarrow NaO^\dagger + M \qquad (5.68)$$

$$NaO^\dagger + O \rightarrow Na(^2P) + O_2 \qquad (5.69)$$

$$Na(^2P) \rightarrow Na(^2S) + hv \,(\lambda = 589.0, 589.6 \text{ nm}) \qquad (5.70)$$

or possibly,[43] in place of 5.69

$$NaO^\dagger + O_3 \rightarrow Na(^2P) + 2O_2^\dagger \qquad (5.71)$$

Two other suggestions have been energy transfer to Na from vibrationally excited O_2, and the formation of NaH followed by

$$NaH + O \rightarrow Na(^2P) + OH \qquad (5.72)$$

However, with regard to reaction 5.72 it is questionable whether NaH is as readily produced as NaO in the D region at night.

The Meinel vibration-rotation bands of $OH(^2\Pi)$ are the strongest feature of the nightglow. The region of maximum intensity of the emission, as measured during rocket flights, is around 90 km (Fig. 5.18). The nature of the process (or processes) responsible for excitation of the Meinel bands was for some time a controversial question. The most favoured mechanism since 1953 has been the chemiluminescent reaction between H atoms and O_3, but the alternative reaction between H and vibrationally excited O_2 has been difficult to rule out altogether (see Section 5.3.1). Evans and Llewellyn have argued that previous estimates of intensity of the OH emission based

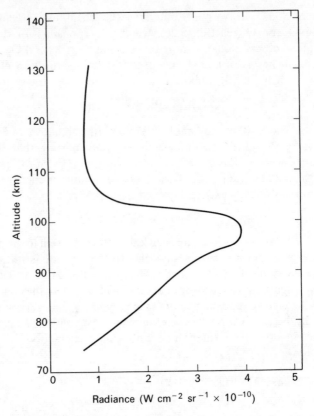

Fig. 5.18 Altitude profile of the OH (8, 3) vibration–rotation band in the nightglow, from a rocket measurement.[40]

on laboratory data for reaction 5.29 have been too high, because of the need to allow for different rates of population of the different vibrational levels. Their calculations lead to a value of 1 MR as the total intensity of the Meinel system.[44]

5.4 The chemistry of excited species

Excited species which are separated from their ground states by forbidden transitions are very likely to lose their energy of excitation by way of chemical reactions, rather than by radiative transitions. Such long-lived or 'metastable' species can therefore be expected to play an important role in atmospheric chemistry. However, excited species which are not especially long-lived may also undergo important chemical reactions provided the product of the rate constant and the reactant concentration is sufficiently large (cf. eqn. 2.10). In this section we are concerned with the collisional quenching processes that remove excited atoms, molecules, or ions, either by simple deactivation or by chemical reaction.

The nightglow emissions from metastable atomic and molecular oxygen provide a good illustration of the role of quenching reactions in the atmosphere. It was noted in Section 5.3.3 that the red line rising from $O(^1D)$ is emitted at high altitudes where the pressure is low, because this state has a radiative lifetime of 110 s and is efficiently quenched by O_2 and N_2. The strong green line of atomic oxygen arises from $O(^1S)$, which has a lifetime of 0.74 s and is also more weakly quenched. Consequently this emission has its maximum intensity at a relatively low altitude, near 100 km. The red line is scarcely detectable from the region of maximum intensity of green emission, even though the lower state of the green transition is $O(^1D)$. The $^1\Delta_g$ state of O_2 has a lifetime considerably greater even than that of $O(^1D)$, but has a very small rate constant for quenching by O_2 or N_2, so that the infrared atmospheric bands are emitted well below 100 km.

Important excited species of the atmosphere are listed in Table 5.4 together with their energies and radiative lifetimes. Most laboratory studies of deactivation rates of excited species have given information only on the rate of the overall quenching process, and not on specific reaction paths, which in general are not well established. The rates of specific chemical reaction paths can be determined by measuring product yields, and much work of this nature is currently in progress. Similarly, rates of particular

Table 5.4 Excited species of importance in the atmosphere.[a]

Ground State	Excited State	Excitation Energy (eV)	Radiative Lifetime (s)
$H(^2S)$	$H(^2P)$	10.2	1.6×10^{-9} [b]
$He(1^1S)$	$He(2^3S)$	19.7	4.5×10^4
$O(^3P)$	$O(^1D)$	1.96	110
	$O(^1S)$	4.17	0.74
$O^+(^4S)$	$O^+(^2D)$	3.31	1.3×10^4
	$O^+(^2P)$	5.0	5
$O_2(X^3\Sigma_g^-)$	$O_2(a^1\Delta_g)$	0.98	2.7×10^3
	$O_2(b^1\Sigma_g^+)$	1.63	12
	$O_2(A^3\Sigma_u^+)$	4.34	~ 1
$O_2^+(X^2\Pi_g)$	$O_2^+(a^4\Pi_u)$	4.04	metastable
$OH(X^2\Pi)_{v=0}$	$OH(X^2\Pi)_{v=9}$	3.48	6×10^{-2}
$N(^4S)$	$N(^2D)$	2.39	9.36×10^4
	$N(^2P)$	3.56	12
$N^+(^3P)$	$N^+(^1D)$	1.89	2.48×10^2
	$N^+(^1S)$	4.04	0.9
$N_2(X^1\Sigma_g^+)$	$N_2(A^3\Sigma_u^+)$	6.17	~ 2
	$N_2(^3\Delta_u)$	7.1	metastable

a Modified from references 22 and 45.
b The effective lifetime is probably considerably longer because of radiation trapping.

energy transfer channels can sometimes be determined by measuring quantum yields of sensitized luminescence. We now consider some individual examples from Table 5.4.

5.4.1   QUENCHING RATES FOR $O(^1D)^*$

Comparatively large amounts of atomic oxygen in the ^1D state are produced by photodissociation of ozone below 100 km and of molecular oxygen at high altitudes. The main quenching agent for $O(^1D)$ in the atmosphere is molecular nitrogen, which becomes vibrationally excited in the process

$$O(^1D) + N_2 \rightarrow N_2^\dagger + O(^3P) \qquad (5.73)$$

The quenching rate for N_2 is compared with those of other quenching species in Table 5.5. The rapid rate of reaction 5.73 enables quenching to

Table 5.5   Quenching coefficients of $O(^1D)$.

Quenching Species	Reported Rate Constant (cm^3 molecule^{-1} s^{-1})	Reference
H_2	1.5×10^{-10}	22
N_2	5×10^{-11}	22
O_2	5×10^{-11}	22
NO	1.5×10^{-10}	22
N_2O	1.8×10^{-10}	22
H_2O	3×10^{-10}	52c
CO	2×10^{-11}	22
CO_2	$(1-2) \times 10^{-10}$	52c
O_3	6×10^{-10}	46

compete effectively with radiation at altitudes up to 250 km. Felder and Young[49] have found that an efficient channel for quenching by N_2 is the transfer of energy from $O(^1D)$ to higher vibrational levels of N_2. This process, followed by 5.74,

$$N_2^\dagger(v' > 0) + e \rightarrow N_2^\dagger(v'' < v') + e + \text{kinetic energy} \qquad (5.74)$$

has been suggested to account for anomalously high electron temperatures in the E region.[22] Similarly, in the analogous reaction with oxygen, where the three possible excited products are:

$$O(^1D) + O_2(^3\Sigma_g^-)_{v''=0} \rightarrow O(^3P) + O_2(^1\Sigma_g^+) \qquad (5.33a)$$

$$\rightarrow O(^3P) + O_2(a^1\Delta_g) \qquad (5.33b)$$

$$\rightarrow O(^3P) + O_2(^3\Sigma_g^-)_{v''>0} \qquad (5.33c)$$

between 20% and 80% of the total quenching occurs via process 5.33c,

---

* The rate coefficients recommended by the Climatic Impact Assessment Program of the U.S. Department of Transportation are listed in the appendix.

producing vibrationally excited O_2.[50] The reaction of $O(^1D)$ with N_2O is believed to be an important source of NO and NO_2 in the stratosphere (see Section 4.3.3 and 7.4).

5.4.2 QUENCHING OF $O(^1S)$

Processes which generate $O(^1S)$ in the atmosphere are dissociative recombination of O_2^+, photoelectron excitation of $O(^3P)$, and the Chapman reaction 5.27. $O(^1S)$ is generally less reactive than $O(^1D)$, and has a shorter lifetime, so that quenching can compete with radiation only up to about 95 km. Quenching rates for various atmospheric gases are listed in Table 5.6;

Table 5.6 Quenching coefficients of $O(^1S)$.

Quenching Species	Reported Rate Constant (cm^3 molecule^{-1} s^{-1})	Reference
H_2	$\sim 2 \times 10^{-15}$	53
O	$1 \times 10^{-13}, 8 \times 10^{-12}$	22, 51
N	1×10^{-12}	51
N_2	$< 5 \times 10^{-17}$	52b
O_2	$4.9 \times 10^{-12} \exp(-860/T)$	52b
NO	8×10^{-10}	52a
CO	1×10^{-14}	52a
H_2O	1×10^{-10}	22, 52a
NH_3	5×10^{-10}	53
N_2O	1.6×10^{-11}	22
NO_2	5×10^{-10}	53
CO_2	$3.3 \times 10^{-11} \exp(-1320/T)$	52b
O_3	6×10^{-10}	52a

it is apparent from this table that N_2 and O_2 are not very effective quenching agents of $O(^1S)$. A surprising feature of Table 5.6 is the rapid quenching by H_2O, O_3 and NO. This fast quenching rate for H_2O suggests that $O(^1S)$ could be an important agent for removing H_2O in the mesosphere. The large difference between the two listed values for the important oxygen atom quenching rate indicates the difficulty of this particular measurement, and also emphasizes the uncertainties that exist in the values of many quenching rate coefficients at the time of writing.

5.4.3 QUENCHING OF METASTABLE MOLECULAR OXYGEN

The most abundant metastable species in the atmosphere is $O_2(a^1\Delta_g)$, which is produced mainly by the photolysis of ozone below 85 km. Other processes that contribute in a minor way include recombination of atomic oxygen (reaction 5.46) and collisional deactivation of $O_2(b^1\Sigma_g^+)$. Photoionization of $O_2(^1\Delta_g)$ is a major source of O_2^+ ions in the D region. Because $O_2(^1\Delta_g)$ has a radiative lifetime of 45 minutes, quenching reactions might be expected to control its concentration. However, as shown by the rate

Table 5.7   Quenching coefficients of $O_2(a^1\Delta_g)$.

Quenching Species	Reported Rate Constant (cm^3 molecule^{-1} s^{-1})	Reference
Ar, He	$\leqslant 10^{-20}$	54
H_2	4×10^{-18}	54
O	$\leqslant 1.3 \times 10^{-16}$	54
N	2.8×10^{-16}	54
N_2	$\leqslant 3 \times 10^{-21}$	52c
O_2	2×10^{-18}	54, 22
NO	5×10^{-17}	54
H_2O	4×10^{-18}	54
NH_3	7×10^{-18}	54
NO_2	5×10^{-18}	54
N_2O	$\leqslant 10^{-19}$	54
CO_2	$< 8 \times 10^{-20}, 4 \times 10^{-18}$	54
O_3	$2 \times 10^{-15}, 2 \times 10^{-14}, 5 \times 10^{-11} \exp(-2830/T)^*$	22, 55, 56
O^-	3×10^{-10}	57
$O_2{}^-$	2×10^{-10}	57

* Temperature range 183–321 K.

constant values in Table 5.7, the $^1\Delta_g$ state has proven remarkably immune from quenching by the major atmospheric gases. Thus oxygen molecules in this state are able to live out their long radiative lifetime below 100 km, and build up to a concentration of about 5×10^9 molecules cm^{-3}. An interesting and somewhat unusual reaction which has been reported for $O_2(^1\Delta_g)$ in the laboratory is the energy pooling process, in which two molecules together emit a single quantum of light:

$$O_2(^1\Delta_g) + O_2(^1\Delta_g) \rightarrow (O_4^*) \rightarrow 2O_2(^3\Sigma_g^-) + h\nu \qquad (5.75)$$

The light emission from reaction 5.75 consists of bands at 634 nm (both ground state molecules left with $v = 0$) and 764 nm (one with $v = 0$, one

Table 5.8   Quenching coefficients of $O_2(b^1\Sigma_g^+)$.

Quenching Species	Reported Rate Constant (cm^3 molecule^{-1} s^{-1})	Reference
He, Ar	1×10^{-17}	54
H_2	$4 \times 10^{-13}, 10^{-12}$	54, 48
N_2	$2 \times 10^{-15}, 5 \times 10^{-13}$	54
O_2	$1.5 \times 10^{-16}, 3 \times 10^{-13}$	54
NO	$6 \times 10^{-14}, 6 \times 10^{-12}$	54
NO_2	2×10^{-14}	54
N_2O	7×10^{-14}	53
NH_3	9×10^{-14}	53
H_2O	5×10^{-12}	54, 48
CO	4×10^{-15}	53
CO_2	$3 \times 10^{-13} - 2 \times 10^{-11}$	54
CH_4	1×10^{-13}	53
O_3	2.5×10^{-11}	52c
SO_2	3×10^{-15}	53

with $v = 1$). Alternatively the products can be one ground state and one $^1\Sigma_g^+$ molecule:

$$O_2(^1\Delta_g) + O_2(^1\Delta_g) \rightarrow O_2(^3\Sigma_g^-) + O_2(b^1\Sigma_g^+) \tag{5.76}$$

where $k_{5.76} = 2.2 \times 10^{-18}$ cm^3 molecule^{-1} s^{-1}.[22]

$O_2(b^1\Sigma_g^+)$ is the upper state of the transition responsible for the atmospheric bands of O_2, and is produced by optical excitation of O_2 and also by energy transfer from $O(^1D)$ to O_2 (reaction 5.48). Measured quenching rates for this state are listed in Table 5.8, from which it appears that the $b^1\Sigma_g^+$ state is generally more reactive than the $a^1\Delta_g$ state, and that molecular nitrogen is the dominant quenching species.

$O_2(A^3\Sigma_u^+)$, the upper state of the Herzberg I system, appears to react rapidly with H_2 according to

$$H_2 + O_2(A^3\Sigma_u^+) \rightarrow 2OH \tag{5.77}$$

but no quantitative data are available.[58]

5.4.4 QUENCHING OF METASTABLE ATOMIC AND MOLECULAR IONS OF OXYGEN

The rates of production of the metastable ions $O^+(^2P)$ and $O^+(^2D)$ have been estimated on the basis of the assumption that they are both produced by photoionization and electron impact on atomic oxygen.[59] The long radiative lifetime of the ^2D state ensures that deactivation occurs by chemical reaction rather than by radiation. At great altitudes deactivation also occurs during inelastic collisions with electrons

$$O^+(^2D) + e \rightarrow O^+(^4S) + e \tag{5.78}$$

but in the F region the important processes are probably a displacement reaction with N_2, and charge transfer with N_2 and O_2.

$$O^+(^2D) + N_2 \rightarrow NO^+ + N \tag{5.79a}$$

$$\rightarrow O + N_2^+ \tag{5.79b}$$

$$O^+(^2D) + O_2 \rightarrow O + O_2^+ \tag{5.80}$$

Rate constants for reactions 5.79 and 5.80 are not well established, but $k_{5.79b} \sim 10^{-9}$ cm^3 molecule^{-1} s^{-1}, and the rate constant of 5.80 is stated to be at least an order of magnitude greater than the corresponding rate constant for ground state $O^+(^4S)$, which is 2×10^{-11} cm^3 molecule^{-1} s^{-1} at 300 K.[60] The charge transfer reaction of $O^+(^2D)$ to N_2 is therefore an important high altitude source of N_2^+ ions.

A considerable fraction of the O_2^+ ions produced in the atmosphere are in the metastable $a^4\Pi_u$ state, but little is known of their subsequent chemistry.[61] Likely deactivation processes are

$$O_2^+(a^4\Pi_u) + O \rightarrow O + O_2^+ \tag{5.81}$$

and
$$O_2^+(a^4\Pi_u) + N_2 \rightarrow O_2 + N_2^+ \tag{5.82a}$$

$$\rightarrow NO^+ + NO \tag{5.82b}$$

$$\rightarrow N_2O^+ + O \tag{5.82c}$$

A consequence of the additional 4.04 eV of energy of the $a^4\Pi_u$ state relative to that of the $X^2\Pi_g$ ground state is that reactions 5.81 and 5.82a, which are endothermic and slow for the ground state species, are rapid for $O_2^+(a^4\Pi_u)$. Thus $k_{5.82a} \sim 2 \times 10^{-10}$ cm^3 molecule^{-1} s^{-1}.[62] Another process that is endothermic for $O_2^+(X^2\Pi_g)$ but exothermic for $O_2^+(a^4\Pi_u)$ is 5.82c; this occurs more slowly than 5.82a.[62]

5.4.5   QUENCHING OF $N(^2D)$

Nitrogen atoms in the ^2D level have a radiative lifetime of 26 hours. Possible production mechanisms have been discussed in Section 4.3.3; they include dissociative recombination of NO^+, electron impact on N_2, and the reaction between N_2^+ and O. Quenching by O_2 is the principal loss process below 250 km:

$$N(^2D) + O_2 \rightarrow NO + O(^3P) \tag{5.42}$$

with a rate coefficient $k_{5.42} = 1.4 \times 10^{-11}$ cm^3 molecule^{-1} s^{-1}.[63] This quenching reaction with O_2 is thought to be responsible for producing a considerable fraction of the total mesospheric nitric oxide. Other measured quenching rates are listed in Table 5.9.

Table 5.9   Quenching rates for $N(^2D)$ and $N(^2P)$.

Quenching Species	Reported Rate Constant (cm^3 molecule^{-1} s^{-1})	Reference
$N(^2D)$		
Ar, He	$<2 \times 10^{-16}$	64
H_2	5×10^{-12}	64
N_2	$<6 \times 10^{-15}, 1 \times 10^{-14}$	64, 65
O_2	$1 \times 10^{-11}, 5 \times 10^{-12}$	22, 65
NO	2×10^{-10}	64
N_2O	2×10^{-12}	22
CO	6×10^{-12}	64
CO_2	6×10^{-13}	64
NH_3	1×10^{-10}	64
$N(^2P)$		
N_2	3×10^{-19}	22

5.4.6 REACTIONS OF $N_2(A^3\Sigma_u^+)$

The metastable species $N_2(A^3\Sigma_u^+)$ gives rise to the Vegard–Kaplan bands $(A^3\Sigma_u^+ \rightarrow X^1\Sigma_g^+)$ of molecular nitrogen, which are observed in the dayglow. Electron impact on N_2 is thought to be the principal excitation process, but other processes that contribute are radiative transitions from the $B^3\Pi_g$ and $C^3\Pi_u$ levels, and the ion–molecule reaction

$$N_2^+ + NO \rightarrow N_2(A^3\Sigma_u^+) + NO^+ \tag{5.83}$$

In the normal atmosphere $N_2(A^3\Sigma_u^+)$ is produced at a rate of about $5 \times 10^9 \, \text{cm}^{-2} \, \text{s}^{-1}$ [61] but Vegard–Kaplan emission does not appear strongly in the dayglow, presumably because the radiative lifetime of approximately $2 \, \text{s}$ [66] permits collisional deactivation to compete effectively with radiation. From the list of measured quenching coefficients in Table 5.10

Table 5.10 Quenching coefficients of $N_2(A^3\Sigma_u^+)$.

Quenching Species	Reported Rate Constant ($\text{cm}^3 \, \text{molecule}^{-1} \, \text{s}^{-1}$)	Reference
H_2	3×10^{-15}	64
N	5×10^{-11}	22
O	$\sim 5 \times 10^{-11}$	22
N_2	10^{-19}	22
$N_2(A^3\Sigma_u^+, v = 0)$	$\sim 2 \times 10^{-11}$	22
O_2	$\sim 10^{-13}$	22
NO	7×10^{-11}	22
N_2O	1×10^{-11}	64
CO	3×10^{-11}	64
CO_2	5×10^{-14}	64

it appears that atomic oxygen must be the main atmospheric quenching species. The energy transferred in the collision (reaction 5.84) is sufficient to populate the $O(^1S)$ level.

$$N_2(A^3\Sigma_u^+) + O(^3P) \rightarrow N_2(X^1\Sigma_g^+) + O(^3P, {}^1D \text{ or } {}^1S) \tag{5.84}$$

Emission of the oxygen green line at 557.7 nm is observed during laboratory studies of this reaction; the NO γ-bands (arising from $NO(A^2\Sigma_g^+)$) similarly occur during quenching by NO.[61] Molecular nitrogen in the $A^3\Sigma_u^+$ state also undergoes an energy pooling process, analogous to that of $O_2(^1\Delta_g)$, to form the higher $C^3\Pi_u$ state.[67]

$$N_2(A^3\Sigma_g^-) + N_2(A^3\Sigma_g^-) \rightarrow N_2(X^1\Sigma_g^+) + N_2(C^3\Pi_u) \tag{5.85}$$

The quenching efficiency of atomic nitrogen for the A state appears to depend markedly on the degree of vibrational excitation of the molecule. Low vibrational levels ($v = 0$ or 1) are efficiently quenched, but an upper limit of 5×10^{-13} cm^3 molecule^{-1} s^{-1} has been estimated for quenching by N when high levels of the A state are populated.[68]

5.4.7   VIBRATIONALLY EXCITED NITROGEN*

In the ground electronic state this is known to be present in the atmosphere as a result of the collisional deactivation of O(^{1}D) by N$_2$ (reaction 5.73). Laboratory studies have shown that the degree of vibrational excitation of N$_2$ has a marked influence on the rates of several important reactions. One such process is the ion–atom interchange

$$O^+ + N_2(v) \rightarrow NO^+ + N \tag{5.86}$$

whose rate is increased by a factor of 20 at a vibrational temperature of 4000 K, by comparison with N$_2$ ($v = 0$).[69]

5.4.8   VIBRATIONALLY EXCITED OH

In the ground electronic state this is, as previously noted, the source of the intense Meinel bands of OH in the infrared. Not much is known about the quenching reactions of OH†,

$$OH^\dagger + O_3 \rightarrow HO_2 + O_2 \tag{5.87a}$$

$$\rightarrow H + 2O_2 \tag{5.87b}$$

$$\rightarrow OH + O_2 + O \tag{5.87c}$$

except that its reaction with ozone[70] depends markedly on the vibrational level of the OH radical.[71] The rate of reaction 5.87a has been measured for OH($v = 9$); $k_{5.87a} = 7.7 \times 10^{-12}$ cm^3 molecule^{-1} s^{-1}. The radiative lifetime of OH($v = 9$) is 6.4×10^{-2} s.[71]

It appears safe to conclude this section with the prediction that, as our present sketchy knowledge of the part played by excited species in atmospheric chemistry is improved, their relative importance will be assessed more highly.

* A discussion of vibrational energy transfer processes in the stratosphere has recently been given by R. L. Taylor, Can. J. Chem., 52, 1436 (1974).

References

1   Hunten, D. M., in 'The Radiating Atmosphere', Ed. McCormac, B. M., D. Reidel Publishing Co., Dordrecht, Holland, 1971.
2   Chamberlain, J. W., 'Physics of the Aurora and Airglow', Academic Press, New York, 1961.
3   Hunten, D. M., Space Sci. Rev., 6, 493 (1967).
4   Nicholls, R. W., Annls Géophys., 20, 144 (1964).
5   Calvert, J. G. and Pitts, J. N. Jnr., 'Photochemistry', John Wiley and Sons Inc., New York, 1966.

6a   Garvin, D., Broida, H. P. and Kostkowski, H. J., *J. chem. Phys.*, **32,** 880 (1960).
6b   Rabinovitch, B. S. and Flowers, M. C., *Q. Rev. chem. Soc.*, **18,** 122 (1964).
7   Karl, G., Kruss, P. and Polanyi, J. C., *J. chem. Phys.*, **46,** 224 (1967).
8   Evans, M. G. and Polanyi, M., *Trans. Faraday Soc.*, **35,** 178 (1939).
9   Pearce, J. B., in *'The Radiating Atmosphere'*, Ed. McCormac, B. M., D. Reidel Publishing Co., Dordrecht, Holland, 1971.
10   Fastie, W. G., Crosswhite, H. M. and Heath, D. E., *J. geophys. Res.*, **69,** 4129 (1964).
11   Rundle, H. W., in *'The Radiating Atmosphere'*, Ed. McCormac, B. M., D. Reidel Publishing Co., Dordrecht, Holland, 1971.
12   Hanson, W. B., *J. geophys. Res.*, **74,** 3720 (1969).
13   Noxon, J. F., *Space Sci. Rev.*, **8,** 92 (1968).
14   Green, A. E. S. and Barth, C. A., *J. geophys. Res.*, **72,** 3975 (1967).
15   Barth, C. A., *Annls Géophys.*, **22,** 198 (1966).
16   Anderson, J. G., *J. geophys. Res.*, **76,** 4634 (1971); **76,** 7820 (1971).
17   Ajello, J. M., *J. chem. Phys.*, **53,** 1156 (1970).
18   Broadfoot, A. L., in *'The Radiating Atmosphere'*, Ed. McCormac, B. M., D. Reidel Publishing Co., Dordrecht, Holland, 1971.
19   Dalgarno, A., McElroy, M. B. and Stewart, A. I., *Planet. Space Sci.*, **14,** 1321 (1966).
20   Schoen, R. I., *Can. J. Chem.*, **47,** 1879 (1969).
21   Wallace, L. and McElroy, M. B., *Planet. Space Sci.*, **14,** 677 (1966).
22   Zipf, E. C., *Can. J. Chem.*, **47,** 1863 (1969).
23   Weill, G. M., in *'Atmospheric Emissions'*, Ed. McCormac, B. M. and Omholt, A., Van Nostrand, New York, 1969.
24   Lawrence, G. M. and McEwan, M. J., *J. geophys. Res.*, **78,** 8314 (1973); Hays, P. B. and Sharp, W. E., *J. geophys. Res.*, **78,** 1153 (1973).
25   Barth, C. A., *Annls Géophys.*, **20,** 182 (1964).
26   Llewellyn, E. J. and Evans, W. F. J., in *'The Radiating Atmosphere'*, Ed. McCormac, B. M., D. Reidel Publishing Co., Dordrecht, Holland, 1971.
27   Nagata, T., Tohmatsu, T. and Ogawa, T., *J. Geomagn. Geoelect.*, **20,** 315 (1968).
28   Evans, W. F. J., Hunten, D. M., Llewellyn, E. J. and Vallance Jones, A., *J. geophys. Res.*, **73,** 2885 (1968).
29   Kaufman, F., *Can. J. Chem.*, **47,** 1917 (1969).
30   Gattinger, R. L., in *'The Radiating Atmosphere'*, Ed. McCormac, B. M., D. Reidel Publishing Co., Dordrecht, Holland, 1971.
31   Kvifte, G., in *'Atmospheric Emissions'*, Eds. McCormac, B. M. and Omholt, A., Van Nostrand, New York, 1969.
32   Gadsden, M., *Annls Géophys.*, **26,** 141 (1970).
33   Mathis, J. S., *Astrophys. J.*, **125,** 318 (1957).
34   Evans, W. F. J. and Llewellyn, E. J., *Radio Sci.*, **7,** 45 (1972).
35   Pick, D. P., Llewellyn, E. J. and Vallance Jones, A., *Can. J. Phys.*, **49,** 897 (1971).
36   Evans, W. F. J. and Llewellyn, E. J., *Annls Géophys.*, **26,** 167 (1970).
37   Young, R. A., *Can. J. Chem.*, **47,** 1927 (1969).
38   Wallace, L. and Hunten, D. M., *J. geophys. Res.*, **73,** 4813 (1968).
39   Barth, C. A. and Schaffner, S., *J. geophys. Res.*, **75,** 4299 (1970); Hicks, G. J. and Chubb, T. A., *J. geophys. Res.*, **75,** 6233 (1970).
40   Noxon, J. F., in *'The Radiating Atmosphere'*, Ed. McCormac, B. M., D. Reidel Publishing Co., Dordrecht, Holland, 1971.
41   Gulledge, I. S., Packer, D. M., Tilford, S. G. and Vanderslice, J. T., *J. geophys. Res.*, **73,** 5535 (1968).
42   Stair, A. T. and Gauvin, H. P., in *'Aurora and Airglow'*, Ed. McCormac, B. M., Reinhold, New York, 1967.
43   Saxena, P., *Annls Géophys.*, **25,** 847 (1969).
44   Evans, W. F. J. and Llewellyn, E. J., *Planet. Space Sci.*, **20,** 625 (1972).
45   McEwan, M. J. and Phillips, L. F., *Acc. Chem. Res.*, **3,** 9 (1969).
46   Webster, H. and Bair, E. J., *J. chem. Phys.*, **56,** 6104 (1972).
47   Biedenkapp, D., Hartshorn, L. G. and Bair, E. J., *Chem. Phys. Lett.*, **5,** 379 (1970).
48   Stuhl, F. and Niki, H., *Chem. Phys. Lett.*, **7,** 473 (1970).
49   Felder, W. and Young, R. A., *J. chem. Phys.*, **57,** 572 (1972).

50   McCullough, D. W. and McGrath, W. D., *Chem. Phys. Lett.*, **8**, 353 (1971).
51   Felder, W. and Young, R. A., *J. chem. Phys.*, **56**, 6028 (1972).
52a London, G., Gilpin, R., Schiff, H. I. and Welge, K. H., *J. chem. Phys.*, **54**, 4512 (1971);
     Welge, K. H., Zia, A., Vietzke, E. and Filseth, S. V., *Chem. Phys. Lett.*, **10**, 13 (1971);
     Filseth, S. V., Stuhl, F. and Welge, K. H., *J. chem. Phys.*, **57**, 4064 (1972).
52b Atkinson, R. and Welge, K. H., *J. chem. Phys.*, **57**, 3689 (1972).
52c Schiff, H. I., *Annls Géophys.*, **28**, 67 (1972).
53   Donovan, R. J. and Husain, D., *Ann. Reports chem. Soc.*, **68**, Section A, 123 (1971).
54   Becker, K. H., Groth, W. and Schurath, U., *Chem. Phys. Lett.*, **8**, 259 (1971).
55   Findlay, F. D. and Snelling, D. R., *J. chem. Phys.*, **54**, 2750 (1971).
56   Becker, K. H., Groth, W. and Schurath, U., *Chem. Phys. Lett.*, **14**, 489 (1972).
57   Fehsenfeld, F. C., Albritton, D. L., Burt, J. A. and Schiff, H. I., *Can. J. Chem.*, **47**,
     1793 (1969).
58   Harteck, P. and Reeves, R. R., *Discuss. Faraday Soc.*, **37**, 82 (1964).
59   Dalgarno, A. and McElroy, M. B., *Planet. Space Sci.*, **14**, 1321 (1966).
60   Turner, B. R., Rutherford, J. A. and Compton, D. K. J., *J. chem. Phys.*, **48**, 1602 (1968);
     Stebbings, R. F., Turner, B. R. and Rutherford, J. A., *J. geophys. Res.*, **71**, 771 (1966).
61   Dalgarno, A., *Annls Géophys.*, **26**, 601 (1970).
62   Ryan, K. R., *J. chem. Phys.*, **51**, 4136 (1969).
63   Slanger, T. G., Wood, B. J. and Black, G., *J. geophys. Res.*, **76**, 8430 (1971).
64   Black, G., Slanger, T. G., St. John, G. A. and Young, R. A., *J. chem. Phys.*, **51**, 116
     (1969).
65   Lin, C. and Kaufman, F., *J. chem. Phys.*, **55**, 3760 (1971).
66   Shemansky, D. E. and Carleton, N. P., *J. chem. Phys.*, **51**, 682 (1969).
67   Stedman, D. H. and Setser, D. W., *J. chem. Phys.*, **50**, 2257 (1969).
68   Weinreb, M. P. and Mannella, G. G., *J. chem. Phys.*, **50**, 3129 (1969).
69   Schmeltekopf, A. L., Fehsenfeld, F. C., Gilman, G. I. and Ferguson, E. E., *Planet.
     Space Sci.*, **15**, 401 (1967).
70   Phillips, L. F. and Schiff, H. I., *J. chem. Phys.*, **37**, 1233 (1962).
71   Potter, A. E., Coltharp, R. N. and Worley, S. D., *J. chem. Phys.*, **54**, 992 (1971).
72   Gauthier, M. and Snelling, D. R., *Chem. Phys. Lett.*, **5**, 93 (1970).

6
The Ionosphere

6.1 Introduction

Towards the end of the nineteenth century, the presence of an electrically conducting zone in the upper atmosphere was postulated to account for the diurnal variation of the earth's magnetic field. In 1901 Marconi succeeded in detecting a radio signal which had been transmitted across the Atlantic from Cornwall to Newfoundland, and a year later Kennelly and Heaviside concluded, independently, that the unexpected success of Marconi's experiment must have been due to deviation of the radio signals by an atmospheric conductive layer located at an altitude of approximately 80 km. For a number of years the properties of this 'ionosphere' were studied by examining the characteristics of the radio signals which it reflected back to earth; indeed, until the advent of rockets this technique, known as the radio-ionosonde, was the only one available for detecting variations in ionospheric electron density with altitude, latitude and time. The designation of the various ionospheric regions arose directly from the use of radio-sounding techniques; thus a very distinct and easily identifiable layer was termed the E layer,* and a second, higher layer was called the F layer. This layer produced multiple echoes, and was subsequently subdivided into F_1 and F_2 regions. The existence of a D region below the E layer was inferred from the observed attenuation of waves reflected from the E and F regions, as a result of passage through the D layer. At times layers of intense ionization were observed at heights close to that of the E region; because of their irregular occurrence these became known as 'sporadic E'.

The initial concept of distinct layers of electrons in the atmosphere was eventually modified to one of more or less well defined regions. In fact there appears to be only one ionospheric layer with a definite peak of electron concentration, namely F_2; the other 'layers' merely correspond to changes in electron density gradient which refract radio signals (see Figs. 6.1(a) and 6.1(b)).

A schematic diagram of the ionospheric regions is given in Fig. 6.2. The D region lies between 60 and 90 km altitude. This region between 60 and 90 km is still referred to as the D region at night, although significant ionization is present only during daylight hours. The E region occurs

* The letter E was used by E. V. Appleton in the 1920's to refer to the electric field of the wave reflected from the first layer he recognized.

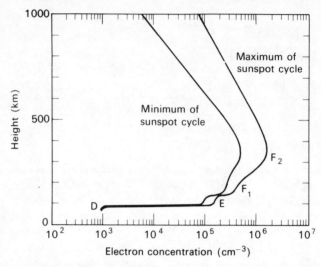

Fig. 6.1a Normal daytime electron distribution at the two extremes of the solar cycle.[1]

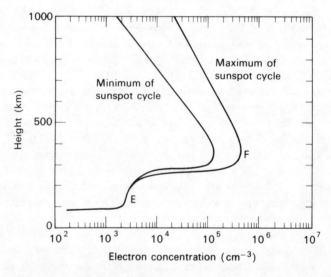

Fig. 6.1b Normal night-time electron distribution at the extremes of the solar cycle.[1]

between 90 and 140 km, and has a daytime electron density about two orders of magnitude greater than that of the D region. The F_1 region has a peak altitude of about 200 km in the daytime and is also essentially absent at night. The maximum electron density of the F_1 region is about 2.5×10^5 electron cm^{-3} near sunspot minimum and 4×10^5 electron cm^{-3} near sunspot maximum. The F_2 region is more erratic, and very sensitive to solar activity. The F_2 peak of electron density, which represents the

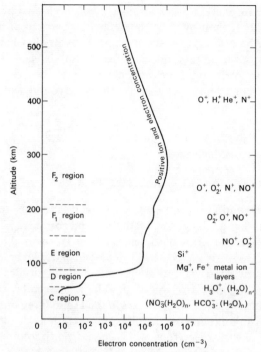

Fig. 6.2 The ionospheric regions.

maximum electron density in the ionosphere, is usually taken as being at 300 km but may vary in altitude between about 200 km and 400 km. A further region below 60 km, the 'C' region, is sometimes postulated, but its existence is difficult to demonstrate conclusively.

Each of the ionospheric regions can be associated with a characteristic type of electron loss process. In the D region, three-body reactions, negative-ion reactions, and electron attachment and detachment are very important, while in the E region the important electron loss process is dissociative recombination, for example

$$NO^+ + e \rightarrow N + O$$

In the high ionosphere (above about 200 or 250 km) electron and ion concentrations are not governed solely by considerations of photochemical equilibrium, because the rates of diffusion of ions and electrons are comparable with the rates of chemical loss processes. Thus the position of the F_2 peak is governed by competition between chemical recombination and diffusion. In the F_1 region, because the principal ions are atomic (see Table 1.1), the characteristic process is an ion–molecule interchange reaction, such as

$$O^+ + N_2 \rightarrow N + NO^+$$

The various ionospheric regions are usually distinguished by their effect on radio waves rather than by their characteristic chemical properties. Nevertheless it can be argued that the chemistry involved, which is our main concern, provides a more valid basis for the subdivision into regions. In this account we shall first discuss the chemistry of the free electrons themselves and then consider the chemistry of the positive and negative ions which are present.

6.2   Electrons in the ionosphere

6.2.1   ELECTRON PRODUCTION AND LOSS PROCESSES

In the middle and lower ionosphere, electrons and ions are produced by radiation from the sun, by galactic radiation and cosmic rays, and by special events such as proton bombardment at high latitudes during polar cap absorption (PCA) events. Generally, charged particle bombardment, or 'corpuscular ionization', is important only at high latitudes because of the magnetic shielding effects of the earth. The relative rates of free electron production from the different processes in the D and E regions are compared in Fig. 6.3, for minimum solar activity and a zenith angle close to 60°. The label $O_2(^1\Delta_g)$ refers to electrons which arise from photoionization of this metastable species, as discussed further in Section 6.3. Once electrons and ions are produced, a variety of processes can occur to transform the primary ions into different chemical species and to reduce both the ion and electron densities; many of these processes are listed in Table 4.2.

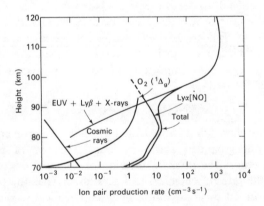

Fig. 6.3   Production of ion pairs in the quiet daytime D and E regions, for a solar zenith angle of about 60°. The results for extreme ultraviolet, Lyman-β, and X-ray ionization are from ref. 2. The photoionization rate of $O_2(^1\Delta_g)$ is from ref. 3 and the photoionization rate of NO by Ly-α from ref. 4.

6.2.2   ELECTRON TEMPERATURE

Primary photoelectrons, produced by ionization of atmospheric gases by solar radiation, commonly have energies much in excess of the mean kinetic energy of the other atmospheric species. Within about a second the photoelectrons lose most of their initial energy by elastic and inelastic collisions, and so reach thermal energies.[5] These thermal electrons then disappear by dissociative recombination with molecular ions and, at lower altitudes, by three-body electron attachment processes. The lifetime of an electron having thermal energy is controlled by the rates of these processes; thus for recombination

$$\tau_e = \frac{1}{k_{ei}[P^+]} \qquad (6.1)$$

where P^+ is the positive ion which is being neutralized, and for electron attachment

$$\tau_e = \frac{1}{k_{ea}[A][M]} \qquad (6.2)$$

where A is the attaching species and M a third body. In the D region at around 80 km, the lifetimes given by equations 6.1 and 6.2 are 10^3–10^4 s and 10–10^2 s, respectively. In the E region the lifetime of thermal electrons is of the order of 10^2 seconds, while in the F region, where the ions are mainly atomic, it is somewhat larger. Thus a photo-electron spends most of its life at thermal energy and our discussion of atmospheric electron chemistry will therefore be centred around the behaviour of thermal electrons.

Below 100 km the electron temperature T_e is the same as the temperature T_g of surrounding atoms and molecules. At higher altitudes, however, T_e considerably exceeds T_g, mainly because of the smaller frequency of collisions between electrons and heavy particles. The temperature profile for the electrons is compared with the gas kinetic temperature and the calculated ion temperature in Fig. 6.4.

6.2.3   IONOSPHERE CONTINUITY EQUATIONS

In any plasma the variation of electron density with time can be described by the continuity equation

$$\frac{d[e]}{dt} = \mathscr{P} - \mathscr{L} + \mathscr{D} \qquad (6.3)$$

where [e] is the electron concentration, $\mathscr{P}$ is the production or ionization rate, $\mathscr{L}$ the electron loss rate due to attachment or recombination processes, and $\mathscr{D}$ the rate of electron loss by diffusion. If there is only one ionizable species present and diffusion may be neglected, equation 6.3 reduces to

$$\frac{d[e]}{dt} = \mathscr{P} - k_{ei}[e][P^+] - k_{ea}[e][A] + k_{cd}[N^-][B] + J_d[N^-] \qquad (6.4)$$

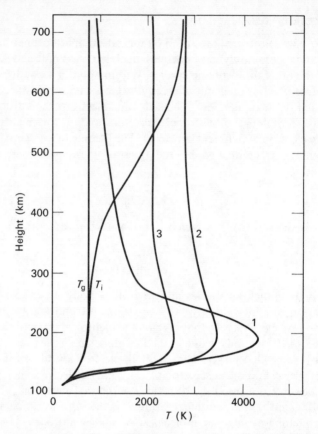

Fig. 6.4 Typical daytime temperature for quiet solar conditions as calculated in ref. 6. (Curves 1, 2 and 3 are calculated electron temperatures such that for 1 heat conduction is ignored; for 2 and 3 heat conduction is included, but in 3 electron–neutral collisions are ignored in the thermal conductivity. T_i is the calculated ion temperature and T_g the neutral atmosphere temperature.)

In equation 6.4, k_{ei}, k_{ea}, k_{cd} and J_d are the rate coefficients for ion–electron recombination, electron attachment, collisional detachment and photodetachment, respectively; [A] is the concentration of attaching species, [B] the concentration of detaching species and [P$^+$], [N$^-$] the concentrations of positive and negative ions, respectively. As both the positive and negative ion concentrations vary with time we can write

$$\frac{d[P^+]}{dt} = \mathscr{P} - k_{ei}[e][P^+] - k_{ii}[P^+][N^-] \tag{6.5}$$

and

$$\frac{d[N^-]}{dt} = k_{ea}[e][A] - k_{ii}[N^-][P^+] - k_{cd}[N^-][B] - J_d[N^-] \tag{6.6}$$

where k_{ii} is the ion–ion recombination coefficient. Within any finite volume the ionosphere is electrically neutral, i.e.

$$[P^+] = [e] + [N^-] \tag{6.7}$$

Below 200 km we can make the simplifying assumption of ignoring diffusion, but the solution of the ionospheric continuity equations 6.4–6.7 still amounts to solution of a set of coupled non-linear differential equations. If the simplifying assumptions of an ionosphere with one ionizing species P, one attaching species A, and one detaching species B are removed, the continuity equations are even more difficult to handle. The ionization process must be summed over all ionizing species present, and the same applies to the processes of ion–electron recombination, electron attachment and detachment, and photo-detachment.

Under normal daytime conditions the rate of disappearance of negative ions by ion–ion recombination ($k_{ii}[P^+][N^-]$) can be neglected in comparison with the rates of attachment ($k_{ea}[e][A]$) and detachment ($k_{cd}[N^-][B]$). Assuming, then, a steady state concentration of negative ions, equation 6.6 reduces to

$$k_{ea}[e][A] = k_{cd}[N^-][B] + J_d[N^-] \tag{6.8}$$

or

$$\frac{[N^-]}{[e]} = \frac{k_{ea}[A]}{(k_{cd}[B] + J_d)} = \lambda \tag{6.9}$$

With the parameter λ defined by equation 6.9, it can be shown for the simple one-species ionosphere[7]

$$\frac{d[e]}{dt} = \frac{\mathscr{P}}{(1 + \lambda)} - (k_{ei} + \lambda k_{ii})[e]^2 - \frac{[e]}{(1 + \lambda)} \cdot \frac{d\lambda}{dt} \tag{6.10}$$

Under most conditions (but not at sunrise or sunset) λ varies slowly with time, and we may assume $d\lambda/dt \sim 0$. Equation 6.10 therefore reduces to

$$\frac{d[e]}{dt} = \frac{\mathscr{P}}{(1 + \lambda)} - (k_{ei} + \lambda k_{ii})[e]^2 \tag{6.11}$$

Further simplifications are possible in equation 6.11, because under normal daytime conditions $d[e]/dt \sim 0$, and the rate of photodetachment is much greater than that of collisional detachment, i.e. $J_d \gg k_{cd}[B]$. Therefore

$$\mathscr{P} \sim (1 + \lambda)(k_{ei} + \lambda k_{ii})[e]^2 \tag{6.12}$$

where

$$\lambda \sim \frac{k_{ea}[A]}{J_d}$$

Reference is often made in the literature to an *effective* electron recombination coefficient, k_{eff}, which can be calculated in two ways, either from equation 6.11,

$$\frac{d[e]}{dt} = \frac{\mathscr{P}}{(1 + \lambda)} - k'_{eff}[e]^2 \tag{6.13}$$

or, alternatively, by using

$$(1 + \lambda)\frac{d[e]}{dt} = \mathscr{P} - k_{eff}[e]^2 \tag{6.14}$$

Hence

$$k'_{eff} = (k_{ei} + \lambda k_{ii}) \tag{6.15}$$

and

$$k_{eff} = (1 + \lambda)(k_{ei} + \lambda k_{ii}) \tag{6.16}$$

Values of k_{eff} can be obtained from measured values of electron concentration and its variation with time, together with calculated values of the electron production rate. Unfortunately, confusion has sometimes arisen because of failure of the experimenter to state which form of k_{eff} has been used.[8] Clearly, from 6.15 and 6.16, the two coefficients k_{eff} and k'_{eff} are equal only when $\lambda \sim 0$, or when $[N^-] \ll [e]$. The results of some attempts at calculating k_{eff} and λ in the D region are shown in Figs. 6.5 and 6.6. The probable reason for the large discrepancy between the two sets of results for k_{eff} in Fig. 6.5 is that the production rates $\mathscr{P}$ used in references 4 and 9 were based on different nitric oxide distributions. The curve derived from reference

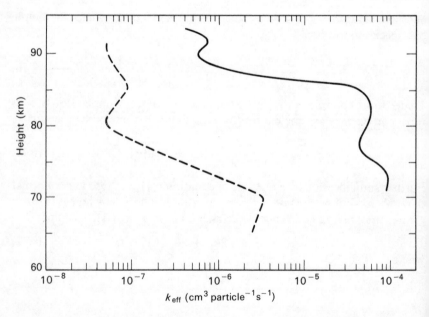

Fig. 6.5 Height variation of k_{eff}, the effective electron loss coefficient as determined in ref. 4 (full line) and ref. 9 (dotted line).

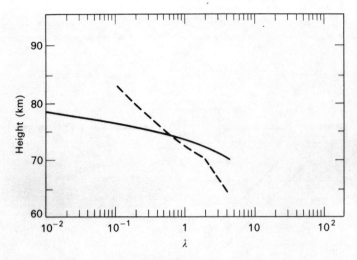

Fig. 6.6 Height variation of $\lambda = [N^-]/[e]$ during the daytime as estimated in ref. 4 (full line) and ref. 9 (dotted line).

4 is preferred because the nitric oxide figures used are considered to be more reliable.

6.2.4 ELECTRON CHEMISTRY

To briefly summarize the results of the previous sections, fast electrons produced by primary ionization processes lose their excess energy fairly rapidly, and are removed more slowly at thermal energies by the processes of attachment to neutral species, recombination with positive ions, and diffusion. Where equilibrium between the various species is attained the electron production and loss rates will balance and a small steady state concentration will be established (equation 6.12). A basic reaction scheme for the production and loss of electrons is outlined in Fig. 6.7.

The important chemical processes influencing the electron concentration in the atmosphere are shown, together with their rate coefficients, in Table 6.1. Electron attachment and electron–ion recombination rates are dependent on the electron energy, but the nature of this dependence is considered to be outside the scope of this book; we are concerned here with electrons at thermal or near thermal energies. The two important electron–ion recombination processes of the atmosphere are *dissociative* recombination and *radiative* recombination. Dissociative recombination may be thought of as a two-step process:

$$XY^+ + e \rightleftharpoons (XY^*) \rightleftharpoons X + Y + \text{kinetic energy} \qquad (6.17)$$

The radiationless electron capture step, in which there is no energy change, leads to the formation of an unstable excited XY molecule. This intermediate

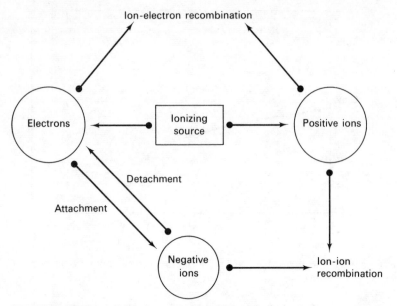

Fig. 6.7   Basic chemical reaction scheme for electrons in the ionosphere.

has a higher probability of dissociation to neutral X and Y than auto-ionization to $XY^+ + e$.

The three-body stabilized recombination process (6.18)

$$XY^+ + e + M \rightarrow XY + M \tag{6.18}$$

is slow compared with two body dissociative recombination, and the three-body rate coefficient decreases rapidly with increasing electron temperature. In the E region, the rate of dissociative recombination exceeds that of three-body recombination by more than a factor of 10^5. A comparison of the rates of the two main electron–ion recombination processes, (radiative recombination and dissociative recombination) shows that radiative recombination becomes the main electron loss process only when the ratio of atomic to molecular ions is 10^4 or greater, conditions which are attained in the magnetosphere, protonosphere, and possibly in the sporadic E layer.

Three-body attachment (reaction 6.19)

$$e + X + M \rightarrow X^- + M \tag{6.19}$$

results from a collision between an electron, an electronegative gas molecule, and a third body which carries away sufficient energy to stabilize the excited negative ion. It is interesting to note that, whereas three-body attachment is the main electron attachment process for low energy electrons in O_2, it is not important in either N_2 or CO_2. On the other hand dissociative attachment,

$$XY + e \rightarrow X^- + Y \tag{6.20}$$

Table 6.1 **Chemical production and loss processes of ionospheric electrons. Unless otherwise stated, rate coefficients are at 300 K. Units of the coefficients for two- and three-particle processes respectively are: cm³ particle^{-1} s^{-1} and cm⁶ particle^{-2} s^{-1}.**

Process		Rate Coefficient	Ref.
Electron–Ion Recombination			
(1) *Radiative Recombination*			
$P^+ + e$	$\to P^* + h\nu^\alpha$	$k_{ei} \sim 4 \times 10^{-12}(T/300)^{-0.7}$	10
(2) *Dissociative Recombination*			
$NO^+ + e$	$\to N + O^\beta$	$k_{ei} = 4.5 \times 10^{-7}(T/300)^{-1.0}$	10
$O_2^+ + e$	$\to O + O^\beta$	$k_{ei} = 2 \times 10^{-7}(T/300)^{-0.7}$	10
$N_2^+ + e$	$\to N + N^\beta$	$k_{ei} = 3 \times 10^{-7}(T/300)^{-0.02}$	10
$CO_2^+ + e$	$\to CO + O$	$k_{ei} = 4 \times 10^{-7}$	10
$(NO)_2^+ + e$	$\to NO + NO$	$k_{ei} = 1.7 \times 10^{-6}$	10
$O_4^+ + e$	$\to O_2 + O_2$	$k_{ei} \sim 2.3 \times 10^{-6}$ at 205 K	10
$H_3O^+ + e$	$\to H + H_2O^{\gamma,\varepsilon}$	$k_{ei} = 1 \times 10^{-6}$ at 540 K	10
$H^+(H_2O)_2 + e$	$\to H + 2H_2O^\gamma$	$k_{ei} = 2 \times 10^{-6}$ (540 K), 2.2 $\times 10^{-6}$ (415 K)	10
$H^+(H_2O)_3 + e$	$\to H + 3H_2O^\gamma$	$k_{ei} = 4 \times 10^{-6}$ (540 K), 3.8 $\times 10^{-6}$ (300 K)	10
$H^+(H_2O)_4 + e$	$\to H + 4H_2O^\gamma$	$k_{ei} = 4.9 \times 10^{-6}$	10
$H^+(H_2O)_5 + e$	$\to H + 5H_2O^\gamma$	$k_{ei} = 6 \times 10^{-6}$ (205 K)	10
$H^+(H_2O)_6 + e$	$\to H + 6H_2O^\gamma$	$k_{ei} \approx 1 \times 10^{-5}$ (205 K)	10
$N_4^+ + e$	$\to N_2 + N_2$	$k_{ei} \sim 2 \times 10^{-6}$	10
Electron Attachment			
(1) *Three-body attachment*			
$e + O_2 + O_2$	$\to O_2^- + O_2$	$k_{ea} = 1.4 \times 10^{-29} (300/T)\exp(-600/T)$ (195 $< T <$ 600 K)	11
$e + O_2 + N_2$	$\to O_2^- + N_2$	$k_{ea} = 1 \times 10^{-31}$ (300 $< T <$ 500 K)	11
$e + O_2 + H_2O$	$\to O_2^- + H_2O^\gamma$	$k_{ea} = 1.4 \times 10^{-29}$ (300 $< T <$ 400 K)	11
$e + O_2 + CO_2$	$\to O_2^- + CO_2$	$k_{ea} = 3.3 \times 10^{-30}$ (300 $< T <$ 525 K)	11
$e + NO + NO$	$\to NO^- + NO^\gamma$	$k_{ea} = 8 \times 10^{-31}$	14a
$e + N_2O + N_2O$	$\to N_2O^- + N_2O^\gamma$	$k_{ea} = 6 \times 10^{-33}$	11
$e + NO_2 + N_2$	$\to NO_2^- + N_2^\gamma$	$k_{ea} = 4 \times 10^{-11}$ cm³ molecule^{-1} sec^{-1} (Independent of N₂ pressure in the range 3–70 torr).	11
(2) *Dissociative Attachment*			
$e + O_3$	$\to O^- + O_2$	$k_{ea} = 5 \times 10^{-12}$	ζ
(3) *Radiative Attachment*			
$e + O$	$\to O^- + h\nu$	$k_{ea} = 1.3 \times 10^{-15}$ (150 $< T <$ 500 K)	11
$e + O_2$	$\to O_2^- + h\nu$	$k_{ea} \sim 10^{-19\pm1}$	11
$e + OH$	$\to OH^- + h\nu$	$k_{ea} \sim 10^{-15\pm1}$	11
Electron Detachment$^\gamma$			
(1) *Collisional Detachment with Neutrals*			
$O_2^- + O_2$	$\to O_2 + O_2 + e$	$k_{cd} = 2.7 \times 10^{-10}$ $(T/300)^{1/2} \exp(-5590/T)$ (375 $< T <$ 600 K)	11

Table 6.1   (Continued).

Process		Rate Coefficient	Ref.
$O_2^- + N_2$	$\rightarrow O_2 + N_2 + e$	$k_{cd} = 1.9 \times 10^{-12}$ $(T/300)^{3/2} \exp(-4990/T)$	11
$O^- + O_2$	$\rightarrow O + O_2 + e$	$k_{cd} = 2.3 \times 10^{-9} \exp(-26\,000/T_i)^{\delta}$ $(4000 < T_i < 20\,000 \text{ K})$	11
$O^- + N_2$	$\rightarrow O + N_2 + e$	$k_{cd} = 2.3 \times 10^{-9} \exp(-26\,000/T_i)^{\delta}$ $(4000 < T_i < 20\,000 \text{ K})$	11
$O_2^- + O_2(^1\Delta_g)$	$\rightarrow O_2 + O_2 + e$	$k_{cd} \sim 2 \times 10^{-10}$	13
$NO^- + CO_2$	$\rightarrow NO + CO_2 + e$	$k_{cd} = 10^{-11}$	14a
$NO^- + N_2O$	$\rightarrow NO + N_2O + e$	$k_{cd} = 6 \times 10^{-12}$	14a

(2) *Associative Detachment* [γ]

Process		Rate Coefficient	Ref.
$H^- + H$	$\rightarrow H_2 + e$	$k_{ad} = 1.3 \times 10^{-9}$	12
$H^- + CO$	$\rightarrow HCO + e$	$k_{ad} \sim 5 \times 10^{-11}$	40
$H^- + NO$	$\rightarrow HNO + e$	$k_{ad} = 4.6 \times 10^{-10}$	40
$H^- + O_2$	$\rightarrow HO_2 + e$	$k_{ad} = 1.2 \times 10^{-9}$	12
$O^- + O$	$\rightarrow O_2 + e$	$k_{ad} = 1.4 \times 10^{-10}$	13
$O^- + N$	$\rightarrow NO + e$	$k_{ad} = 2.0 \times 10^{-10}$	12
$O^- + H_2$	$\rightarrow H_2O + e$	$k_{ad} = 6.0 \times 10^{-10}$	12
$O^- + NO$	$\rightarrow NO_2 + e$	$k_{ad} = 1.6 \times 10^{-10}$	12
$O^- + N_2$	$\rightarrow N_2O + e$	$k_{ad} < 5 \times 10^{-13}$	40
$O^- + CO$	$\rightarrow CO_2 + e$	$k_{ad} = 4.4 \times 10^{-10}$	12
$O^- + CO_2$	$\rightarrow CO_3 + e$	$k_{ad} < 1 \times 10^{-13}$	12
$O^- + SO_2$	$\rightarrow SO_3 + e$	$k_{ad} = 7 \times 10^{-10}$	12
$C^- + CO$	$\rightarrow C_2O + e$	$k_{ad} = 4.1 \times 10^{-10}$	40
$C^- + N_2O$	$\rightarrow CO + N_2 + e$	$k_{ad} = 9 \times 10^{-10}$	40
$C^- + CO_2$	$\rightarrow 2CO + e$	$k_{ad} = 4.7 \times 10^{-11}$	40
$O^- + O_2(^1\Delta_g)$	$\rightarrow O_3 + e$	$k_{ad} \sim 3 \times 10^{-10}$	13
$O_2^- + O$	$\rightarrow O_3 + e$	$k_{ad} = 3 \times 10^{-10}$	12
$O_2^- + N$	$\rightarrow NO_2 + e$	$k_{ad} = 5.0 \times 10^{-10}$	12
$OH^- + H$	$\rightarrow H_2O + e$	$k_{ad} = 1.8 \times 10^{-9}$	40
$OH^- + O$	$\rightarrow HO_2 + e$	$k_{ad} = 2.0 \times 10^{-10}$	12
$OH^- + N$	$\rightarrow HNO + e$	$k_{ad} < 1 \times 10^{-11}$	12
$CN^- + H$	$\rightarrow HCN + e$	$k_{ad} = 8 \times 10^{-10}$	12
$S^- + H_2$	$\rightarrow H_2S + e$	$k_{ad} < 1 \times 10^{-15}$	40
$S^- + O_2$	$\rightarrow SO_2 + e$	$k_{ad} = 3 \times 10^{-11}$	12
$SO_3^- + H_2O$	$\rightarrow H_2SO_4 + e$	$k_{ad} < 1 \times 10^{-12}$	40
$HS^- + H$	$\rightarrow H_2S + e$	$k_{ad} = 1.3 \times 10^{-9}$	40
$O_3^- + O$	$\rightarrow 2O_2 + e$	$k_{ad} = 1 \times 10^{-10}$	14b
$Cl^- + H$	$\rightarrow HCl + e$	$k_{ad} = 1 \times 10^{-9}$	40

(3) *Photodetachment*

Process		Rate Coefficient	Ref.
$O^- + h\nu$	$\rightarrow O + e$	see Fig. 6.8(a)	15
$O_2^- + h\nu$	$\rightarrow O_2 + e$	see Fig. 6.8(b)	16
$OH^- + h\nu$	$\rightarrow OH + e$	see Fig. 6.8(c)	15
$O_3^- + h\nu$	$\rightarrow O_3 + e$	see Fig. 6.8(a)	17

α P^+ represents a positive atomic ion. The recombination coefficient is not particularly sensitive to the identity of the positive ion, varying by less than a factor of 2 in going from H^+ to K^+.

β   Rates chosen for $T_e = T_i = T_g$.

γ   Products not established.

δ T_i is the calculated ion temperature.

ε   Some workers believe the products to be $H_2 + OH$.

ζ   D. Stelman, J. L. Moruzzi and A. V. Phelps, *Journal of Chemical Physics* **56**, 4183 (1972).

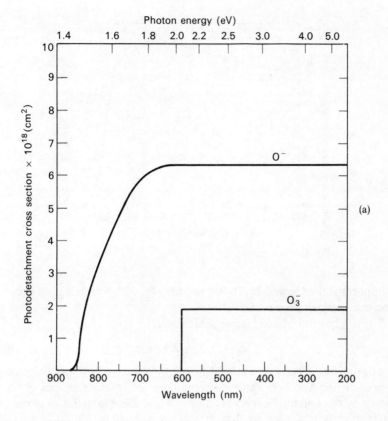

(a)

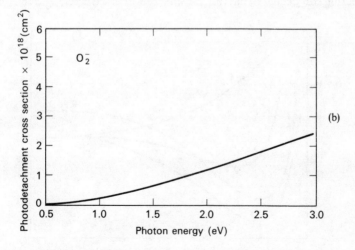

(b)

Fig. 6.8   Photodetachment cross sections for O^-, O_3^-, O_2^- and OH^- as a function of photon energy. (Fig. 6.8a from ref. 17, 6.8b from ref. 16 and 6.8c from ref. 15.)

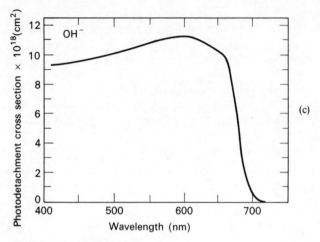

Fig. 6.8 **(Continued)**

is important for O_3 and N_2O, but not for O_2, with which it only occurs at electron energies above 4.5 eV.[11]

The mechanism of associative detachment

$$X + Y^- \rightarrow XY + e \qquad (6.21)$$

can be understood with the aid of Fig. 6.9.[12] When the two species X and Y^- approach one another, three different types of interaction can occur, as shown by the potential curves labelled 1, 2 and 3 in Fig. 6.9. For curve 1 the interaction is repulsive so that at low energies only reflection or elastic scattering can occur. In curve 2 the interaction is attractive and the potential

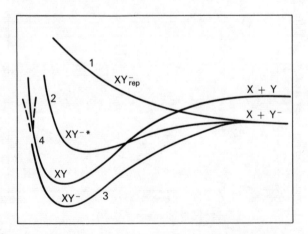

Fig. 6.9 Schematic representation of potential energy curves involved in associative detachment.

curve for XY⁻* cuts the XY potential curve, curve 4. Associative detachment may therefore occur as a result of a non-radiative transition from curve 2 to curve 4. In this second case the XY⁻* potential curve lies above that of the XY curve and the auto-detachment process

$$XY^{-*} \rightarrow XY + e \qquad (6.22)$$

is exothermic. Non-radiative transitions are commonly very rapid, in which case the probability of auto-detachment is large compared with that for dissociation of XY⁻*. For curve 3, the XY⁻ ion is more stable than the molecule XY and associative detachment is unlikely to occur. This case is uncommon; most molecules do not have stable negative ions. The occurrence in the atmosphere of the associative detachment reaction

$$O_2^- + O \rightarrow O_3 + e \qquad (6.23)$$

has an important bearing on radio propagation characteristics of the ionosphere. Because of reaction 6.23, the ratio of electrons to negative ions is higher than it would otherwise be. Electrons in the D region attenuate radio-frequency electromagnetic waves, but negative ions do not, a fact well demonstrated by the day and night variation of radio reception. At night the electron density of the D region drops to near zero, not only because of the large decrease in photo-ionization but also because the rate of reaction 6.23 is less than its daytime value as a result of the lower O atom concentration.

6.2.5 ELECTRON CONCENTRATIONS

The electron concentration in the lower ionosphere has been measured by a number of different techniques, involving both ground-based observations and rocket probes. The results of some rocket measurements are shown in Figs. 6.10a and 6.10b for day and night respectively. The effect of changing production rate with time is indicated in Fig. 6.11, which shows the variation of the electron concentration at a number of heights with the solar zenith angle χ.

A notable feature of many rocket probe measurements of the electron concentration of the D region is the large gradient in electron density observed between 80 and 90 km (Fig. 6.10a). A rapid change in electron density in the D region implies, of course, a rapid change in either production or loss rates. Reid[4] analysed the processes affecting the concentrations of electrons in the D region and concluded that the change in electron density was mainly due to a change in the rate of dissociative recombination. Rocket sampling of the D region has shown an abrupt disappearance of ions derived from water molecules around 85 km (see Section 6.6.1) and it is reasonable to suppose that this disappearance is related to the sharp ledge in the electron density profiles at around the same altitude. Because of the small amount of water vapour in the mesosphere, it is necessary for the

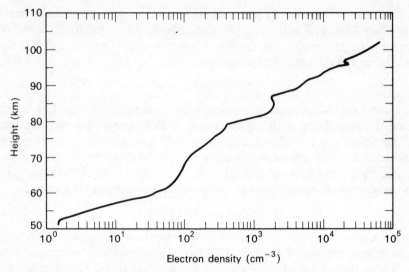

Fig. 6.10a　Rocket measurement of height distribution of electrons in the daytime, for a solar zenith angle near 60°.[18]

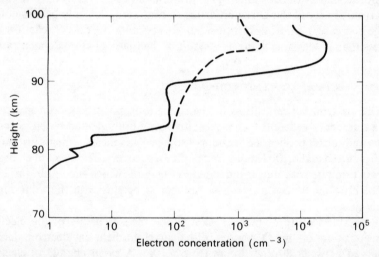

Fig. 6.10b　Rocket measurements of the height distribution of electrons at night in the lower ionosphere. The full curve is from ref. 19, the dashed curve from ref. 20.

dissociative recombination rate of water cluster ions to be extremely large for a change in water vapour concentration to affect the electron density. This requirement is not contradicted by present experimental knowledge.

At higher altitudes, as the concentration of molecular ions decreases, the chemical lifetime of thermal-energy electrons increases. In the E and F regions the main molecular ions are NO^+ and O_2^+. In the F region, however, the electron temperature is about 1000 K or higher (Fig. 6.4) and the rate of

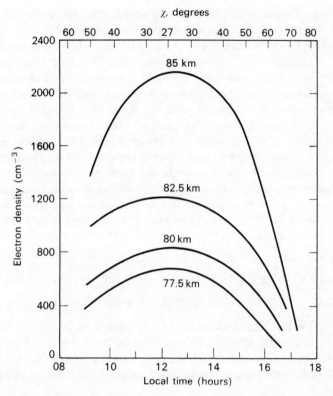

Fig. 6.11 Time variation of electron concentration at specified heights, from ground based observations in Crete.[21]

dissociative recombination is lower because k_{ei}, the dissociative recombination rate coefficient, is generally a decreasing function of temperature.

6.3 Daytime production of ions

6.3.1 THE LOWER IONOSPHERE

The major production processes of ions in the lower ionosphere under quiet daytime conditions are shown in Fig. 6.3. It is evident that, except at very low altitudes, the most important process is photoionization. Lyman-α (121.6 nm) radiation from the sun provides the largest single contribution to the ionization between 75 and 90 km, even though it has sufficient energy (10.19 eV) to ionize only minor atmospheric constituents such as nitric oxide, metastable excited species, and some metal atoms. The importance of nitric oxide as a source of ions in the D region, despite the low concentration of NO compared with O_2, N_2 and to a lesser extent $O_2(^1\Delta_g)$, results from the high intensity of the solar Lyman-α line, which amounts to about 5×10^{-3} W m^{-2} at the top of the earth's atmosphere (cf. Section 2.2).

The ionization threshold of $O_2(^1\Delta_g)$ is 111.8 nm, whereas for ground state $O_2(^3\Sigma_g^-)$ it is 102.7 nm. Radiation of wavelength less than 111.8 nm is able to ionize significant amounts of $O_2(^1\Delta_g)$ because the ionization cross section of this species is large in regions which correspond to windows in the absorption spectrum of ground state O_2. Radiation of higher energy than 102.7 nm is largely absorbed by ground state O_2, which becomes the primary source of O_2^+ at higher altitudes. Appreciable concentrations of $O_2(^1\Delta_g)$ are present in the D and lower E regions, and its photoionization provides a significant contribution to the total photoionization rate. The rate of production of O_2^+ ions from $O_2(^1\Delta_g)$, shown in Fig. 6.3, was calculated by Paulsen *et al.*[3] with allowance for the absorption by carbon dioxide of radiation between 102.7 and 111.8 nm. The production rate $\mathscr{P}(O_2^+)$ in ion cm^{-3} s^{-1} is given by[3]

$$\mathscr{P}(O_2^+)$$
$$= [O_2(^1\Delta_g)](0.549 \times 10^{-9} \exp(-2.406 \times 10^{-20}[O_2]) + 2.614 \times 10^{-9}$$
$$\times \exp(-8.508 \times 10^{-20}[O_2])) \qquad (6.24)$$

where $[O_2]$ is the column density of O_2 in molecules per cm^2 column and $[O_2(^1\Delta_g)]$ is the metastable concentration in molecule cm^{-3}. When the absorption by CO_2 is considered, the rate of production of O_2^+ via $O_2(^1\Delta_g)$ is found to be significantly less than the original estimates of Hunten and McElroy.[22]

The relatively hard 0.2–0.8 nm X-rays penetrate deep into the mesosphere and these, together with cosmic rays, ionize all atmospheric constituents.

A comparison of the ion production rates of Fig. 6.3 with the known electron loss rates highlights what is perhaps the major problem of D region ion chemistry, and also puts in perspective the reliability of current estimates of the NO concentration and other significant variables. Current estimates of ion production rates are about an order of magnitude too large to account for the observed electron densities, and do not satisfactorily reproduce the experimental observations of relative positive ion composition at 80 km.[23]

6.3.2 THE UPPER IONOSPHERE

In the daytime the primary source of ions and electrons in the E and F regions is soft X-ray and extreme ultraviolet radiation from the sun, at wavelengths below about 102.7 nm. Important components of this ionizing radiation are the hydrogen Lyman-β line (102.6 nm), the C III line at 97.7 nm, the He I and II lines at 58.4 and 30.4 nm, and soft X-rays of wavelength less than 1 nm. Molecular oxygen is ionized by Lyman-β and shorter wavelength radiation, atomic oxygen is ionized below 91.04 nm, molecular nitrogen below 79.58 nm and atomic nitrogen below 85.31 nm. Atomic nitrogen is not, however, an important source of ions because its concentration is small in comparison with those of N_2, O_2 and O around 200 km. This is a consequence of the lack of an efficient process for photodissociating molecular nitrogen. The principal ions formed directly by ionizing radiation in the earth's atmosphere are therefore N_2^+ and O_2^+ with smaller amounts of O^+ and, at lower altitudes, NO^+. Some metal atoms, for example Na and Mg, are also likely to be ionized directly.

The intensity of solar radiation falling on the upper atmosphere is markedly dependent on the degree of surface activity of the sun. In Table 6.2 we

Table 6.2 Solar fluxes for solar minimum[24] conditions and solar medium[25] conditions.

Wavelength Range (nm)	F_λ^a (*Solar min.*) (10^9 photons cm^2 s^{-1})	F_λ^b (*Solar med.*) (10^9 photons cm^2 s^{-1})
131.0–102.7 (excluding H Ly-α)	80	~17.5
121.6 (H Ly-α)	270	~300
102.7–91.1 (excluding H Ly-β and C III)	7.1	3.7
102.57 (H Ly-β)	2.3	3.5
97.7 (C III)	4.0	4.4
91.1–80.0	13.4	8.3
80.0–63.0	5.6	2.4
63.0–46.0	9.6	4.7
46.0–37.0	2.0	0.63
37.0–28.0	9.2	10.3
28.0–20.5	3.5	4.5^c
20.5–15.3	~7.5	4.6^c
15.3–10		0.4^c

a   The integrated solar flux in the specified wavelength ranges for solar minimum conditions are less reliable than the later solar medium fluxes. The integrated values shown in this column are thought to be too large.[25]
b $F_{10.7\,cm} = 144$
c $F_{10.7\,cm} = 177$

compare the fluxes for solar minimum conditions ($F_{10.7}^* \sim 72 \times 10^{-22}$ W m^{-2} Hz^{-1})[24] with those for medium solar activity ($F_{10.7}^* \sim 130–170 \times 10^{-22}$ W m^{-2} Hz^{-1}).[25] The solar flux varies according to the position in the 11 year solar cycle, and also with the 27 day period of the sun's rotation. Figure 6.12 shows the variation during one solar rotation of individual components of the XUV (extreme ultraviolet)† radiation, together with the

---

* This refers to the 10.7 cm wavelength of the solar coronal emission, which provides a convenient indication of the level of solar activity. Measurements of the solar flux at 10.7 cm (2800 MHz) are made daily at many locations (cf. Section 2.2.1).
† The alternative abbreviation EUV is sometimes used to describe very short wavelength ultra-violet radiation ($\lambda < 100$ nm).

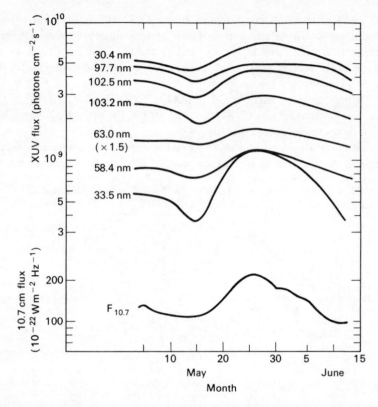

Fig. 6.12 Variation of extreme ultraviolet (XUV) solar fluxes derived from satellite measurements during a solar rotation.[25]

variation of the solar flux monitored at 2800 MHz. Total ion production rates, including primary and secondary ionization of N_2, O_2 and O arising from solar XUV radiation (102.7 nm > λ > 10.0 nm), are shown in Fig. 6.13.

6.4 Production of ions at night

Removal of the main ionizing source, solar radiation, drastically reduces the ionization rate but does not eliminate it altogether. It was formerly thought that there were no ionizing sources in the ionosphere at night, but observations of electron concentrations in the E and lower F regions have shown that the electron density never drops below 10^3 cm^{-3} and may sometimes exceed 10^4 cm^{-3}. The observed persistence of many atmospheric ions such as N_2^+ and O_2^+ at night also implies the presence of an ionizing source of sufficient energy to produce up to ten ion pairs per cm^3 per second in the regions below 140 km and above 180 km.[27,28]

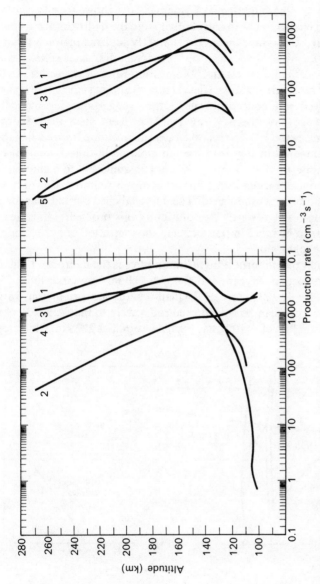

Fig. 6.13 Estimated total photoionization rates in the E and F regions from XUV solar radiation (102.7 nm > λ > 10 nm) for solar minimum conditions.[26] Primary production is shown in the left figure and secondary production is shown in the right figure. Curve 1 is total ion production rate; curve 2, O_2^+; curve 3, O^+; curve 4, N_2^+ and curve 5, N^+.

The occurrence of the Lyman-α line in the nightglow, as a result of resonant scattering of solar Lyman-α by hydrogen atoms in the outermost layers of the atmosphere, points to one possible ionizing source in the D region. Measurements of the night-time Lyman-α flux show wide variations, associated with changes in both solar activity and the hydrogen atom density in the exosphere. An average flux at the top of the ionizing region is typically $4\,\text{kR}\,(\sim 5 \times 10^{-6}\,\text{W m}^{-2}\,\text{sec}^{-1}\,\text{sr}^{-1})^{29}$ and for a nitric oxide concentration of around $3 \times 10^{7}\,\text{cm}^{-3}$ at 80 km,[30] ionization rates of the order of 10^{-2} ion pairs $\text{cm}^{-3}\,\text{sec}^{-1}$ are possible. Clearly this rate is not sufficient to account for the observed ion concentrations. Other suggestions for night-time ionization sources have been X-ray emission from stars (e.g. Scorpius XR 1) and cosmic rays.[31,32] The relative rates of ionization from each of these sources are compared in Fig. 6.14 for the quiet night-time D region. It is evident that none of these ionization rates is sufficient to maintain the observed ion concentrations and a further source is required. The nature of this source has not been established. The curve labelled corpuscular ionization may go some way towards the solution of this problem; ionization by particles appear to be more important than the ionization of nitric oxide by Lyman-α in the night-glow.

In the E region the important night-time source is the hydrogen Lyman-β line ionizing molecular oxygen, and the two helium resonance lines, He I (58.4 nm) and He II (30.4 nm). H Lyman-β radiation originating in the hydrogen geo-corona can be transported radiatively to the dark side of the earth. A Lyman-β flux of ~ 10 R has been measured at 227 km, which leads

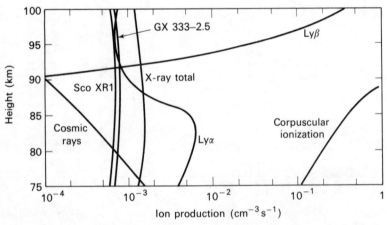

Fig. 6.14   Ionization rates in the quiet night-time D and lower E regions, estimated for a latitude of 30°N and a local time of 0500 hours.[32] The curves labelled Sco XR1 and GX 333–2.5 refer to ionization by X-ray stars of these designations. The curves labelled Lyα and Lyβ refer to ionization by scattered solar Lyman α and Lyman β radiation. The curve labelled corpuscular ionization refers to the estimated rate of ion pair production from particle radiation.[33]

to a calculated production rate of 0.43 O_2^+ ions cm^{-3} s^{-1} at around 100 km.[34] The helium line intensities are between 4 and 12 R. These three ultraviolet lines, all produced by resonant scattering, may be sufficient to maintain the night-time E and F_1 regions.[34]

Data obtained from satellites have shown that the electron temperature exceeds the ion temperature at night, an observation which implies the existence of a night-time heat source. Particle bombardment of 4 keV protons on the atmosphere (at a flux of 5×10^7 protons cm^{-2} s^{-1}) can produce 10^{-8} W cm^{-2} (column) s^{-1} of neutral heating, and about 10^{-10} W cm^{-2} (column) s^{-1} of electron heating, with the largest energy deposition occurring near 200 km.[35] The normal night-time ionosphere may therefore prove to be maintained by a combination of heating and ionization in the F region with downward diffusion of protons from the protonosphere, ionization by UV resonance lines in the nightglow, and ionization by corpuscular radiation.

6.5   Ion concentrations and reaction rates

Our understanding of atmospheric ion chemistry is based on direct measurements of the relevant ionospheric parameters. The measured quantities include ion and neutral particle densities and the solar flux. The results of these investigations define the areas within which laboratory investigations are required in order to provide explanations for the field observations. In recent years much progress has been made on both field and laboratory studies, and the subject of atmospheric ion chemistry has changed very rapidly. In 1965 the Handbook of Geophysics and Space Environments[36] listed 15 ion–molecule reactions thought to be important in the ionosphere. Of these 15, only 5 had been studied in the laboratory, generally with poor agreement among the results obtained by different investigators. Within five years all of the listed reactions had been studied, and the rate coefficients of the important reactions had been established to within a factor of two. The first rocket flights recording the observation of hydrated positive ions were reported in 1965. In 1967, Ferguson[37] commented, in a review of ionospheric chemistry, that whereas the rate coefficients of the important ion-neutral reactions for the E and F regions were known to within a factor of two, in the D region many of the important reactions were as yet unmeasured. In 1971, the same author was able to review in some detail[38] the complicated reaction schemes which control the D region positive ion chemistry, and was able to compare predictions based on laboratory data with observations from rocket flights. The first rocket flights with detailed monitoring of negative ion concentrations in the D region were also reported in 1970–71, and results found so far have agreed with predictions based on laboratory data. Many of the reactions of positive ions of importance to ionospheric chemistry are listed, with their rate coefficients, in Table 6.3.

Table 6.3 Positive ion–molecule reactions of relevance to planetary ionospheres[a]

Reaction		Rate Coefficient[b]	Reference
$H^+ + O$	$\rightarrow O^+ + H$	3.8×10^{-10}	39
$H^+ + NO$	$\rightarrow NO^+ + H$	1.9×10^{-9}	39
$H^+ + CO_2$	$\rightarrow COH^+ + O$	3×10^{-9}	40
$H_2^+ + H_2$	$\rightarrow H_3^+ + H$	2×10^{-9}	40
$H_2^+ + N_2$	$\rightarrow N_2H^+ + H$	2×10^{-9}	40
$H_2^+ + Ar$	$\rightarrow ArH^+ + H$	1.2×10^{-9}	40
$H_3^+ + N_2$	$\rightarrow N_2H^+ + H_2$	1.5×10^{-9}	40
$H_3^+ + CO$	$\rightarrow COH^+ + H_2$	1.4×10^{-9}	40
$H_3^+ + NO$	$\rightarrow NOH^+ + H_2$	1.4×10^{-9}	40
$H_3^+ + CO_2$	$\rightarrow CO_2H^+ + H_2$	1.9×10^{-9}	40
$H_3^+ + NO_2$	$\rightarrow NO^+ + OH + H_2$	7×10^{-10}	40
$H_3^+ + N_2O$	$\rightarrow N_2OH^+ + H_2$	1.8×10^{-9}	40
$H_3^+ + CH_4$	$\rightarrow CH_5^+ + H_2$	1.6×10^{-9}	40
$H_3^+ + H_2O$	$\rightarrow H_3O^+ + H_2$	$\sim 3 \times 10^{-9}$	40
$H_3^+ + NH_3$	$\rightarrow NH_4^+ + H_2$	$\sim 3.6 \times 10^{-9}$	40
$He^+ + O_2$	$\left.\begin{array}{l}\rightarrow He + O + O^+ \\ \rightarrow O_2^+ + He\end{array}\right\}$	1×10^{-9}	40
$He^+ + N_2$	$\rightarrow He + N + N^+$ $\rightarrow He + N_2^+$	$(12 \pm 5) \times 10^{-10}$ Branching $N^+ \sim 65\%$ $N_2^+ \sim 35\%$	41
$He^+ + NO$	$\rightarrow N^+ + O + He$	2×10^{-9}	40
$He^+ + CO$	$\rightarrow C^+ + O + He$	2×10^{-9}	40
$O^+ + H$	$\rightarrow H^+ + O$	$\sim 8 \times 10^{-10}$	42
$O^+ + H_2$	$\rightarrow OH^+ + H$	2×10^{-9}	47
$O^+ + O_2$	$\rightarrow O + O_2^+$	2×10^{-11}	41
$O^+ + NO$	$\rightarrow NO^+ + O$	$<1.3 \times 10^{-12}$	43
$O^+ + N_2$	$\rightarrow NO^+ + N$	1.3×10^{-12}	41
$O^+ + NO_2$	$\rightarrow NO_2^+ + O$	1.6×10^{-9}	44
$O^+ + N_2O$	$\rightarrow N_2O^+ + O$	2.2×10^{-10}	40
$O^+ + N_2O$	$\rightarrow NO^+ + NO$	2.3×10^{-10}	40
$O^+ + CO_2$	$\rightarrow O_2^+ + CO$	1.2×10^{-9}	41
$O^+ + H_2O$	$\rightarrow H_2O^+ + O$	2.3×10^{-9}	40
$O^+(^2D) + N_2$	$\rightarrow N_2^+ + O$	$\sim 1 \times 10^{-9}$	40
$O_2^+ + N$	$\rightarrow NO^+ + O$	1.8×10^{-10}	41
$O_2^+ + NO$	$\rightarrow NO^+ + O_2$	6.3×10^{-10}	43
$O_2^+ + N_2$	$\rightarrow NO^+ + NO$	$<1 \times 10^{-15}$	41
$O_2^+ + NO_2$	$\rightarrow NO_2^+ + O_2$	6.6×10^{-10}	40
$O_2^+ + NH_3$	$\rightarrow NH_3^+ + O_2$	2.4×10^{-9}	40
$N^+ + H_2$	$\rightarrow NH^+ + H$	7×10^{-10}	47
$N^+ + O_2$	$\rightarrow NO^+ + O$	3.5×10^{-10}	41
$N^+ + O_2$	$\rightarrow N + O_2^+$	4.5×10^{-10}	41
$N^+ + CO$	$\rightarrow CO^+ + N$	5×10^{-10}	41
$N^+ + CO_2$	$\rightarrow CO_2^+ + N$	1.3×10^{-9}	41
$N_2^+ + O$	$\rightarrow N_2 + O^+$	$<1 \times 10^{-11}$	41
$N_2^+ + O$	$\rightarrow NO^+ + N$	1.4×10^{-10}	45
$N_2^+ + H_2$	$\rightarrow N_2H^+ + H$	1.5×10^{-9}	40
$N_2^+ + O_2$	$\rightarrow N_2 + O_2^+$	6.6×10^{-11}	40
$N_2^+ + N$	$\rightarrow N^+ + N_2$	$<1 \times 10^{-11}$	40
$N_2^+ + NO$	$\rightarrow NO^+ + N_2$	3.3×10^{-10}	45
$N_2^+ + CO$	$\rightarrow CO^+ + N_2$	7×10^{-11}	41
$N_2^+ + CO_2$	$\rightarrow CO_2^+ + N_2$	9×10^{-10}	41
$N_2^+ + H_2O$	$\rightarrow H_2O^+ + N_2$	2.2×10^{-9}	40
$N_2^+ + H_2O$	$\rightarrow N_2H^+ + OH$	2×10^{-9}	40
$NO^+ + O_3$	$\rightarrow NO_2^+ + O_2$	$<10^{-14}$	40
$N_2O^+ + H_2$	$\rightarrow N_2OH^+ + H$	4×10^{-10}	47
$N_2H^+ + CO_2$	$\rightarrow CO_2H^+ + N_2$	9.2×10^{-10}	40
$N_2H^+ + N_2O$	$\rightarrow N_2OH^+ + N_2$	8×10^{-10}	40
$N_2H^+ + CH_4$	$\rightarrow CH_5^+ + N_2$	9×10^{-10}	40
$N_2H^+ + H_2O$	$\rightarrow H_3O^+ + N_2$	$\sim 5 \times 10^{-10}$	40
$C^+ + O_2$	$\rightarrow CO^+ + O$	1×10^{-9}	41

Table 6.3 (Continued).

Reaction		Rate Coefficient[b]	Reference
$C^+ + CO_2$	$\rightarrow CO^+ + CO$	1.9×10^{-9}	41
$C^+ + H_2O$	$\rightarrow COH^+ + H$	2×10^{-9}	40
$CO^+ + H_2$	$\rightarrow COH^+ + H$	2×10^{-9}	47
$CO^+ + O$	$\rightarrow O^+ + CO$	1.4×10^{-10}	39
$CO^+ + N$	$\rightarrow$ Products	$<2 \times 10^{-11}$	40
$CO^+ + NO$	$\rightarrow NO^+ + CO$	3.3×10^{-10}	39
$CO^+ + O_2$	$\rightarrow O_2^+ + CO$	2×10^{-10}	41
$CO^+ + CO_2$	$\rightarrow CO_2^+ + CO$	1.1×10^{-9}	41
$CO^+ + H_2O$	$\rightarrow$ Products	2.2×10^{-9}	40
$S^+ + NO$	$\rightarrow NO^+ + S$	4.2×10^{-10}	40
$S^+ + O_2$	$\rightarrow SO^+ + O$	1.6×10^{-11}	40
$SO_2^+ + O_2$	$\rightarrow O_2^+ + SO_2$	2.8×10^{-10}	40
$H_2O^+ + O_2$	$\rightarrow O_2^+ + H_2O$	$\sim 2 \times 10^{-10}$	48
$H_2O^+ + H_2O$	$\rightarrow H_3O^+ + OH$	1.8×10^{-9}	48
$Ar^+ + H_2$	$\rightarrow ArH^+ + H$	7×10^{-10}	40
$Ar^+ + N_2$	$\rightarrow N_2^+ + Ar$	5×10^{-11}	40
$Ar^+ + CO$	$\rightarrow CO^+ + Ar$	9×10^{-11}	40
$Ar^+ + NO$	$\rightarrow NO^+ + Ar$	3×10^{-10}	40
$Ar^+ + O_2$	$\rightarrow O_2^+ + Ar$	5×10^{-11}	40
$Ar^+ + H_2O$	$\left.\begin{array}{l} \rightarrow H_2O^+ + Ar \\ \rightarrow ArH^+ + OH \end{array}\right\}$	1.4×10^{-9}	40
$Ar^+ + CO_2$	$\rightarrow CO_2^+ + Ar$	7×10^{-10}	40
$Ar^+ + CH_4$	$\rightarrow CH_3^+ + H + Ar$	6.5×10^{-10}	40
	$\rightarrow CH_2^+ + H_2 + Ar$	1.4×10^{-10}	40
$CO_2^+ + H$	$\left.\begin{array}{l} \rightarrow HCO^+ + O \\ \rightarrow H^+ + CO_2 \end{array}\right\}$	$\sim 6 \times 10^{-10}$	40
$CO_2^+ + H_2$	$\rightarrow CO_2H^+ + H$	1.4×10^{-9}	47
$CO_2^+ + N$	$\rightarrow$ Products	$\leqslant 10^{-11}$	45
$CO_2^+ + O$	$\left.\begin{array}{l} \rightarrow O_2^+ + CO \\ \rightarrow O^+ + CO_2 \end{array}\right\}$	2.6×10^{-10}	45
$CO_2^+ + NO$	$\rightarrow NO^+ + CO_2$	1.2×10^{-10}	45
$CO_2^+ + O_2$	$\rightarrow O_2^+ + CO_2$	5×10^{-11}	45
$NO_2^+ + NO$	$\rightarrow NO^+ + NO_2$	2.9×10^{-10}	40

Binary Reactions of Clusters

$N_4^+ + O_2$	$\rightarrow O_2^+ + 2N_2$	4×10^{-10}	40
$N_4^+ + H_2O$	$\rightarrow H_2O^+ + 2N_2$	1.9×10^{-9}	48
$NO^+.NO + NH_3$	$\rightarrow NO^+.NH_3 + NO$	1.3×10^{-9}	49
$NO^+.NO + H_2O$	$\rightleftarrows NO^+.H_2O + NO$	$k_f = 1.4 \times 10^{-9}$	49
		$k_r = 9 \times 10^{-14}$	49
$NO^+.NH_3 + NH_3$	$\rightarrow NH_4^+ + ONNH_2$	9×10^{-10}	50
$NO^+.CO_2 + H_2O$	$\rightarrow NO^+.H_2O + CO_2$	$\sim 1 \times 10^{-9}$	40
$NO^+.H_2O + NH_3$	$\rightarrow NH_4^+ + HNO_2$	1×10^{-9}	40
$NO^+.(H_2O)_3 + H_2O$	$\rightarrow H^+(H_2O)_3 + HNO_2$	3×10^{-10}	51
$O_2^+.N_2 + N_2$	$\rightarrow O_2^+ + N_2 + N_2$	2×10^{-11}	c
$O_2^+.N_2 + O_2$	$\rightarrow O_4^+ + N_2$	$>5 \times 10^{-11}$	40
$O_2^+.N_2 + H_2O$	$\rightarrow O_2^+.H_2O + N_2$	4×10^{-9}	c
$O_2^+.H_2O + H_2O$	$\rightarrow H_3O^+.OH + O_2$	1×10^{-9}	40
$O_2^+.(H_2O)_2 + H_2O$	$\left.\begin{array}{l} \rightarrow H^+.(H_2O)_2(OH) + O_2 \\ \rightarrow H^+.(H_2O)_2 + OH + O_2 \end{array}\right\}$	6.3×10^{-11}	48
$NH_3^+ + NH_3$	$\rightarrow NH_4^+ + NH_2$	1.7×10^{-9}	40
$H_3O^+ + NH_3$	$\rightarrow NH_4^+ + H_2O$	2.1×10^{-9}	40
$H_3O^+(H_2O) + NH_3$	$\rightarrow$ products	2.6×10^{-9}	40
$H_3O^+(H_2O)_2 + NH_3$	$\rightarrow$ products	1.6×10^{-9}	40
$H_3O^+(H_2O)_3 + NH_3$	$\rightarrow$ products	2.1×10^{-9}	40
$NH_4^+(H_2O) + NH_3$	$\rightarrow NH_4^+(NH_3) + H_2O$	1.2×10^{-9}	40
$NH_4^+(H_2O)_2 + NH_3$	$\rightarrow NH_4^+(NH_3)(H_2O) + H_2O$	$\geqslant 9 \times 10^{-10}$	40
$NO_2^+.H_2O + NH_3$	$\rightarrow NH_4^+ + HNO_3$	1.1×10^{-9}	40

Table 6.3   (Continued).

Reaction		Rate Coefficient[b]	Reference
Three-body reactions of non cluster ions			
$H^+ + H_2 + H_2$	$\rightarrow H_3^+ + H_2$	3×10^{-29}	d
$N^+ + N_2 + N_2$	$\rightarrow N_3^+ + N_2$	5×10^{-29}	46
$N_2^+ + N_2 + N_2$	$\rightarrow N_2.N_2^+ + N_2$	8×10^{-29}	46
$O_2^+ + N_2 + N_2$	$\rightarrow O_2.N_2^+ + N_2$	8×10^{-31}	c
$O_2^+ + O_2 + O_2$	$\rightarrow O_4^+ + O_2$	2.6×10^{-30}	d
$NO^+ + NO + NO$	$\rightleftarrows NO^+.NO + NO$	$k_f = 5 \times 10^{-30}$	49
		$k_r = 9 \times 10^{-16}$	49
$NO^+ + NO + H_2O$	$\rightarrow NO^+.H_2O + NO$	1.6×10^{-28}	49
$NO^+ + N_2 + M$	$\rightarrow NO^+.N_2 + M$	$\sim 3.5 \times 10^{-31}$	52
$NO^+ + NH_3 + NO$	$\rightarrow NO^+.NH_3 + NO$	5×10^{-28}	50
$NH_4^+ + NH_3 + NO$	$\rightarrow NH_4^+.NH_3 + NO$	1×10^{-27}	50
$H_3O^+ + H_2O + N_2$	$\rightleftarrows H^+.(H_2O)_2 + N_2$	$k_f = 3.4 \times 10^{-27}$	46
		$k_r = 7 \times 10^{-26}$	46
Three-body reactions of cluster ions			
$H^+.(H_2O)_2 + H_2O + O_2$	$\rightarrow H^+.(H_2O)_3 + O_2$	2×10^{-27}	48
$H^+.(H_2O)_2 + H_2O + N_2$	$\rightleftarrows H^+.(H_2O)_3 + N_2$	$k_f = 2.3 \times 10^{-27}$	48
		$k_r = 7 \times 10^{-18}$	48
$H^+.(H_2O)_3 + H_2O + NO$	$\rightleftharpoons H^+.(H_2O)_4 + NO$	$k_f = 3.5 \times 10^{-27}$	49
		$k_r = 1.3 \times 10^{-14}$	49
$H^+.(H_2O)_3 + H_2O + N_2$	$\rightleftharpoons H^+.(H_2O)_4 + N_2$	$k_f = 2.4 \times 10^{-27}$	46
		$k_r = 4 \times 10^{-14}$	46
$H^+.(H_2O)_3 + H_2O + O_2$	$\rightleftharpoons H^+.(H_2O)_4 + O_2$	$k_f = 2 \times 10^{-27}$	48
		$k_r = 4 \times 10^{-14}$	48
$H^+.(H_2O)_4 + H_2O + NO$	$\rightleftharpoons H^+.(H_2O)_5 + NO$	$k_f/k_r = 3.7 \times 10^{-16}$ cm^3	49
$H^+.(H_2O)_4 + H_2O + O_2$	$\rightleftharpoons H^+.(H_2O)_5 + O_2$	$k_f = 0.9 \times 10^{-27}$	48
		$k_r = 6 \times 10^{-12}$	48
$NO^+.NO + NO + NO$	$\rightleftharpoons NO^+.(NO)_2 + NO$	$k_f/k_r = 1 \times 10^{-18}$ cm^3	49
$NO^+.H_2O + H_2O + NO$	$\rightleftharpoons NO^+.(H_2O)_2 + NO$	$k_f = 1.1 \times 10^{-27}$	49
		$k_r = 1.4 \times 10^{-14}$	49
$NO^+.(H_2O)_2 + H_2O + NO$	$\rightleftharpoons NO^+.(H_2O)_3 + NO$	$k_f = 1.9 \times 10^{-27}$	49
		$k_r = 1.9 \times 10^{-12}$	49
$NH_4^+.NH_3 + NH_3 + NO$	$\rightleftharpoons NH_4^+.(NH_3)_2 + NO$	$k_f = 2.7 \times 10^{-27}$	50
		$k_r = 2.4 \times 10^{-15}$	50
$NH_4^+.(NH_3)_2 + NH_3 + NH_3$	$\rightleftharpoons NH_4^+.(NH_3)_3 + NH_3$	$k_f = 2.4 \times 10^{-27}$	50
		$k_r = 1.2 \times 10^{-12}$	50

a   See also Tables 6.4, 6.5 and 6.6 for important hydration schemes of O_2^+ and NO^+ and association reactions of NO^+.
b   Rate coefficients are at 300 K. Units are cm^3 particle^{-1} s^{-1} for binary processes and cm^6 particle^{-2} s^{-1} for three-body processes.
c   C. J. Howard, V. M. Bierbaum, H. W. Rundle, and F. Kaufman, *J. chem. Phys.*, **57**, 3491 (1972).
d   E. Graham *et al.*, *J. chem. Phys.*, **59**, 4648 (1973); J. D. Payzant *et al.*, *J. chem. Phys.*, **59**, 5615 (1973).

In Section 6.3 we noted that the major primary ions in the E region and at higher altitudes are N_2^+, O_2^+ and O^+, formed by direct photoionization of N_2, O_2 and O. In the D region we must also consider Lyman-α ionization of NO and 102.7–111.8 nm ionization of $O_2(^1\Delta_g)$. Once these primary ions are formed, rapid reactions with neutral atoms and molecules occur, with the possibility of forming several alternative sets of products from any given primary ion. The reactions which follow primary ionization are discussed in the next sections for each ionospheric region in turn. Throughout this discussion it should be borne in mind that ion chemistry is only one of the many processes which govern the observed ionospheric composition. The

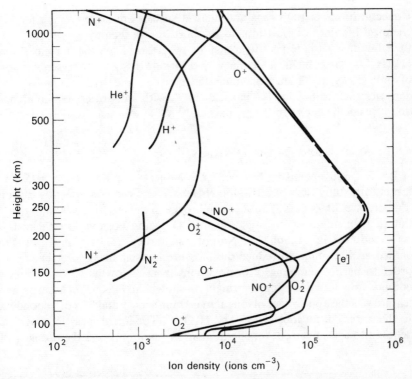

Fig. 6.15 Positive ion composition of a solar minimum, daytime ionosphere, from ref. 53. The ion distributions shown are based on results from two mass spectrometer experiments. The data has been normalized to the electron density distribution [e] measured during the same general period.

task of distinguishing the effects of ion chemistry from those of other processes, such as ion transport, is itself a difficult assignment. An overall picture of positive ion densities, as measured by a mass spectrometer above 90 km is given in Fig. 6.15.

6.6 D region ion chemistry

There are four main reasons why the D region is the most complex of the ionospheric regions, namely:

(a) the large variety of the available neutral molecules, notably NO, $O_2(^1\Delta_g)$, CO_2 and H_2O, leads to the possibility of complex reaction paths between the primary and product ions;

(b) the pressure is sufficient to allow three-body processes to be important;

(c) there is the possibility of forming significant amounts of negative ions;

(d) the presence of metal ions.

Photoequilibrium is usually assumed to exist for ions in the D region.

Concentrations of both ions and neutral species are large enough for the chemical lifetimes of most ions to be short during the day, compared with their lifetimes with respect to transport processes. Even the 'terminal' ion $H^+ (H_2O)_2$ has a chemical lifetime with respect to dissociative recombination of only 500 s at 82 km (assuming the electron and hydronium hydrate densities each to be 2×10^3 particle cm^{-3}, with a dissociative recombination rate coefficient of 10^{-6} cm^3 particle^{-1} s^{-1}).

6.6.1　POSITIVE IONS IN THE D REGION

Positive ion chemistry has been the subject of more active field and laboratory study than negative ion chemistry and can consequently call on a greater wealth of data. The successful rocket flights of Narcisi and Bailey[54] initiated an intensive investigation into D region positive ion chemistry. As mentioned in Section 3.3, Narcisi and Bailey used a liquid nitrogen chilled zeolite pump to reduce the pressure in their mass spectrometer in order to permit sampling of positive ions from as low as 50 km. The ion composition found during one flight is shown in Fig. 6.16. A surprising feature was the appearance of ions at mass numbers 19 and 37, corresponding to the water cluster ions H_3O^+ and $H_5O_2{}^+$. (The ion at mass 18, H_2O^+ is believed to arise from rocket contamination.) Not only was the presence of

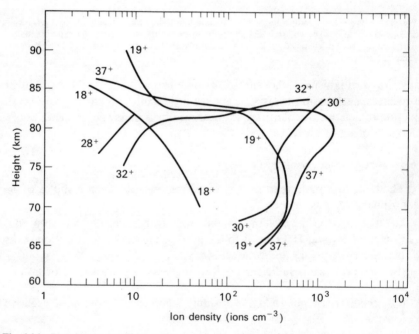

Fig. 6.16　D region positive ion profiles reported by Narcisi and Bailey.[54]

water cluster ions confirmed by subsequent flights, but ions of the type H^+ $(H_2O)_n$ were found to dominate the ion composition of the D region below about 82 km under all conditions: day or night, sunrise or sunset, during an aurora, a meteor shower, a sporadic E event, and a total solar eclipse.[55] The water cluster ion found in greatest abundance during the rocket flights of Narcisi was $H_5O_2^+$ at mass 37, followed by H_3O^+ at mass 19, and $H_7O_3^+$ at mass 55. As fragmentation of the heavier hydrates could have occurred during sampling, the relative concentrations of these hydrates found by the mass spectrometers is not a true indication of their respective concentrations. Krankowsky *et al.*[56] report the appearance of $H_9O_4^+$ at mass 73; presumably they used less drastic sampling conditions than Narcisi and co-workers. NO^+, the ion which was previously expected to be predominant in the D region, is generally less abundant than $H_5O_2^+$ in the daytime at mid-latitudes, but its relative abundance increases at higher altitudes. No NO^+ ions were found below 85 km at night or during twilight. Only above ~ 82 km (see Fig. 6.25), does NO^+ become the principal ion. At first rocket contamination was considered to be a possible source of the water cluster ions, but the reproducibility of the ion signals obtained during repeated rocket flights by different workers, on both the up and down legs of the flights, have established the clusters as genuine constituents of the D region. It would, of course, be difficult to explain how neutral water molecules could convert to $H_5O_2^+$ in a sufficiently short time to be sampled by the rocket from which they had been out-gassed.

The ions detected at 82 km during several rocket flights through the D region are shown in Fig. 6.17, and include those at mass numbers 19 (H_3O^+), 21 ($H_3{}^{18}O^+$), 24–25–26 (Mg^+), 30 (NO^+), 32 (S^+, O_2^+), 34 (S^+), 37 ($H_5O_2^+$), 39 ($H_5{}^{16}O.{}^{18}O^+$), 52 (Cr^+), 55 ($H_7O_3^+$), 54–56 (Fe^+), and 58–60 (Ni^+).[55] Other peaks detected, and their possible assignments, are 41 ($Na^+.(H_2O)$), 46 (NO_2^+), 48 ($NO^+.(H_2O)$ with possibly small amounts of SO^+), and 50 ($O_2^+.(H_2O)$ with perhaps a contribution from ${}^{34}SO^+$). The origin and behaviour of sulphur ions are not well understood. Narcisi's assignment of the ions at mass 32 and 34 amu to sulphur was based on the abundance ratio of the naturally occurring sulphur isotopes.[55] The fact that they are not observed above 82 km suggests that their origin is terrestrial.*
Metal ion chemistry is discussed in Section 6.7.

A further striking and unexpected feature of the rocket-borne mass spectrometer findings was the observation of a drastic transition near the mesopause (~ 85–86 km). Water cluster ions and sulphur ions disappear

---

* Recent laboratory evidence indicates that these ions at mass 32 and 34 cannot be S^+ because of the rapid interchange reaction between S^+ and O_2. (E. E. Ferguson, private communication).

$$S^+ + O_2 \rightarrow SO^+ + O$$
$$k = 1.6 \times 10^{-11} \text{ cm}^3 \text{ molecule}^{-1} \text{ s}^{-1}$$

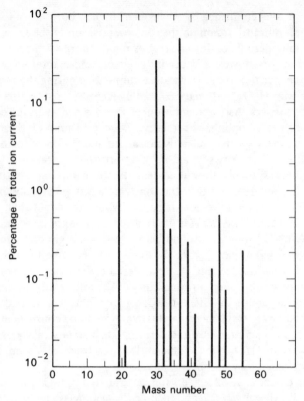

Fig. 6.17 Positive ions detected at 82 km over Fort Churchill, Canada, near sunset. (After Narcisi[55])

completely and, just as abruptly, metal ions appear, all within about 2 km. The findings of about 20 rocket flights indicate that metal ions, and ions of some oxides of these metals, are present in layers between about 82 and 120 km. A fairly reproducible metal ion layer, mainly Mg^+, Fe^+, plus smaller amounts of Na^+, Al^+, Ca^+ and Ni^+, occurs with a peak ion density around 93 km. The peak has a halfwidth of between 5 and 10 km. Higher altitude layers are mainly composed of ions of Si, and/or Mg, and/or Fe (see Fig. 6.25). It is generally accepted that meteor ablation is the origin of most of these metal ions. During conditions of a high influx of meteors, and in the presence of the sporadic E phenomenon (Section 6.7.2), the metal ion composition of the lower ionosphere is considerably enhanced.[55]

6.6.2 THEORY OF FORMATION OF WATER MOLECULE CLUSTER IONS

The problem is to find a mechanism for converting the primary ions O_2^+ and NO^+, as produced by photoionization, into the observed water cluster ions. Any proposed model should account not only for the large concentrations of hydrated hydronium ions, but also for their dramatic decrease around the mesopause. Ferguson and Fehsenfeld[57,58] have proposed such a reaction sequence, and Good et al.[48] have provided independent verification of their scheme. Essentially it is considered that the positive ions O_2^+ and NO^+ will become hydrated by the process

$$P^+ + H_2O + M \rightarrow P^+.(H_2O) + M \qquad (6.25)$$

followed by successive three-body hydration steps to produce larger clusters $P^+.(H_2O)_n$, where P^+ represents either O_2^+ or NO^+. At some point in the chain of clustering reactions a bimolecular reaction with water (6.26) becomes exothermic and predominates over further clustering steps.

$$P^+.(H_2O)_n + H_2O \rightarrow H^+.(H_2O)_n + P + OH \qquad (6.26)$$

The detailed mechanism for the $O_2^+–H_2O$ reaction sequence is given in Table 6.4. It is somewhat modified from the original scheme suggested by Fehsenfeld and Ferguson[57] because of the subsequent discovery that the reaction

$$O_4^+ + O \rightarrow O_2^+ + O_3 \qquad (6.27)$$

is fast.[43] Early computed altitude profiles of $H_5O_2^+$, based on the O_2^+ production rates of Hunten and McElroy[22] (see Section 6.3.1), gave quite good agreement with observed profiles, and successfully reproduced the abrupt reduction in $H_5O_2^+$ concentrations around 82 km.[59] However, when the lower O_2^+ production rates of Paulsen et al.[3] are adopted, the calculated $H_5O_2^+$ profile falls well short of Narcisi and Bailey's observed profile (Fig. 6.18). Thus the O_2^+ scheme of Table 6.4 is inadequate to account for the observed concentrations of $H_5O_2^+$ at low altitudes.

For NO^+ the detailed mechanism for converting NO^+ into $H_5O_2^+$, based on laboratory measurements, is given in Table 6.5.[43] Some problems associated with the NO^+ scheme have become apparent. First, the major product of the $NO^+–H_2O$ reaction sequence is at mass 55, $(H_3O^+.(H_2O)_2)$ and not the observed mass 37 $(H_3O^+.(H_2O))$. The experimentally observed

Table 6.4 D region scheme for converting O_2^+ into $H_3O^+.H_2O$ [43]

Reaction	Rate Coefficient
$O_2^+ + H_2O + N_2 \rightarrow O_2^+.(H_2O) + N_2$	2.8×10^{-28}
$O_2^+ + H_2O + O_2 \rightarrow O_2^+.(H_2O) + O_2$	1.9×10^{-28}
$O_2^+ + O_2 + O_2 \rightarrow O_2.O_2^+ + O_2$	2.4×10^{-30}
$O_2.O_2^+ + O \rightarrow O_2^+ + O_3$	$(3 \pm 2) \times 10^{-10}$
$O_2.O_2^+ + H_2O \rightarrow O_2^+.(H_2O) + O_2$	$(1.7 \pm 0.4) \times 10^{-9}$
$O_2^+.(H_2O) + H_2O \rightarrow H_3O^+.(OH) + O_2$	$(1.4 \pm 0.5) \times 10^{-9}$
$\rightarrow H_3O^+ + OH + O_2$	$\leqslant 3 \times 10^{-10}$
$H_3O^+.(OH) + H_2O \rightarrow H^+.(H_2O)_2 + OH$	3.2×10^{-9}

Rate coefficients are quoted for temperatures ~ 300 K.
Units are cm^6 particle^{-2} s^{-1} for three-body reactions and cm^3 particle^{-1} s^{-1} for binary reactions.

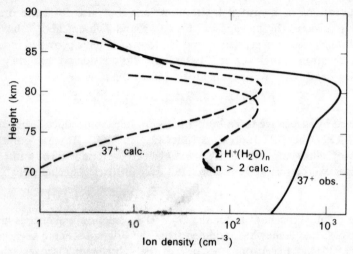

Fig. 6.18 Calculated and observed $H_3O^+.H_2O$ concentrations in the D region. (After Ferguson[38])

mass 37 peak may be too large because the water cluster ion $H_3O^+.(H_2O)_2$ is fragmented to some extent in the shock wave of the supersonic rocket and in the electric field used to attract ions into the mass spectrometer.[55] However, it is unlikely that fragmentation processes alone could make up the deficiency between observed and predicted signals at mass 55 and mass 37. A more serious objection to the NO^+–H_2O scheme is due to Reid[23] who has shown that reactions 6.28 and 6.29

$$NO^+.(H_2O)_2 + H_2O + M \rightarrow NO^+.(H_2O)_3 + M \qquad (6.28)$$

$$NO^+.(H_2O)_3 + H_2O \rightarrow H^+.(H_2O)_3 + HNO_2 \qquad (6.29)$$

are much too slow to allow the complete conversion of NO^+ hydrates to hydronium ions.

To summarize the position regarding the two main reaction sequences stemming from O_2^+ and NO^+, shown schematically in Fig. 6.19, we can say that at present there appears to be no completely satisfactory explanation

Table 6.5 D region scheme for converting NO^+ into $H_3O^+.(H_2O)_n$ where n = 1 or 2[43]

Reaction		Rate Coefficient k_f	k_{rev}
$NO^+ + H_2O + M$	$\rightleftharpoons NO^+.(H_2O) + M$	1.6×10^{-28}	1.0×10^{-14}
$NO^+.(H_2O) + H_2O + M$	$\rightleftharpoons NO^+.(H_2O)_2 + M$	1.0×10^{-27}	1.6×10^{-14}
$NO^+.(H_2O)_2 + H_2O + M$	$\rightleftharpoons NO^+.(H_2O)_3 + M$	1.7×10^{-27}	1.5×10^{-12}
$NO^+.(H_2O)_3 + H_2O$	$\rightarrow H^+.(H_2O)_3 + HNO_2$	7×10^{-11}	

Rate coefficients are quoted for temperatures ~ 300 K.
Units are cm^6 particle^{-2} s^{-1} for three-body reactions and cm^3 particle^{-1} s^{-1} for binary reactions.

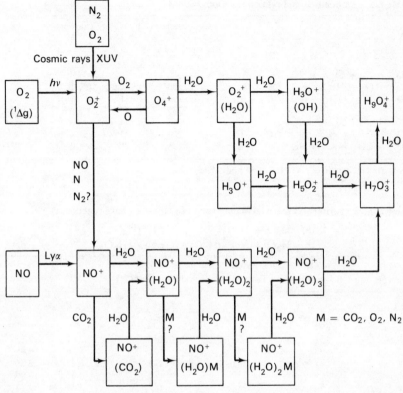

Fig. 6.19 Diagrammatic reaction scheme for D region positive ions, after Ferguson.[38]

for the observed positive ion profiles of Narcisi and Bailey. What is lacking is a fast reaction sequence for converting NO^+ ions into water cluster ions below 82 km. It is possible that the most important initial association reaction of the NO^+ ion, is not with the water molecule, i.e. not

$$NO^+ + H_2O + M \rightleftharpoons NO^+.(H_2O) + M \qquad (6.30)$$

but with a different species such as CO_2 or H which might have a faster association rate. Several other schemes have been suggested[38] but as yet none has provided a satisfactory solution to the problem. Some of the known association reactions of NO^+, and their rate coefficients are given in Table 6.6.

The subject of cluster ions in general is an interesting field for study. Gaseous ion clusters have been observed to form with many neutral molecules besides H_2O; some examples are $NH_3.NH_4^+$, $CO_2.CO_2^+$, $HNO_3.(NO_3^-)$ and $CO_2.O_2^-$. Usually the cluster molecules are quite weakly bonded to the ion. Some experimental measurements of ΔH, $\Delta G°$ and ΔS for the reaction

$$H_2O + H^+.(H_2O)_{n-1} \rightleftharpoons H^+.(H_2O)_n \qquad (6.31)$$

Table 6.6   Association and related reactions of NO^+.

Reaction	Temperature (K)	Rate Coefficient	Ref.
Association			
$NO^+ + N_2 + M \rightarrow NO^+.N_2 + M$	300	3.5×10^{-31}	52
$NO^+ + O_2 + He \rightarrow NO^+.O_2 + He$	200	$<6 \times 10^{-34}$	43
$NO^+ + CO_2 + He \rightarrow NO^+.CO_2 + He$	197	$(1 \pm 3) \times 10^{-29}$	43
	235	$(7 \pm 3.5) \times 10^{-30}$	43
	290	$(4 \pm 2) \times 10^{-30}$	43
$NO^+ + CO_2 + N_2 \rightarrow NO^+.CO_2 + N_2$	200	$(2.5 \pm 1.5) \times 10^{-29}$	43
$NO^+ + CO_2 + Ar \rightarrow NO^+.CO_2 + Ar$	196	$(3 \pm 1) \times 10^{-29}$	43

Units are cm^6 particle^{-2} s^{-1} for three body reactions.

Table 6.7   Experimental thermodynamic values for the gas phase reactions:[61,51]

$$\text{(a)} \quad H_2O + H^+.(H_2O)_{n-1} \rightleftharpoons H^+(H_2O)_n$$
$$\text{(b)} \quad H_2O + NO^+.(H_2O)_{n-1} \rightleftharpoons NO^+(H_2O)_n$$

$n-1, n$	$-\Delta H$ (kJ mol^{-1})	$-\Delta G^\circ$ (kJ mol^{-1})	$-\Delta S$ (J deg^{-1} mol^{-1})	A^*	B^*
(a)					
0, 1	690				
1, 2	151	105	139	7.41	9.55
2, 3	93	57	121	4.76	9.05
3, 4	71	35.6	118	3.63	8.86
4, 5	64	23	136	3.33	10.00
5, 6	54	16.3	127	2.84	9.46
6, 7	49	11.7	123	2.53	9.24
7, 8	43	9.2	113	2.17	8.50
(b)					
0, 1	78.7		98.7		
1, 2	65.3		97.5		

* The constants A and B refer to the equation $\log_{10} K_{(n-1,n)} = (A/T) \times 10^3 - B$, where K is the equilibrium constant. Water vapour pressures are in torr.

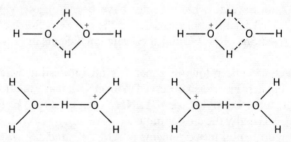

Fig. 6.20   Suggested valence bond structures for the $(H_4O_2)^+$ and $(H_5O_2)^+$ ions, after Good et al.[46]

are summarized in Table 6.7. They show a general trend towards decreasing stability as the size of the cluster ion increases. A comparison of the ΔH data with the proton affinity for water has suggested that in lower hydrates all water molecules are equivalent and that the best descriptive notation is $H^+.(H_2O)_2$ rather than $H_3O^+.(H_2O)$.[61] Suggested valence bond structures for the cluster ions $H_3O^+.OH$ and $H^+(H_2O)_2$ are shown in Fig. 6.20.[48] The weak bond energy of these cluster ions (e.g. the bond energy of $H_5O_2^+-H_2O$ is only ~ 93 kJ) raises the possibility of their photodissociation by visible solar radiation (process 6.32), but laboratory studies have shown that photodissociation is not an efficient process.[38]

$$H_7O_3^+ + h\nu \rightarrow H_5O_2^+ + H_2O \qquad (6.32)$$

6.6.3 D REGION NEGATIVE IONS

The history of D region negative ions is even more recent than that of the positive ions. Few *in situ* mass spectrometric measurements have been reported since the first observations by Johnson *et al.* in 1958.[62] Johnson and co-workers identified the most abundant negative ion in the E region as NO_2^- at mass 46. Two subsequent reports of negative ion analysis have appeared, by Narcisi *et al.*[63] and Arnold *et al.*[64] These measurements were made at night, in the second case during a weak aurora. The results are shown in Tables 6.8(a) and 6.8(b). In the E region between 90 and 114 km Narcisi *et al.* report low concentrations of negative ions of mass 16, 35 and 46 amu, which probably correspond to the ions O^-, Cl^- and NO_2^-. A subsequent flight detected the heavier ions (>78 amu) of Table 6.8(a). Peaks at 62, 80, 98, 116, 134 and 152 amu were found, with an error of ± 0.015 M for any particular mass number M. These heavier ions are spaced 18 amu apart indicating they may be hydrates of some basic ion; perhaps $CO_3^-.(H_2O)$ at 78 amu or, considered more likely, $NO_3^-.(H_2O)$ at 80 amu. An altitude profile of the more important ions observed by Arnold *et al.*[64] on the upward leg of their rocket flight is given in Fig. 6.21.

The experimental information available at present, although limited, indicates that the occurrence of negative ions is almost entirely limited to the region below 92 km, with the heavier negative ions being confined to the low temperature region of the mesopause (80–90 km). Rocket contamination has not yet been ruled out as the origin of some peaks, e.g. Cl^- at 35 amu.[63] The major point of divergence of the two sets of results is that whereas Arnold and co-workers[64] detect the ions HCO_3^- and $NO_2^-.H_2O$, they do not find the extensive hydration of NO_3^- reported by Narcisi *et al.*[63] These differences may in part be experimental artefacts and in part due to genuine differences in negative ion composition. Clearly further rocket observations are required.

The main D region negative ion production mechanism is that of electron

Table 6.8(a)   D region negative ion results from Narcisi *et al.*[63]

	Mass No. (amu)	Counts per 10 msec	Altitude (km)
Ascent	35	5	78.6
	37	3	78.6
	61 ± 1	4	79
	63 ± 1	4	79
	32	2	81.3
	35	15	81.4
	37	6	81.4
	61 ± 1	17	81.7
	63 ± 1	17	81.7
	76 ± 1	9	81.9
	61 ± 1	2	84.3
	63 ± 1	2	84.3
	$76 - 78$	13	84.4
	35	2	86.4
	76 ± 1	6	87
	>78	9	87
	>78	rising	89.3
Descent	>78	rising	89.3
	>78	rising	86.9
	>78	rising	84.3
	32	2	82.4
	35	7	82.3
	37	2	82.3
	61 ± 1	9	81.9
	63 ± 1	7	81.9

Table 6.8(b)   Negative ions observed in the D region after Arnold *et al.*[64,73]

Mass No. (amu)	Negative Ion (Tentative Identification)	Maximum counts per mass peak
32	O_2^-	33
35	Cl^-	210
	$O_2H_3^-$	
37	Cl^-	64
60	CO_3^-	407
61	HCO_3^-	119
62	NO_3^-	77
68	$O_2^-.(H_2O)_2$	16
76	CO_4^-	22
78	$CO_3^-.H_2O$	18
93 ± 1	$NO_2^-.(HNO_2)$	73
	$CO_4^-.(H_2O)$	
111 ± 1	$NO_2^-.(HNO_2).(H_2O)$	70
	$CO_4^-.(H_2O)_2$	
125 ± 1	$NO_3^-.(HNO_3)$	50

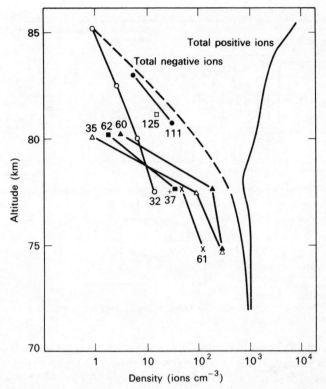

Fig. 6.21   Altitude profile of negative ions measured during a rocket flight above Andoya, Norway. The dashed portion of the total negative ion density curve is obtained from the sum of the individual constituents. (After Arnold *et al.*[64])

attachment to O_2 to form O_2^-;

$$e + 2O_2 \rightarrow O_2^- + O_2 \tag{6.33}$$

$$k_{ea} = 1.4 \times 10^{-29}(300/T)\exp(-600/T) \; [11]$$

All other established negative ions in the D region can be generated from the O_2^- ions via ion–neutral reactions. A schematic outline of the processes involved, excluding hydration reactions, is shown in Fig. 6.22. In this scheme NO_3^- and HCO_3^- are 'terminal' negative ions, in that they do not react further to form different negative ions but instead are lost by hydration and by positive ion recombination. Many other such schemes have appeared in the literature and no doubt others will follow. The one shown here takes into account the most recent laboratory data at the time of writing.

There are three important reactions removing O_2^-, namely

$$O_2^- + O \rightarrow O_3 + e \tag{6.34}$$

$$O_2^- + O_3 \rightarrow O_3^- + O_2 \tag{6.35}$$

and $\qquad O_2^- + O_2 + M \rightarrow O_4^- + M \tag{6.36}$

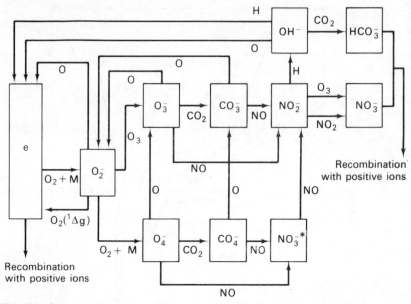

Fig. 6.22　Schematic flow diagram for D region negative ion chemistry, not including water cluster ions. (After Ferguson[65])

The 'secondary' negative ions O_3^- and O_4^- undergo a complex series of reactions (shown in Fig. 6.22) which are thought to lead ultimately to the stable negative ions of the D region, NO_3^- and HCO_3^-.[65] The importance of the minor neutral species O_3, CO_2, NO and H in negative ion chemistry is evident. These minor species offer exothermic reaction paths for ions produced early in the reaction sequence; it is unfortunate that their concentrations are not better established. Generally the CO_2 mole fraction of 3×10^{-4} at ground level is adopted for the D region. With this assumption it is not to be expected that O_3^- produced in reaction 6.37

$$O_3 + O_2^- \rightarrow O_3^- + O_2 \qquad (6.37)$$
$$k_{6.37} = 3.5 \times 10^{-10} \text{ cm}^3 \text{ particle}^{-1} \text{ s}^{-1}$$

will become a major D region negative ion because of its rapid reaction with CO_2:

$$O_3^- + CO_2 \rightarrow CO_3^- + O_2 \qquad (6.38)$$
$$k_{6.38} = 4 \times 10^{-10} \text{ cm}^3 \text{ particle}^{-1} \text{ s}^{-1}$$

The lifetime of O_3^- with respect to reaction 6.38 is only 0.01 s at 80 km. Similarly, because of the expected rapid reaction of the hydrated ion $O_2^-.(H_2O)_2$ with CO_2:

$$O_2^-.(H_2O)_2 + CO_2 \rightarrow CO_4^- + 2H_2O \qquad (6.39)$$
$$k_{6.39} \sim 10^{-9} \text{ cm}^3 \text{ particle}^{-1} \text{ s}^{-1}$$

Table 6.9   Negative ion-neutral reactions of atmospheric importance.[a]

Reaction		*Rate Coefficient*	*Ion Energy* (At 300 K unless otherwise stated)	*Ref.*
Binary reactions of non-cluster ions				
$H^- + O_2$	$\rightarrow O^- + OH$	$<1 \times 10^{-11}$		40
	$\rightarrow O_2^- + H$	$<1 \times 10^{-11}$		40
	$\rightarrow OH^- + O$	$<1 \times 10^{-11}$		40
$H^- + N_2O$	$\rightarrow OH^- + N_2$	1.1×10^{-9}		40
$H^- + NO_2$	$\rightarrow NO_2^- + H$	2.9×10^{-9}		40
$H^- + H_2O$	$\rightarrow OH^- + H_2$	3.8×10^{-9}		40
$O^- + N$	$\rightarrow NO + e$	2.2×10^{-10}		40
$O^- + H_2$	$\rightarrow OH^- + H$	3.3×10^{-11}		40
$O^- + O_3$	$\rightarrow O_3^- + O$	6×10^{-10}		67
$O^- + H_2O$	$\rightarrow OH^- + OH$	1.4×10^{-9}		40
$O^- + NO_2$	$\rightarrow NO_2^- + O$	1.2×10^{-9}		66
$O^- + NO_2$	$\rightarrow O_2^- + NO$	1.8×10^{-11}	~ 1 eV	66
$O^- + N_2O$	$\rightarrow NO^- + NO$	2×10^{-10}		40
$O^- + CH_4$	$\rightarrow OH^- + CH_3$	1×10^{-10}		40
$O_2^- + H$	$\rightarrow$ products	1.5×10^{-9}		40
$O_2^- + H_2$	$\rightarrow$ products	$<1 \times 10^{-12}$		40
$O_2^- + O_3$	$\rightarrow O_3^- + O_2$	3.5×10^{-10}		67
$O_2^- + NO_2$	$\rightarrow NO_2^- + O_2$	8×10^{-10}		66
$O_2^- + HCl$	$\rightarrow Cl^- + HO_2$	1.6×10^{-9}		65
$O_2^- + N_2O$	$\rightarrow O_3^- + N_2$	$<1 \times 10^{-12}$		40
$O_2^- + SO_2$	$\rightarrow SO_2^- + O_2$	4.8×10^{-10}		40
$O_3^- + H$	$\rightarrow OH^- + O_2$	8.4×10^{-10}		40
$O_3^- + NO$	$\rightarrow NO_2^- + O_2$	1×10^{-11}		67[b]
$O_3^- + CO_2$	$\rightarrow CO_3^- + O_2$	4×10^{-10}		66
$O_3^- + NO_2$	$\rightarrow$ products	2.8×10^{-10}		40
$O_3^- + N_2$	$\rightarrow$ products	$<1 \times 10^{-15}$		66
$O_3^- + SiO$	$\rightarrow SiO_3^- + O$	fast		66
$O_3^- + CO$	$\rightarrow$ products	slow		66
$O_3^- + SO_2$	$\rightarrow SO_3^- + O_2$	1.7×10^{-9}		40
$OH^- + NO_2$	$\rightarrow NO_2^- + OH$	1.2×10^{-9}		66
$C^- + H_2$	$\rightarrow$ products	$<1 \times 10^{-13}$		40
$C^- + O_2$	$\rightarrow O^- + CO$	4×10^{-10}		40
$CO_3^- + O$	$\rightarrow O_2^- + CO_2$	8.0×10^{-11}		67
$CO_3^- + NO$	$\rightarrow NO_2^- + CO_2$	9.0×10^{-12}		66
$CO_3^- + NO_2$	$\rightarrow$ products	2×10^{-10}		40
$CO_3^- + SO_2$	$\rightarrow SO_3^- + CO_2$	2.3×10^{-10}		40
$NH_2^- + NO_2$	$\rightarrow NO_2^- + NH_2$	1.0×10^{-9}		66
$NO^- + HCl$	$\rightarrow Cl^- + HNO$	1.6×10^{-9}		65
$NO^- + O_2$	$\rightarrow O_2^- + NO$	5×10^{-10}		40
$NO^- + NO_2$	$\rightarrow NO_2^- + NO$	7.4×10^{-10}		40
$NO^- + N_2O$	$\rightarrow NO_2^- + N_2$	2.8×10^{-14}		40
$NO_2^- + H$	$\rightarrow OH^- + NO$	3×10^{-10}		68
$NO_2^- + NO_2$	$\rightarrow NO_3^- + NO$	$\sim 4 \times 10^{-12}$		69
$NO_2^- + HCl$	$\rightarrow Cl^- + HNO_2$	1.4×10^{-9}		65
$NO_2^- + HBr$	$\rightarrow Br^- + HNO_2$	1.9×10^{-9}		65
$NO_2^- + O_3$	$\rightarrow NO_3^- + O_2$	1.8×10^{-11}		66
$NO_2^- + O$	$\rightarrow$ products	$<10^{-11}$		40
$NO_2^- + N$	$\rightarrow$ products	$<10^{-11}$		40
$NO_3^- + O$	$\rightarrow$ products	$<10^{-11}$		40
NO_3^{-*}[c] $+ NO$	$\rightarrow NO_2^- + NO_2$	$\sim 1.5 \times 10^{-11}$		69
$NO_3^- + NO$	$\rightarrow NO_2^- + NO_2$	$\sim 3 \times 10^{-15}$		14
$NO_3^- + HCl$	$\rightarrow Cl^- + HNO_3$	$<10^{-12}$		65
$NO_3^- + HBr$	$\rightarrow Br^- + HNO_3$	6.3×10^{-10}		65
$N_2O^- + O_2$	$\rightarrow O_3^- + N_2$	fast	80 K	66
$F^- + NO_2$	$\rightarrow NO_2^- + F$	$<2.5 \times 10^{-11}$		66
$Cl^- + NO_2$	$\rightarrow NO_2^- + Cl$	$<6 \times 10^{-12}$		66

Table 6.9   (Continued).

Reaction		Rate Coefficient	Ion Energy (At 300 K unless otherwise stated)	Ref.
Binary reactions of cluster ions				
$O_4^- + O$	$\rightarrow O_3^- + O_2$	4.0×10^{-10}		72
	$\rightarrow O^- + 2O_2$			
$O_4^- + O_2$	$\rightarrow O_2^- + 2O_2$	2×10^{-14}		71
$O_4^- + NO$	$\rightarrow NO_3^{-*e} + O_2$	2.5×10^{-10}		69
$O_4^- + N_2$	$\rightarrow O_2^-.N_2 + O_2$	$<1.0 \times 10^{-11}$		70
$O_4^- + N_2O$	$\rightarrow O_2^-.N_2O + O_2$	$<1.0 \times 10^{-12}$		70
$O_4^- + CO$	$\rightarrow CO_3^- + O_2$	$<2.0 \times 10^{-11}$		70
$O_4^- + CO_2$	$\rightarrow CO_4^- + O_2$	4.3×10^{-10}		70
$O_4^- + H_2O$	$\rightarrow O_2^-.H_2O + O_2$	1.5×10^{-9}		71
$CO_4^- + O$	$\rightarrow O_3^- + CO_2$	1.5×10^{-10}		72
	$\rightarrow CO_3^- + O_2$			
$CO_4^- + NO$	$\rightarrow NO_3^{-*e} + CO_2$	4.8×10^{-11}		69
$O_2^-.H_2O + O_2$	$\rightarrow O_4^- + H_2O$	2.5×10^{-15}		40
$O_2^-.H_2O + CO_2$	$\rightarrow CO_4^- + H_2O$	5.8×10^{-10}		70
$O_2^-.H_2O + NO$	$\rightarrow NO_3^- + H_2O$	3.1×10^{-10}		70
$O_2^-.H_2O + O_3$	$\rightarrow$ products	3×10^{-10}		40
$O_2^-.(H_2O)_2 + O_2$	$\rightarrow O_2^-(H_2O) + H_2O + O_2$	1.1×10^{-14}		71
$O_2^-.(H_2O)_2 + O_3$	$\rightarrow$ products	3.4×10^{-10}		40
$O_2^-.(H_2O)_3 + O_2$	$\rightarrow O_2^-(H_2O)_2 + H_2O + O_2$	$3.5 \times 10^3 \text{ s}^{-1 \text{ d}}$		71
$O_3^-.H_2O + CO_2$	$\rightarrow$ products	3×10^{-10}		40
$O_3^-.(H_2O)_2 + CO_2$	$\rightarrow$ products	2×10^{-10}		40
$NO_2^-(H_2O) + O_2$	$\rightarrow NO_2^- + H_2O + O_2$	1.6×10^{-15}		75
$NO_2^-(H_2O)_2 + O_2$	$\rightarrow NO_2^-(H_2O) + H_2O + O_2$	5.8×10^{-14}		75
$NO_3^-(H_2O) + O_2$	$\rightarrow NO_3^- + H_2O + O_2$	1.4×10^{-14}		40
$CO_3^-.(H_2O) + NO$	$\rightarrow$ products	1.8×10^{-11}		40
$CO_3^-.(H_2O) + NO_2$				
	$\rightarrow$ products	1.5×10^{-10}		40
$NO_2^-.(H_2O) + SO_2$				
	$\rightarrow NO_2^-.(SO_2) + H_2O$	1.5×10^{-9}		40
$SO_4^- + NO_2$	$\rightarrow NO_2^- + SO_2 + O_2$	2.5×10^{-10}		40
	$\rightarrow NO_3^- + SO_3$	1×10^{-10}		40
Three-body reactions of negative ions				
$O^- + CO_2 + CO_2 \rightarrow CO_3^- + CO_2$		9.0×10^{-29}		66
$O^- + CO_2 + He \rightarrow CO_3^- + He$		1.5×10^{-28}		66
$O^- + N_2 + He \rightarrow N_2O^- + He$		$\sim1 \times 10^{-30}$	80	66
$O^- + O_2 + O_2 \rightarrow O_3^- + O_2$		1.4×10^{-30}		71
$O^- + H_2O + O_2 \rightarrow O^-.H_2O + O_2$		1×10^{-28}		74
$O_2^- + N_2 + He \rightarrow O_2^-.N_2 + He$		4×10^{-32}		76
$O_2^- + O_2 + O_2 \rightarrow O_4^- + O_2$		5×10^{-31}		71
$O_2^- + O_2 + He \rightarrow O_4^- + He$		3.4×10^{-31}	200	76
$O_2^- + CO_2 + O_2 \rightarrow CO_4^- + O_2$		2.0×10^{-29}		76
$O_2^- + CO_2 + CO_2 \rightarrow CO_4^- + CO_2$		9.0×10^{-30}		76
$O_2^- + CO_2 + He \rightarrow CO_4^- + He$		4.7×10^{-29}	200	76
$O_2^- + H_2O + O_2 \rightarrow O_2^-.H_2O + O_2$		1.6×10^{-28}		71
$OH^- + CO_2 + M \rightarrow HCO_3^- + M$		10^{-29}		68
$O_3^- + H_2O + O_2 \rightarrow O_3^-.H_2O + O_2$		1.5×10^{-28}		71
$O_3^- + N_2 + N_2 \rightarrow$ products		$<1.5 \times 10^{-31}$		66
$NO_2^- + H_2O + M \rightarrow NO_2^-.(H_2O) + M$		$1.2 \times 10^{-28}(NO)$	$(K^e = 5.2 \times 10^{-14})$	75
		$8.4 \times 10^{-29}(O_2)$		
$NO_2^-(H_2O) + H_2O + M$				
	$\rightarrow NO_2^-(H_2O)_2 + M$	$3.8 \times 10^{-29}(O_2)$	$(K^e = 6.6 \times 10^{-16})$	75
$NO_3^- + H_2O + M \rightarrow NO_3^-(H_2O) + M$		$7.5 \times 10^{-29}(O_2)$	$(K^e = 5.3 \times 10^{-15})$	75
$NO_3^- + HCl + Ar \rightarrow NO_3^-.HCl + Ar$		5×10^{-28}		65

Table 6.9   (Continued).

Reaction	Rate Coefficient	Ion Energy (At 300 K unless otherwise stated)	Ref.
$O_2^-.H_2O + H_2O + O_2$ $\rightleftharpoons O_2^-.(H_2O)_2 + O_2$	5.4×10^{-28}		71
$O_2^-.(H_2O)_2 + H_2O + O_2$ $\rightleftharpoons O_2^-(H_2O)_3 + O_2$	2.2×10^{-11} [d]		71

a   Excluding electron detachment processes; see Table 6.1.
b   Although the products listed in reference 47 are $NO_3^- + O$ it is now believed they are more likely to be $NO_2^- + O_2$ (E. E. Ferguson, private communication).

c   It is believed the conformation of the NO_3^{-*} ion is $(O—O—N—O)^-$ rather than ⟨structure⟩ expected

from reactions of NO_2^- with O_3 and NO_2.[69]
d   Valid at pressures above 3 torr. Apparently independent of the third body.
e K = equilibrium constant in cm molecule units.

Table 6.10   Stabilities of some negative ions.

Ion	Electron Affinity of Neutral Species (kJ)	Dissociation Energy (kJ)		Ref.
$OH^-.(H_2O)$		$OH^-—H_2O$	106.6	77
$OH^-.(H_2O)_2$		$OH^-.(H_2O)—H_2O$	68.6	77
$OH^-.(H_2O)_3$		$OH^-.(H_2O)_2—H_2O$	63.1	77
$OH^-.(H_2O)_4$		$OH^-.(H_2O)_3—H_2O$	59.4	77
$OH^-.(H_2O)_5$		$OH^-.(H_2O)_4—H_2O$	59.0	77
O_2^-	42.5	$O^-—O$	394	74
O_4^-	96.5	$O_2^-—O_2$	52.1	74
$O_2^-.H_2O$	115.8	$O_2^-—H_2O$	77.2	74
$O_2^-.(H_2O)_2$	193	$O_2^-—2H_2O$	154.4	74
$O_2^-.(H_2O)_2$		$O_2^-.H_2O—H_2O$	71.9	77
$O_2^-.(H_2O)_3$		$O_2^-(H_2O)_2—H_2O$	64.4	77
$O_2^-.(H_2O)_5$	357	$O_2^-—5H_2O$	318.4	74
$O_2^-.NO$	> 328	$O_2^-—NO$	>241	74
NO_2^-	230[‡]			
$NO_2^-.(H_2O)$		$NO_2^-—H_2O$	59.8	77
$NO_2^-.(H_2O)_2$		$NO_2^-.(H_2O)—H_2O$	54.0	77
$NO_2^-.(H_2O)_3$		$NO_2^-(H_2O)_2—H_2O$	43.5	77
CO_4^-	115.8	$O_2^-—CO_2$	77.2	74
$CO_4^-.H_2O$		$CO_4^-—H_2O$	~61	77
$CO_4^-.(H_2O)_2$		$CO_4^-.H_2O—H_2O$	~44	77
NO_3^-	376			65

‡   D. B. Dunkin, F. C. Fehsenfeld and E. E. Ferguson, *Chemical Physics Letters*, **15**, 257 (1972).

it was initially suggested that hydrated ions of O_2^- should not attain appreciable concentrations. However, laboratory measurements have since shown that $O_2^-.(H_2O)_2$ may be reformed by reaction 6.40;

$$CO_4^-.(H_2O) + H_2O \rightarrow O_2^-.(H_2O)_2 + CO_2 \tag{6.40}$$

and indeed an ion of mass 68 was observed in the rocket flight of Arnold and co-workers (Table 6.8(b)).

Although the process is not included in the scheme of Fig. 6.22, it is to be expected that all long-lived ions will become hydrated, to form the species actually observed in the laboratory and during rocket flights. The mechanism of hydration has not been established; thus, for example, the hydration of NO_3^- could occur either by the direct process 6.41, or by the two-step process shown in reactions 6.42 and 6.43.

$$NO_3^- + H_2O + M \rightarrow NO_3^-.H_2O + M \tag{6.41}$$

$$NO_3^- + O_2 + M \rightarrow NO_3^-.O_2 + M \tag{6.42}$$

$$NO_3^-.O_2 + H_2O \rightarrow NO_3^-.H_2O + O_2 \tag{6.43}$$

In Table 6.9 we list the rate coefficients of the reactions shown in the scheme of Fig. 6.22 (where rates are known) together with the rate coefficients of other negative ion reactions of interest to the D region. The stabilities of some negative cluster ions are compared in Table 6.10.

6.7   E region ion chemistry

As mentioned in Section 6.3.2, the main E region ionizing sources are the Lyman-β (102.6 nm) and C III (97.7 nm) emission lines, which ionize only O_2, the Lyman continuum from 91.1 to ~ 80.0 nm which ionizes both atomic and molecular oxygen, and the XUV and X-ray radiation below 28 nm which ionizes all atmospheric constituents. The major primary ions are N_2^+ and O_2^+, with smaller amounts of O^+ and NO^+. Rapid reactions of N_2^+ with both O_2 and O (6.44 and 6.45) prevent N_2^+ from being a major E region ion despite its relatively large production rate. The dominant ions of the E region are therefore O_2^+ and NO^+.

$$N_2^+ + O_2 \rightarrow O_2^+ + N_2 \tag{6.44}$$

$$k_{6.44} = 1 \times 10^{-10} \text{ cm}^3 \text{ molecule}^{-1} \text{ s}^{-1}$$

$$N_2^+ + O \rightarrow NO^+ + N \tag{6.45}$$

$$k_{6.45} = 1.4 \times 10^{-10} \text{ cm}^3 \text{ molecule}^{-1} \text{ s}^{-1}$$

Because electrons and ions are removed solely by dissociative recombination processes (e.g. 6.46 and 6.47):

$$NO^+ + e \rightarrow N + O \tag{6.46}$$

$$O_2^+ + e \rightarrow O + O \tag{6.47}$$

the chemistry of the E region is more straightforward than that of the D region. In the daytime ion, electron and neutral particle densities are generally sufficient to maintain photochemical equilibrium, but transport processes have been shown to be essential to an understanding of the nighttime E region.[78] The chemistry of E region metal ions is discussed separately in the section on Sporadic E (Section 6.7.2).

The E region is one of the most studied parts of the ionosphere, and many rocket investigations have been made of its ion composition. The results of two such investigations, one during the day and the other at night, are shown in Figs. 6.23 and 6.24. The water peak at mass 18 is attributed to rocket contamination. Of the two major ions observed at night, NO^+ and O_2^+, the NO^+ ion has the larger concentration in the E region. During the day either ion may predominate.[79] Keneshea et al.[80] coordinated a series of rocket flights to make simultaneous measurements of neutral winds, total charged particle densities, neutral densities and positive ion composition at both sunrise and sunset. The ion composition as measured during a flight at sunset (solar zenith angle 88.5°) is shown in Fig. 6.25. The water cluster ions at 19^+ and 37^+ are seen to disappear near the mesopause. The main metal ion layer appears in the lower E region, around 93 km, and is largely composed of Mg^+ and Fe^+. Another very narrow layer appears near 110 km and is mainly Si^+ (mass 28). Narcisi[55] reports that this layer has been measured in more than six flights in the vicinity of 110 km and is therefore a common feature of the daytime ionosphere. A second rocket flight, 48

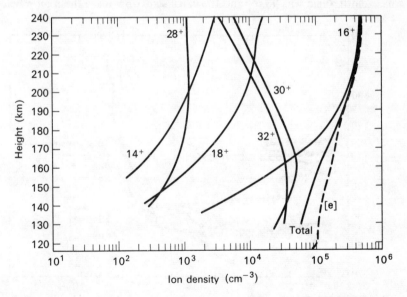

Fig. 6.23 Measurement of the daytime, mid-latitude ion composition of the E and F regions, after Johnson.[79] The curve labelled Total, gives the total positive ion concentration; similarly [e] gives the electron concentration.

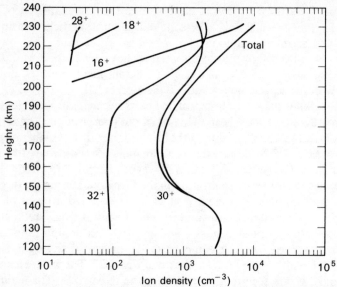

Fig. 6.24   Measurement of the night-time, mid-latitude ion composition of the E and F regions, after Johnson.[79]

minutes after the results of Fig. 6.25 were obtained, showed a substantial reduction in the NO^+ density below 110 km, although the metallic ion densities around 93 km had not changed markedly. During this flight the solar zenith angle was 98.6°, and the earth's shadow filled the region below 74 km.

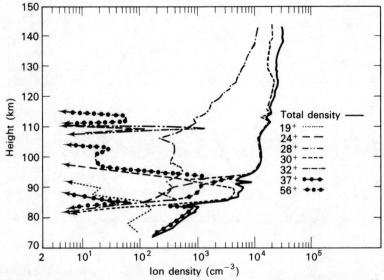

Fig. 6.25   Major positive ions observed at sunset during a rocket flight of Narcisi *et al.*[55] Solar zenith angle = 88.5°.

6.7.1 MODELS FOR THE E REGION

Of the multitude of reactions listed in Tables 6.1 and 6.3, only seven are required to account for the observed chemistry of the E region. These are:

$$N_2^+ + O_2 \rightarrow O_2^+ + N_2 \qquad (6.44)$$

$$k_{6.44} = 1 \times 10^{-10} \text{ cm}^3 \text{ molecule}^{-1} \text{ s}^{-1}$$

$$N_2^+ + O \rightarrow NO^+ + N \qquad (6.45)$$

$$k_{6.45} = 1.4 \times 10^{-10} \text{ cm}^3 \text{ molecule}^{-1} \text{ s}^{-1}$$

$$NO^+ + e \rightarrow N + O \qquad (6.46)$$

$$k_{6.46} = 4.5 \times 10^{-7} \, (T/300)^{-1.0}$$

$$O_2^+ + e \rightarrow O + O \qquad (6.47)$$

$$k_{6.47} = 2 \times 10^{-7} \, (T/300)^{-0.7}$$

$$O^+ + O_2 \rightarrow O_2^+ + O \qquad (6.48)$$

$$k_{6.48} = 2 \times 10^{-11} \text{ cm}^3 \text{ molecule}^{-1} \text{ s}^{-1}$$

$$O^+ + N_2 \rightarrow NO^+ + N \qquad (6.49)$$

$$k_{6.49} = 2 \times 10^{-12} \text{ cm}^3 \text{ molecule}^{-1} \text{ s}^{-1}$$

$$O_2^+ + NO \rightarrow NO^+ + O_2 \qquad (6.50)$$

$$k_{6.50} = 6.3 \times 10^{-10} \text{ cm}^3 \text{ molecule}^{-1} \text{ s}^{-1}$$

With the sequence of reactions 6.44 through 6.47 as a basis for an E region model, and using the NO altitude profile of Barth,[81] Keneshea et al.[80] calculated the electron, O_2^+ and NO^+ concentrations on a diurnal basis. Their calculated concentrations, allowing for chemical loss processes only, are compared with their experimental results for NO^+, O_2^+ and the total ion density (at sunset, solar zenith angle = 88.5°) in Fig. 6.26. The agreement is good at higher altitudes but marked discrepancies occur below 100 km. When the effect of transport by neutral winds* was included in the model, the agreement between calculated and observed profiles improved (Fig. 6.27). Thus the effect of transport processes cannot be ignored in the night-time E region. The relative importance of the two processes removing NO^+, namely chemical loss and transport, can be estimated by comparing the lifetimes of the NO^+ ion with respect to each of the loss processes. The full effect of the neutral winds measured by Narcisi et al.[55] occurs in a time of the order of 1000 s; in the daytime the lifetime of NO^+ in an ambient electron density of 1.5×10^5 electrons cm^{-3} is less than 25 s so that transport is not significant. At night, however, the electron density falls by about two orders of magnitude, thereby increasing the lifetime of NO^+ with respect to dissociative recombination to ~2500 s. Hence transport of NO^+ is an

---

* Measured by photographing a chemical released from the rocket during its flight.

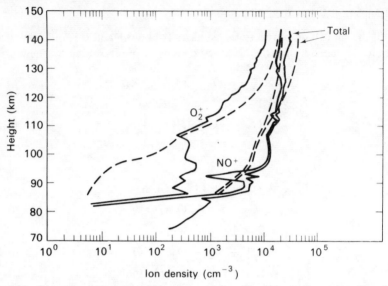

Fig. 6.26 Comparison of calculated (chemistry only) profiles with experimental profiles of $[NO^+]$, $[O_2^+]$ and total ion densities.[80] The dashed curves are the model calculations, the solid curves are the experimental results. Curves labelled Total represent the total positive ion density. The solar zenith angle $\chi = 88.5°$.

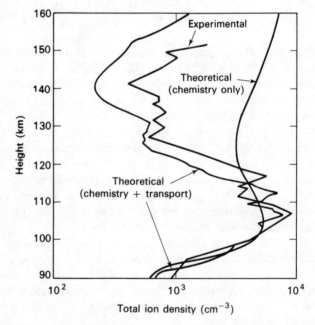

Fig. 6.27 Comparison of calculated (chemistry plus transport) and measured total positive ion composition for a solar zenith angle $\chi = 98.8°$.[80] (Note that the solar zenith angle is different from that in Fig. 6.26.)

important feature of the night-time E region. Stubbe[78] has compared the relative rates of chemical and transport losses of NO^+, and has concluded that solutions of the O_2^+ and NO^+ continuity equations for the night-time ionosphere are meaningless if diffusion and neutral winds are ignored.

6.7.2 SPORADIC E AND METAL ION CHEMISTRY

Effects associated with local regions of increased ionization in the E region are referred to as Sporadic E phenomena, a term which embraces several distinct types of scattering and reflection effects. There are in addition three distinct categories of sporadic E effects, namely equatorial, mid-latitude and auroral, which arise in different ways. In this section we are concerned mainly with the mid-latitude phenomenon, which we shall find involves the chemistry of gaseous metal ions. Indeed, the narrow intense regions of increased ionization near 105 km, which have been referred to for many years as Sporadic E, can now be identified as layers of increased metal ion density.

Mass spectrometric observations of metal ions were first reported by Istomin,[82] and their presence has since been confirmed by many rocket flights. A feature of the investigations of Narcisi[55] and Young et al.[83] was the observation of narrow, localized layers of metal ions (Fig. 6.25). Young and co-workers demonstrated that an intense sporadic E layer centred at 106 km was composed of metal ions with peak concentration of about 10^5 ion cm^{-3}, which is two orders of magnitude greater than the ambient ion concentration in the vicinity of the layer. The layer in this particular case measured less than 2 km vertically and more than 60 km horizontally. A significant increase in metal ion concentration has been observed during a meteor shower,[84] and it is reasonable to conclude that most metal ions have their origin in meteor ablation in the atmosphere. A typical sporadic E layer lies between 95 and 115 km altitude, is 1–3 km thick, and has a sharp ion density profile which may show fine structure (Fig. 6.28). The horizontal extent of a layer may be several hundred kilometres, with patches of exceptionally high ion densities, large ion density gradients, or both. An interesting side effect of the presence of a large concentration of metal ions is the small NO^+ concentration within the layer, by comparison with the regions immediately above and below. Within a metal ion layer there is an increased electron density, which enhances the dissociative recombination rates of the molecular ions O_2^+ and NO^+ (6.46 and 6.47), thereby reducing their concentrations. Atomic ions, on the other hand, have very low electron recombination rates; thus their concentration is not markedly influenced by the increased electron density.

The question arises as to how the metal atoms produced by meteor ablation are first ionized, then arranged in layers, and finally removed from the layers into the surrounding atmosphere. Although some metal ions are

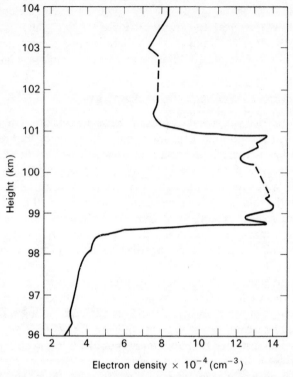

Fig. 6.28   Detail of a typical daytime electron density profile of a sporadic E layer showing a narrow steep sided layer between 99 and 101 km. The layer was measured during a Nike Apache rocket flight.[85,86]

produced directly through meteor ablation, by far the largest fraction are the result of photoionization and charge transfer processes.[87]

$$M + h\nu \rightarrow M^+ + e \qquad (6.51)$$

$$M + O_2^+ \text{ (or NO}^+) \rightarrow M^+ + O_2 \text{ (or NO)} \qquad (6.52)$$

The ionization potentials of most atmospheric metal atoms are relatively low; consequently, charge transfer is exothermic and occurs readily (see Table 6.11). Metal ions can also be formed from the metal oxides MO and MO_2 by photoionization and charge transfer (reactions 6.53–6.56) followed by ion–atom interchange (reaction 6.57), which is exothermic for many ionospheric metal oxides.

$$MO + h\nu \rightarrow MO^+ + e \qquad (6.53)$$

$$MO_2 + h\nu \rightarrow MO_2^+ + e \qquad (6.54)$$

$$MO + O_2^+ \text{ (or NO}^+) \rightarrow MO^+ + O_2 \text{ (or NO)} \qquad (6.55)$$

$$MO_2 + O_2^+ \text{ (or NO}^+) \rightarrow MO_2^+ + O_2 \text{ (or NO)} \qquad (6.56)$$

and $\qquad\qquad MO^+ + O \rightarrow M^+ + O_2 \qquad (6.57)$

Table 6.11   Rate coefficients for charge transfer to neutral metal atoms from Rutherford et al.[88]

Reaction		Extrapolated Thermal Energy Rate Coefficient[a] (cm^3 molecule^{-1} s^{-1})
$N^+ + Na$	$\rightarrow N + Na^+$	small
$O^+ + Na$	$\rightarrow O + Na^+$	small
$N_2^+ + Na$	$\rightarrow N_2 + Na^+$	1.9×10^{-9}
$O_2^+(X^2\Pi_g) + Na$	$\rightarrow O_2 + Na^+$	1.4×10^{-9}
$O_2^+(a^4\Pi_u) + Na$	$\rightarrow O_2 + Na^+$	2.0×10^{-9}
$H_2O^+ + Na$	$\rightarrow H_2O + Na^+$	2.7×10^{-9}
$N_2O^+ + Na$	$\rightarrow N_2O + Na^+$	2.0×10^{-9}
$N^+ + Mg$	$\rightarrow N + Mg^+$	1.2×10^{-9}
$O^+ + Mg$	$\rightarrow O + Mg^+$	small
$N_2^+ + Mg$	$\rightarrow N_2 + Mg^+$	$\sim 7 \times 10^{-10}$
$O_2^+ + Mg$	$\rightarrow O_2 + Mg^+$	1.2×10^{-9}
$O_2^+(a^4\Pi_u) + Mg$	$\rightarrow O_2 + Mg^+$	$> 3 \times 10^{-9}$
$NO^+ + Mg$	$\rightarrow NO + Mg^+$	8.1×10^{-10}
$H_2O^+ + Mg$	$\rightarrow H_2O + Mg^+$	2.2×10^{-9}
$N_2O^+ + Mg$	$\rightarrow N_2O + Mg^+$	2.2×10^{-9}
$N^+ + Ca$	$\rightarrow N + Ca^+$	1.1×10^{-9}
$O^+ + Ca$	$\rightarrow O + Ca^+$	7.6×10^{-10}
$N_2^+ + Ca$	$\rightarrow N_2 + Ca^+$	1.8×10^{-9}
$O_2^+(X^2\Pi_g) + Ca$	$\rightarrow O_2 + Ca^+$	1.8×10^{-9}
$O_2^+(a^4\Pi_u) + Ca$	$\rightarrow O_2 + Ca^+$	3.5×10^{-9}
$NO^+ + Ca$	$\rightarrow NO + Ca^+$	4.0×10^{-9}
$H_2O^+ + Ca$	$\rightarrow H_2O + Ca^+$	4.0×10^{-9}
$H_3O^+ + Ca$	$\rightarrow H_2O + H + Ca^+$	4.4×10^{-9}
$N_2O^+ + Ca$	$\rightarrow N_2O + Ca^+$	3.7×10^{-9}
$H^+ + Fe$	$\rightarrow H + Fe^+$	7.4×10^{-9}
$N^+ + Fe$	$\rightarrow N + Fe^+$	1.5×10^{-9}
$O^+ + Fe$	$\rightarrow O + Fe^+$	2.9×10^{-9}
$N_2^+ + Fe$	$\rightarrow N_2 + Fe^+$	4.3×10^{-10}
$NO^+ + Fe$	$\rightarrow NO + Fe^+$	9.2×10^{-10}
$O_2^+ + Fe$	$\rightarrow O_2 + Fe^+$	1.1×10^{-9}
$H_2O^+ + Fe$	$\rightarrow H_2O + Fe^+$	1.5×10^{-9}

a   These rate coefficients were extrapolated to 300 K from higher energy results in the range 1–500 eV.

Some relevant rate coefficients are given in Table 6.12. The rates of processes 6.54 and 6.56 have not been measured, and their relative importance cannot be precisely assessed at present. Charge transfer is expected to increase in importance, relative to photoionization, in the lower E region, because radiation having sufficient energy to ionize M, MO and MO_2 (~ 7 eV) does not penetrate below 100 km except in the few atmospheric 'windows' between 100 and 130 nm (see Chapter 2). The relative efficiencies of photoionization and charge transfer can be estimated from Fig. 6.29; the solid curve represents the ionization rate coefficient due to photoionization, assuming an ionization cross section of 1×10^{-18} cm^2 for all metal and metal oxide species. The crosses are the ionization coefficients $J_{6.51}$ calculated by Swider[87] for the metals shown. The importance of charge transfer relative

Table 6.12 Thermal energy atmospheric metal–ion reactions.[89,90]

Reaction		Rate Coefficient*	Comments
Singly charged species			
$N_2^+ + Na$	$\rightarrow Na^+ + N_2$	5.8×10^{-10}	Table 6.11 gives 1.9×10^{-9}
$NO^+ + Na$	$\rightarrow Na^+ + NO$	7.0×10^{-11}	
$O_2^+ + Na$	$\rightarrow Na^+ + O_2$	6.7×10^{-10}	Table 6.11 gives 1.4×10^{-9}
$O_2^+ + Na$	$\rightarrow NaO^+ + O$	7.7×10^{-11}	
$Mg^+ + O_3$	$\rightarrow MgO^+ + O_2$	2.3×10^{-10}	
$Ca^+ + O_3$	$\rightarrow CaO^+ + O_2$	1.6×10^{-10}	
$Fe^+ + O_3$	$\rightarrow FeO^+ + O_2$	1.5×10^{-10}	
$Na^+ + O_3$	$\rightarrow NaO^+ + O_2$	$<1 \times 10^{-11}$	
$K^+ + O_3$	$\rightarrow KO^+ + O_2$	$<1 \times 10^{-11}$	
$MgO^+ + O$	$\rightarrow Mg^+ + O_2$	$\sim 1 \times 10^{-10}$	
$SiO^+ + O$	$\rightarrow Si^+ + O_2$	$\sim 2 \times 10^{-10}$	
$SiO^+ + N$	$\rightarrow Si^+ + NO$	$\sim 2 \times 10^{-10}$	
	$\rightarrow NO^+ + Si$	$\sim 1 \times 10^{-10}$	
$Na^+, K^+, Ba^+ + O_2(NO)$	$\rightarrow$ products	$<1 \times 10^{-13}$	For ion kinetic energies up to ~ 5 eV
$Mg^+ + O_2 + Ar$	$\rightarrow MgO_2^+ + Ar$	$\sim 2.5 \times 10^{-30}$	
$Ca^+ + O_2 + Ar$	$\rightarrow CaO_2^+ + Ar$	$\sim 6.6 \times 10^{-30}$	
$Ca^+ + O_2 + He$	$\rightarrow CaO_2^+ + He$	$\sim 2 \times 10^{-30}$	
$Fe^+ + O_2 + Ar$	$\rightarrow FeO_2^+ + Ar$	$\sim 1.0 \times 10^{-30}$	
$Na^+ + O_2 + Ar$	$\rightarrow NaO_2^+ + Ar$	$<2 \times 10^{-31}$	
$K^+ + O_2 + Ar$	$\rightarrow KO_2^+ + Ar$	$<2 \times 10^{-31}$	
$Na^+ + 2CO_2$	$\rightleftharpoons Na^+.CO_2 + CO_2$	2×10^{-29}	$k_{rev} = 1 \times 10^{-14}$
$Na^+.CO_2 + 2CO_2$	$\rightleftharpoons Na^+(CO_2)_2 + CO_2$	2×10^{-29}	$k_{rev} = 5 \times 10^{-13}$
$Na^+ + 2O_2$	$\rightleftharpoons Na^+.O_2 + O_2$	5×10^{-32}	$k_{rev} = 8 \times 10^{-13}$
$K^+ + 2CO_2$	$\rightleftharpoons K^+.CO_2 + CO_2$	4×10^{-30}	$k_{rev} = 2.5 \times 10^{-13}$
$Na^+ + H_2O + He$	$\rightarrow Na^+.H_2O + He$	4.7×10^{-30}	
$Na^+ + 2H_2O$	$\rightarrow Na^+.H_2O + H_2O$	1.0×10^{-28}	
$K^+ + H_2O + He$	$\rightarrow K^+.H_2O + He$	2.6×10^{-30}	
$K^+ + 2H_2O$	$\rightarrow K^+.H_2O + H_2O$	4.5×10^{-29}	
$Ba^+ + CO_2 + He$	$\rightarrow Ba^+.CO_2 + He$	2.8×10^{-30}	
$Ca^+ + CO + He$	$\rightarrow Ca^+.CO + He$	2.7×10^{-30}	
$Ca^+ + O_2 + He$	$\rightarrow Ca^+.O_2 + He$	$\sim 2 \times 10^{-30}$	
$Ba^+ + CO_2 + He$	$\rightarrow Ba^+.CO_2 + He$	2.8×10^{-30}	
Doubly charged species			
$Mg^{2+} + Ar + He$	$\rightarrow Mg^{2+}.Ar + He$	3.1×10^{-30}	
$Mg^{2+} + N_2 + He$	$\rightarrow Mg^{2+}.N_2 + He$	1.9×10^{-29}	
$Mg^{2+} + CO + He$	$\rightarrow Mg^{2+}.CO + He$	4.7×10^{-29}	
$Mg^{2+} + CO_2 + He$	$\rightarrow Mg^{2+}.CO_2 + He$	3.1×10^{-27}	
$Ca^{2+} + Ar + He$	$\rightarrow Ca^{2+}.Ar + He$	$\sim 1 \times 10^{-30}$	
$Ca^{2+} + N_2 + He$	$\rightarrow Ca^{2+}.N_2 + He$	6.2×10^{-30}	
$Ca^{2+} + O_2 + He$	$\rightarrow Ca^{2+}.O_2 + He$	8.9×10^{-30}	
$Ca^{2+} + CO + He$	$\rightarrow Ca^{2+}.CO + He$	2.0×10^{-29}	
$Ca^{2+} + CO_2 + He$	$\rightarrow Ca^{2+}.CO_2 + He$	1.1×10^{-27}	
$Ca^{2+} + N_2O + He$	$\rightarrow Ca^{2+}.N_2O + He$	2.5×10^{-27}	
$Ca^{2+} + H_2O + He$	$\rightarrow Ca^{2+}.H_2O + He$	$\sim 5 \times 10^{-28}$	
$Ba^{2+} + N_2 + He$	$\rightarrow Ba^{2+}.N_2 + He$	$\sim 1.6 \times 10^{-30}$	
$Ba^{2+} + O_2 + He$	$\rightarrow Ba^{2+}.O_2 + He$	3×10^{-30}	

Table 6.12 (Continued).

Reaction		Rate Coefficient*	Comments
$Ba^{2+} + CO + He$	$\rightarrow Ba^{2+}.CO + He$	$\sim 5 \times 10^{-30}$	
$Ba^{2+} + CO_2 + He$	$\rightarrow Ba^{2+}.CO_2 + He$	1.1×10^{-28}	
$Ba^{2+} + N_2O + He$	$\rightarrow Ba^{2+}.N_2O + He$	$\sim 1.9 \times 10^{-28}$	
$Ba^{2+} + H_2O + He$	$\rightarrow Ba^{2+}.H_2O + He$	1.1×10^{-28}	

* cm^3 molecule^{-1} s^{-1} for bimolecular reactions and cm^6 molecule^{-2} s^{-1} for three-body processes.

to photoionization is estimated by comparing the photoionization rate with the product of the molecular ion concentration ($\sim 10^3$ cm^{-3} at 80 km to 10^5 cm^{-3} at 110 km for the normal daytime ionosphere) and the appropriate charge transfer rate coefficient given in Table 6.11.

How the metal ions become distributed in layers is a problem that has not been satisfactorily resolved, but it is recognized that physical rather than chemical processes are responsible. Layzer[86] argues that a layer of enhanced

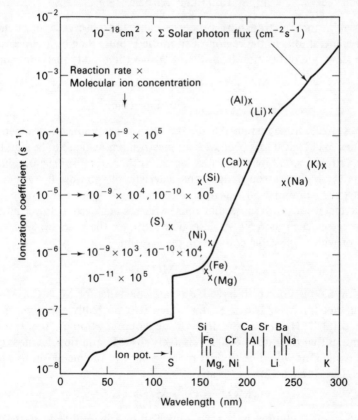

Fig. 6.29 Ionization coefficients for important species of meteoric origin. (After Swider[87]. See text for explanation of symbols.)

ionization could be produced by a local increase in ionization rate, a local decrease in the ion loss rate, or a redistribution of existing ionization. Of these possibilities only redistribution of the existing ionization is likely, and vertical transport appears to be the only plausible mechanism for this redistribution and subsequent formation of the thin layers. Various physical transport processes such as wind shear (where horizontal E–W winds are presumed to take the ionized component of the atmosphere across the earth's magnetic field lines and so generate an electric field which may localize pockets of increased charge density) and other special effects have been invoked, with some success in predicting the existence and shape of layers.[86]

If dynamic processes are to control the distribution of metal ions then the lifetime of the ions with respect to chemical loss processes such as radiative recombination (equation 6.58) must be fairly large.

$$M^+ + e \rightarrow M + h\nu \qquad (6.58)$$

$$k_{ei} \sim 10^{-12} \text{ cm}^3 \text{ particle}^{-1} \text{ s}^{-1}$$

The significance of transport processes, and the observed persistence of positive metal ions in the region of sporadic E must then be a consequence of the slow rate of reaction 6.58. Below 90 km three-body reactions such as

$$M^+ + O_2 + M \rightarrow MO_2^+ + M \qquad (6.59)$$

$$k_{6.59} \sim 10^{-30 \pm 1} \text{ cm}^6 \text{ molecule}^{-2} \text{ s}^{-1}$$

are most likely to be responsible for the observed rapid decline in metal ion densities. If chemical loss processes for electron removal are to be significant ion–electron recombination must occur with a rate sufficiently fast to deplete the high electron density characteristic of sporadic E. One way in which the recombination rate might be increased is by the conversion of atomic metal ions into molecular ions, which is achieved at lower altitudes by processes such as 6.59. At E region altitudes the reaction with ozone (6.60) provides a possible conversion mechanism

$$M^+ + O_3 \rightarrow MO^+ + O_2 \qquad (6.60)$$

Reaction 6.60 is known to have large rate constants for M = Ca, Mg and Fe, but not for M = Na or K, for which it is probably endothermic (see Table 6.12).[89] However, the low ozone concentration in the E region determines that the rate of 6.60 is relatively slow. In addition, as most of the metal oxide ions would be rapidly reduced back to the metal ion by atomic oxygen (see Table 6.12)

$$MO^+ + O \rightarrow M^+ + O_2 \qquad (6.61)$$

the effect of oxidation by ozone must be largely nullified. If the ratio of $[MO^+]$ to $[M^+]$ is controlled by reactions 6.60 and 6.61, then

Table 6.13 Reactions of Mg and associated ions and their rate coefficients used by Anderson and Barth.[91a]

Reaction		Rate Coefficient[a]	Reaction Number
$Mg^+ + O_2 + M$	$\rightarrow MgO_2^+ + M$	$k_{6.62} = 2.5 \times 10^{-30}$	(6.62)
$Mg^+ + O_3$	$\rightarrow MgO^+ + O_2$	$k_{6.63} = 2.3 \times 10^{-10}$	(6.63)
$MgO_2^+ + O$	$\rightarrow MgO^+ + O_2$	$k_{6.64} = 1 \times 10^{-10}$ (est)	(6.64)
$Mg + O_2^+$	$\rightarrow MgO^+ + O$	$k_{6.65} < 5 \times 10^{-10}$	(6.65)
$Mg + NO^+$	$\rightarrow MgO^+ + N$	$k_{6.66} < 5 \times 10^{-10}$	(6.66)
$MgO^+ + O$	$\rightarrow Mg^+ + O_2$	$k_{6.67} = 1 \times 10^{-10}$	(6.67)
$Mg + O_2^+$	$\rightarrow Mg^+ + O_2$	$k_{6.68} = 1.2 \times 10^{-9}$	(6.68)
$Mg + NO^+$	$\rightarrow Mg^+ + NO$	$k_{6.69} = 1.0 \times 10^{-9}$	(6.69)
$Mg + h\nu$	$\rightarrow Mg^+ + e$	$J_{6.70} = 4 \times 10^{-7}$ at zero optical depth	(6.70)
$Mg^+ + e$	$\rightarrow Mg + h\nu$	$k_{6.71} = 1 \times 10^{-12}$ (est)	(6.71)
$MgO^+ + e$	$\rightarrow Mg + O$	$k_{6.72} = 1 \times 10^{-7}$ (est)	(6.72)
$MgO_2^+ + e$	$\rightarrow Mg + O_2$	$k_{6.73} = 3 \times 10^{-7}$ (est)	(6.73)

a Units for termolecular processes are cm^6 molecule^{-2} s^{-1} and for bimolecular are cm^3 molecule^{-1} s^{-1}.

$[MO^+]/[M^+] \sim [O_3]/[O] \sim 10^{-7}$ at around 110 km.[89] Thus the possibility of appreciable concentrations of MO^+ ions existing in the E region can probably be ruled out.

The chemistry of atmospheric magnesium was the subject of a study by Anderson and Barth,[91a] who found by dayglow experiments, using a rocketborne spectrometer, that $[Mg^+] \geq 22[Mg]$. Gadsden[91b] has pointed out that the observed Mg^+ radiances were not correctly related to column densities and therefore the $[Mg^+]/[Mg]$ ratio of Anderson and Barth is likely to be in error. With the reaction scheme shown in Table 6.13, and assuming that the time for removal of Mg^+ by diffusion is significantly larger than the time required for the equilibrium between Mg^+ and Mg to be set up, the following steady state concentrations at 110 km are established:

$$[MgO_2^+] = \frac{k_{6.62}[O_2][N_2][Mg^+]}{k_{6.64}[O]} \tag{6.74}$$

$$[MgO^+] = (k_{6.63}[Mg^+][O_3] + k_{6.64}[MgO_2^+][O] + k_{6.65}[Mg][O_2^+] + k_{6.66}[Mg][NO^+])/k_{6.67}[O] \tag{6.75}$$

$$[Mg^+] = (k_{6.67}[O][MgO^+] + k_{6.68}[O_2^+][Mg] + k_{6.69}[NO^+][Mg] + J_{6.70}[Mg])/k_{6.62}[O_2][N_2] \tag{6.76}$$

and

$$[Mg] = \frac{k_{6.71}[Mg^+][e] + k_{6.72}[MgO^+][e] + k_{6.73}[MgO_2^+][e]}{(k_{6.65} + k_{6.68})[O_2^+] + (k_{6.66} + k_{6.69})[NO^+] + J_{6.70}} \tag{6.77}$$

Anderson and Barth chose an ozone density of 10^4 molecule cm^{-3}, an atomic oxygen density of 3×10^{11} atom cm^{-3}, an electron density of

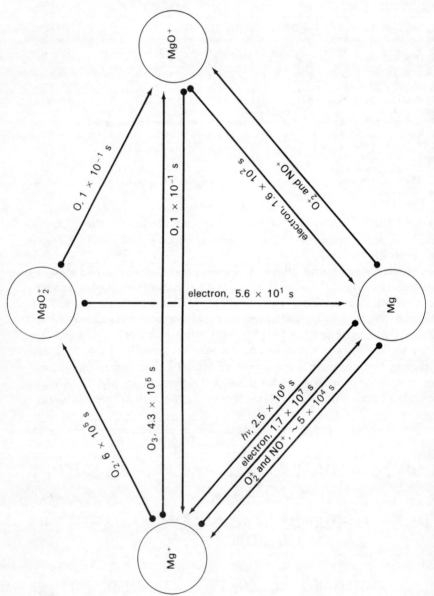

Fig. 6.30 Schematic diagram of the atmospheric chemistry of magnesium and derived ions at 110 km.[91a] Lifetimes with

6×10^4 cm^{-3} and N_2 and O_2 densities of 1.6×10^{12} and 3.5×10^{11} molecule cm^{-3}, respectively, at 110 km. As the values of $[NO^+]$ and $[O_2^+]$ are lower, within a layer, than the ambient concentrations, Anderson and Barth chose values less by a factor of 5 than the predicted values of 1.5×10^4 and 8×10^2 ion cm^{-3}, respectively. Their reaction scheme is summarized in Fig. 6.30, together with the calculated lifetimes of the chemical process indicated. Several significant features emerge from this scheme: (1) The lifetimes of MgO_2^+ and MgO^+ are less than one second and that of Mg about one hour. (2) The lifetime of Mg^+ is determined by the three-body reaction with N_2 and O_2, and is of the order of a day. Therefore there is time for Mg^+ to accumulate (by transport processes) and maintain an ionization layer. (3) The rates of destruction of MgO_2^+ and MgO^+ are determined by their rates of reaction with atomic oxygen, and not by dissociative recombination. (4) Neutral magnesium is destroyed mainly by charge transfer with NO^+ and O_2^+ although its chemical lifetime is sufficiently long for transport processes to modify its distribution. It should be noted that the original source of metal atoms (meteor ablation) and the ultimate sink (downward diffusion) have not been included in Fig. 6.30.

The relative concentrations of MgO_2^+ and Mg^+ are established by eqn. 6.74, so that $[MgO_2^+]/[Mg^+] = 1.4 \times 10^{-7}$ at 110 km. If the rates of the ion interchange reactions, 6.65 and 6.66, are 10^{-9} cm^3 particle^{-1} s^{-1} or smaller, then eqn. 6.75 becomes

$$[MgO^+] = \frac{k_{6.64}[MgO_2^+][O]}{k_{6.67}[O]} \qquad (6.78)$$

or $[MgO^+] \sim [MgO_2^+]$. Further, the ratio

$$\frac{k_{6.73}[MgO_2^+][e]}{k_{6.71}[Mg^+][e]} \sim 10^{-2}$$

and $k_{6.73}[MgO_2^+][e] \sim k_{6.72}[MgO^+][e]$. If, in addition, charge transfer predominates over the photoionization reaction 6.70 then eqn. 6.77 reduces to

$$\frac{[Mg^+]}{[Mg]} \sim \frac{k_{6.68}[O_2^+] + k_{6.69}[NO^+]}{k_{6.71}[e]} \qquad (6.79)$$

If $[Mg^+]/[Mg] \geqslant 22$ as Anderson and Barth estimated, then for eqn. 6.79 to be valid $k_{6.68} \geqslant 4 \times 10^{-11}$ and $k_{6.69} \geqslant 4 \times 10^{-10}$ cm^3 molecule^{-1} s^{-1}. These values are consistent with laboratory data (Table 6.11). Alternatively, if the ratio is closer to 200[91b], again for eqn. 6.79 to be valid, the coefficients $k_{6.68}$ and $k_{6.69}$ are consistent with the measured values of 1.2×10^{-9} and 8×10^{-10}, respectively.

The loss processes of alkali metal ions are more difficult to interpret because of the slow rates of termolecular reactions with O_2 to form NaO_2^+ or KO_2^+, and the slowness of the bimolecular reactions with O_3 to form

NaO^+ or KO^+ (see Table 6.12). Gadsden[92] has investigated the relative importance of chemical and physical processes governing the concentrations of metals, particularly sodium, and has concluded that diffusive loss of atomic sodium is more important than ionization loss (Section 5.3.2).

The overall picture of sporadic E, and of E region metal ion chemistry, is one in which physical rather than chemical processes govern the distribution of metal atoms and the formation and deionization of ion layers. The source of metal atoms is meteor ablation $(\sim 10^{-11} \text{ g year}^{-1})$[92] and the sink is downward diffusion. Metal ions in the D region will be hydrated or otherwise complexed by processes of the type

$$Na^+ + H_2O + M \rightarrow Na^+.H_2O + M \qquad (6.80)$$

$$Na^+ + 2CO_2 \rightarrow Na^+.CO_2 + CO_2 \qquad (6.81)$$

$$Na^+.CO_2 + H_2O \rightarrow Na^+.H_2O + CO_2 \qquad (6.82)$$

$$k_{6.82} \sim 10^{-9} \text{ cm}^3 \text{ molecule}^{-1} \text{ s}^{-1} \text{ }[89]$$

and will ultimately be neutralized by dissociative recombination.

6.8 F region chemistry

There is no major distinction between the chemistry of the F_1 and F_2 regions. In the F region, in contrast to the D and E regions, the principal ions are atomic, and therefore ion interchange processes, such as 6.49, control both the chemistry of the region and the rate of electron loss. As the altitude increases diffusion processes become more important until eventually, well above the F_2 peak, they control the ion distribution even in the daytime. The main distinction between the F_1 and F_2 regions lies in the relative importance of ambipolar diffusion. Generally the chemistry of the F_1 region is simpler than that of the other two regions we have considered, although the higher ambient temperatures (Fig. 6.31) constitute a complicating factor.

6.8.1 THE F_1 REGION

Measured and calculated neutral species and ion distributions in the F_1 region are compared in Figs. 6.32 and 6.33 respectively. There is poor agreement between the measured and calculated neutral densities derived from models based on diffusive equilibrium, particularly in the case of atomic oxygen. However, as pointed out in Section 4.3.1(ii), mass spectrometric measurements of atomic oxygen may be uncertain. Usually densities derived from observations of satellite drag tend to be higher than corresponding mass spectrometric determinations. The major ions formed by photoionization are O^+, O_2^+ and N_2^+; their estimated rates of production are shown in Fig. 6.34. These ions, and also N^+ and He^+, are very reactive.

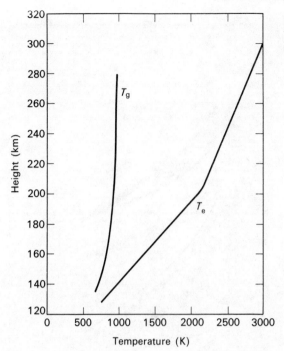

Fig. 6.31 Measured molecular (T_g) and electron (T_e) temperatures in the F region.[93]

Two other ions, NO^+ and H^+, are observed to be present but do not modify the distribution of neutral species in the F_1 region to any great extent.

N_2^+ *ions.* Figure 6.34 shows the rate of production of N_2^+ ions by photo-ionization of N_2. The loss of N_2^+ occurs by reaction 6.44 (mainly at lower altitudes) by reaction 6.45 (mainly at higher altitudes), and by dissociative recombination (6.83).

$$N_2^+ + O_2 \rightarrow O_2^+ + N_2 \tag{6.44}$$

$$k_{6.44} = 1 \times 10^{-10} \text{ cm}^3 \text{ molecule}^{-1} \text{ s}^{-1} \text{ at 300 K}^{41}$$

$$N_2^+ + O \rightarrow NO^+ + N \tag{6.45}$$

$$k_{6.45} = 1.4 \times 10^{-10} \text{ cm}^3 \text{ molecule}^{-1} \text{ s}^{-1} \text{ at 300 K}^{39}$$

$$N_2^+ + e \rightarrow N + N \tag{6.83}$$

$$k_{6.83} = 3 \times 10^{-7} (T/300)^{-0.02 \ 10}$$

The variation of $k_{6.44}$ with temperature has been measured between 300 and 500 K (where it decreases by approximately 50% from its 300 K value) and also for N_2^+ ion temperatures up to 2000 K.[95] The temperature dependence of reaction 6.45 has not been measured at the time of writing. The efficiency of these three loss processes for N_2^+ is such that, in spite of a high rate of

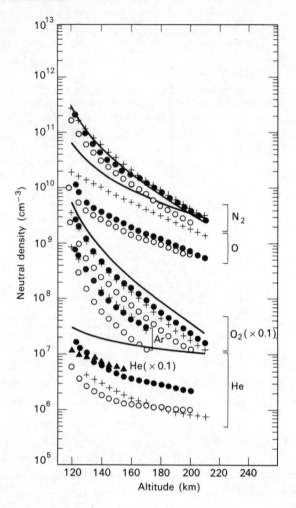

Fig. 6.32 Measured E and F$_1$ neutral densities of Krankowsky *et al.*[94] (shown by individual points) compared with the summer model ($T = 900$ K) of the US Standard Atmosphere Supplements, 1966, (solid lines).

production of N$_2^+$, its concentration in the 120–220 km region is only $\sim 5 \times 10^3$ ion cm^{-3} in the daytime.

NO$^+$ *ions*. Photoionization of nitric oxide in the F region is negligible compared with other sources of NO$^+$ ions because of the very low NO concentration. The two main sources of NO$^+$ are reactions 6.45 and 6.49:

$$O^+ + N_2 \rightarrow NO^+ + N \qquad (6.49)$$

$$k_{6.49} = 1.2 \times 10^{-12} \text{ cm}^3 \text{ molecule}^{-1} \text{ s}^{-1} \text{ at 300 K}$$

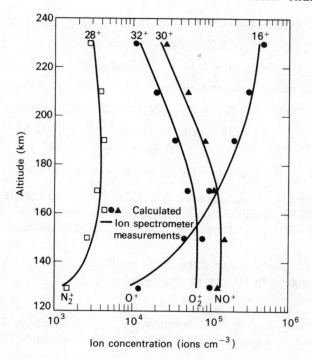

Fig. 6.33 Measured and calculated profiles of the E and F_1 region ions, after Pharo *et al.*[93]

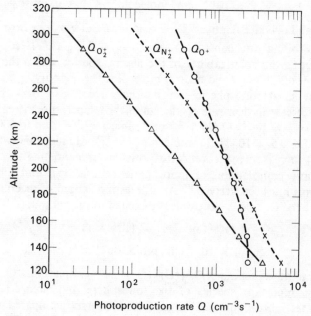

Fig. 6.34 Calculated photoproduction rates of O_2^+, N_2^+ and O^+, after Pharo *et al.*[93]

Loss of NO^+ is by dissociative recombination,

$$NO^+ + e \rightarrow N + O \tag{6.46}$$

where $\qquad k_{6.46} = 4.5 \times 10^{-7} (T/300)^{-1.0}$ [10]

Two other processes, 6.84 and 6.85,

$$N^+ + O_2 \rightarrow NO^+ + O \tag{6.84}$$

where $\quad k_{6.84} = 3.5 \times 10^{-10}$ cm^3 molecule^{-1} s^{-1} at 300 K[41]

and

$$O_2^+ + N \rightarrow NO^+ + O \tag{6.85}$$

where $\quad k_{6.85} = 1.8 \times 10^{-10}$ cm^3 molecule^{-1} s^{-1} at 300 K[41]

may supplement 6.45 and 6.49 as sources of NO^+, but their relative importance cannot be established until both neutral particle densities and the temperature coefficients of their rate constants are better known.

O^+ *ions.* The main source of the dominant F region ion, O^+, is photo-ionization (Fig. 6.34). The charge transfer process

$$N_2^+ + O \rightarrow O^+ + N_2 \tag{6.86}$$

is unobservably slow in the laboratory ($k_{6.86} < 10^{-11}$ cm^3 molecule^{-1} s^{-1}).[95] The O^+ ions are removed by interchange reactions with N_2 (6.49) and O_2 (6.48).

$$O^+ + O_2 \rightarrow O_2^+ + O \tag{6.48}$$

$$k_{6.48} = 2 \times 10^{-11}\ cm^3\ \text{molecule}^{-1}\ s^{-1}\ \text{at 300 K}[41]$$

Both 6.48 and 6.49 are relatively slow reactions, i.e. $k \ll k_L$ where k_L is the Langevin orbiting rate constant given by $k_L = 2\pi e \sqrt{\alpha/\mu}$. (Here e is the ionic charge, α the polarizability of the neutral species, and μ the reduced mass of the colliding ion and neutral molecule. The quantity k_L has a value $\sim 10^{-9}$ for most ionospheric ion–molecule reactions.) Both processes exhibit a decrease in their rate coefficient as the temperature increases, such that $k_{6.48} = 2 \times 10^{-11} (T/300)^{-1/2}$ cm^3 molecule^{-1} s^{-1} up to 600 K, and $k_{6.49}$ at 600 K is 5×10^{-13} cm^3 molecule^{-1} s^{-1}.[95] At much higher temperatures $k_{6.49}$ increases; a theoretical expression for predicting the value of $k_{6.49}$ for any combination of ionospheric translational and vibrational temperatures has been derived.[96] An interesting variant of reaction 6.49 is 6.87, where the O^+ ion is electronically excited to the ^2D state,

$$O^+(^2D) + N_2 \rightarrow N_2^+ + O \tag{6.87}$$

$$k_{6.87} \sim 10^{-9}\ cm^3\ \text{molecule}^{-1}\ s^{-1}\ [95]$$

The long radiative lifetime of $O^+(^2D)$ (1.3×10^4 s) implies that $O^+(^2D)$ is a potentially important species of the ionosphere, but unfortunately the relative rates of reactions 6.87 and 6.49 cannot be compared until the $O^+(^2D)$ particle densities are known. Because the atomic ion O^+ does not associate readily with an electron

$$O^+ + e \to O + h\nu \tag{6.88}$$

$$k_{6.88} \sim 4 \times 10^{-12} \, (T/300)^{-0.7 \ 10}$$

processes 6.48 and 6.49, forming the molecular ions O_2^+ and NO^+, effectively control the electron loss rate in the F_1 region. Reactions 6.48 and 6.49 are followed by the faster dissociative recombination processes 6.46 and 6.47.

$$NO^+ + e \to N + O \tag{6.46}$$

$$k_{6.46} = 4.5 \times 10^{-7} \, (T/300)^{-1.0} \, \text{cm}^3 \, \text{particle}^{-1} \, \text{s}^{-1 \ 10}$$

$$O_2^+ + e \to O + O \tag{6.47}$$

$$k_{6.47} = 2 \times 10^{-7} \, (T/300)^{-0.7} \, \text{cm}^3 \, \text{particle}^{-1} \, \text{s}^{-1 \ 10}$$

O_2^+ *ions.*   Important sources of O_2^+ in the F_1 region are photoionization (Fig. 6.34) and the ion-neutral reactions 6.44, 6.48 and 6.89;

$$N_2^+ + O_2 \to O_2^+ + N_2 \tag{6.44}$$

$$N^+ + O_2 \to O_2^+ + N \tag{6.89}$$

$$k_{6.89} \sim 3 \times 10^{-10} \, \text{cm}^3 \, \text{molecule}^{-1} \, \text{s}^{-1 \ 95}$$

The variation with temperature of the rate coefficients $k_{6.44}$ and $k_{6.48}$ has been described; the temperature dependence of $k_{6.89}$ has yet to be measured. Removal of the O_2^+ formed in reaction 6.89 is by dissociative recombination (reaction 6.47).

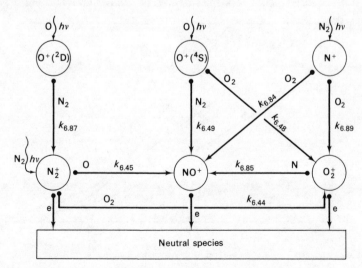

Fig. 6.35   Schematic outline of the important chemical processes responsible for the F_1 and lower F_2 regions.

Daytime models for the distributions of the important F_1 region ions have been based, with some success, on the chemical processes outlined for conditions of photochemical equilibrium, as summarized in Fig. 6.35. The major processes for neutral species are photodissociation and diffusion, which predominate over mass transport and turbulent mixing. At present the limiting factor in the model calculations is the ionospheric composition data (for both neutral and charged species), rather than the laboratory measurements of reaction rates.

6.8.2   THE F_2 REGION

As noted at the beginning of Section 6.8, there is no major distinction between the chemistry of the F_1 and F_2 regions. In the F_2 region solar radiation at wavelengths less than 80 nm is mainly responsible for ionization of the dominant neutral constituents, atomic oxygen and molecular nitrogen, whose distribution with altitude is shown in Fig. 6.36 (from a model by Jacchia[97]). Measured altitude profiles of the major ions in the F_2 region, as observed with rocket-borne mass spectrometers, are shown in Figs. 6.37 and 6.38. Both flights indicated that the predominant ions O^+, H^+, N^+ and He^+ are monatomic, but the results differed by an order of magnitude with regard to the concentrations of He^+ and H^+. Although the two sets of measurements were made from the same rocket base (Wallops Island, Virginia) they were made at different times of the year, March 1966 in the case of Fig. 6.37 and August 1966 for Fig. 6.38, and thus the differing results may reflect genuine seasonal variations. Comparable seasonal variations have been observed from orbiting satellites at around 600–800 km.[100]

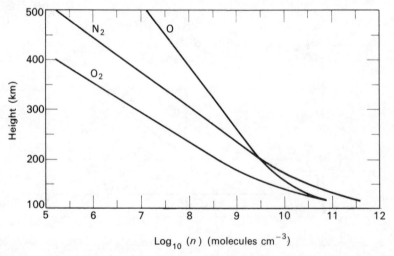

Fig. 6.36   Calculated concentrations of main neutral constituents in the F region, for an exospheric temperature of 900 K.[97]

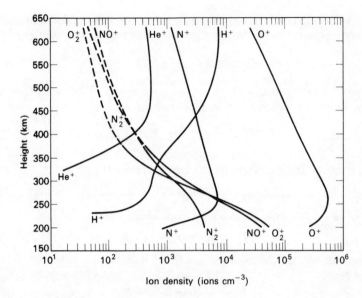

Fig. 6.37   Measured ion distributions of Brinton *et al.*[98] in the F$_2$ region above Wallops Island, Virginia, at 1300 EST. on March 2nd, 1966.

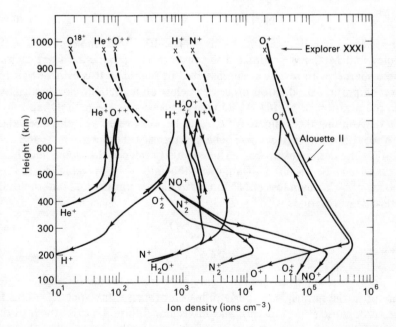

Fig. 6.38   Measured ion distribution of Hoffman *et al.*[99] in the F$_2$ region at 1328 EST on August 15th 1966, above Wallops Island, Virginia. 'Explorer 31' and 'Alouette 2' refer to satellite measurements.

In discussing the chemistry of the F_2 region we must include, in addition to the processes discussed for the F_1 region in Section 6.8.1, reactions of N^+, H^+ and He^+.

N^+ *ions.* Several mechanisms for the production of N^+ ions have been suggested,[101] e.g. process 6.90 at higher altitudes:

$$He^+ + N_2 \rightarrow He + N + N^+ \tag{6.90}$$

where $k_{6.90} \sim 1 \times 10^{-9}$ cm^3 molecule^{-1} s^{-1} [95] (but see also under He^+ ions), photoionization of atomic nitrogen below 300 km:

$$N + h\nu \rightarrow N^+ + e \tag{6.91}$$

and photoionization of molecular nitrogen below about 400 km:

$$N_2 + h\nu \rightarrow N + N^+ + e \tag{6.92}$$

The most important chemical sources of N^+ ions are charge transfer between N and N_2^+:

$$N_2^+ + N \rightarrow N^+ + N_2 \tag{6.93}$$

$$k_{6.93} < 10^{-11} \text{ cm}^3 \text{ molecule}^{-1} \text{ s}^{-1} \text{ [101]}$$

and charge transfer of O_2^+ with $N(^2D)$:

$$N(^2D) + O_2^+ \rightarrow N^+ + O_2 \tag{6.94}$$

where $k_{6.94}$ has been estimated as $\sim 4 \times 10^{-10}$ cm^3 molecule^{-1} s^{-1}.[101] Bailey and Moffett[101] estimated the rates of N^+ production from each of these sources with results as summarized in Fig. 6.39. It is evident that the most important production processes below about 350 km are dissociative photoionization of N_2 (6.92) and charge transfer between $N(^2D)$ and O_2^+ (6.94), assuming the estimate for $k_{6.94}$ to be somewhere near the true value.

Above 350 km dissociative charge transfer between He^+ and N_2 (6.90) is the dominant process. There is also evidence that some additional process may be required to augment the rate of N^+ production around 300 km.[101] The main loss of N^+ ions is by charge transfer and ion–molecule interchange reactions with O_2:

$$N^+ + O_2 \rightarrow NO^+ + O \tag{6.84}$$

and $\qquad\qquad N^+ + O_2 \rightarrow O_2^+ + N \tag{6.89}$

H^+ *ions.* Atomic hydrogen has its major source in the dissociation of water vapour in the mesosphere. Eddy diffusion processes transport H atoms to the turbopause (105–110 km), and molecular diffusion carries them to the base of the exosphere, from which they eventually escape altogether (cf. Section 4.3.2 (iv)). Although photoionization of atomic hydrogen is undoubtedly the primary ionization process at very high altitudes, the main

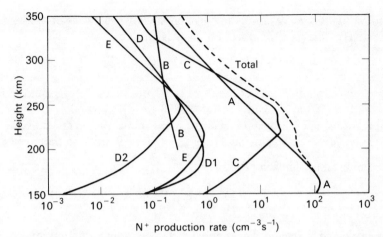

Fig. 6.39 Relative rates of production of N^+ ions in the F region, after Bailey and Moffett.[101]

Curve A, $N_2 + h\nu \rightarrow N + N^+ + e$;
curve B, $He^+ + N_2 \rightarrow He + N^+ + N$;
curve C, $N(^2D) + O_2^+ \rightarrow N^+ + O_2$;
curve D_1, $N(^4S) + h\nu \rightarrow N^+ + e$;
curve D_2, $N(^2D) + h\nu \rightarrow N^+ + e$;
curve E, $N_2^+ + N \rightarrow N_2 + N^+$.

production and loss process for H^+ is believed to be the accidentally resonant charge transfer reaction,

$$O^+(^4S) + H(^2S) \rightarrow H^+ + O(^3P_J) \qquad (6.95)$$

$$\Delta E = \begin{cases} -0.01 \text{ eV}, J = 0 \\ +0.00 \text{ eV}, J = 1 \\ +0.002 \text{ eV}, J = 2 \end{cases}$$

The value of the rate coefficient $k_{6.95}$ is estimated to be about 8×10^{-10} cm^3 particle^{-1} s^{-1}, from extrapolation of high energy measurements.[42] The rate constant of the reverse reaction

$$H^+ + O(^3P) \rightarrow O^+ + H \qquad (6.96)$$

has been measured by Fehsenfeld and Ferguson[39] and is $k_{6.96} = 3.8 \times 10^{-10}$ cm^3 molecule^{-1} s^{-1} at 300 K. The large rate coefficients of the forward and reverse reactions keep H^+ and O^+ in equilibrium with each other in the middle and lower ionosphere. The equilibrium constant K for the conversion of H^+ to O^+ can be estimated from

$$K = \frac{Q(O^+).Q(H)}{Q(O).Q(H^+)} \exp(-\Delta E/kT) \qquad (6.97)$$

where Q represents the electronic partition function of the species, and ΔE is the endothermicity of the reaction (0.02 eV). For O^+, H and H^+ the partition functions are simply the ground state degeneracies of 4, 2 and 1, respectively. The partition function for O is given by

$$Q(O) = \sum_J (2J + 1)\exp(-E_J/kT) \tag{6.98}$$

where the summation is over the three low-lying states, $^3P_2$ (ground state); $^3P_1$ (at 158.270 cm^{-1}) and $^3P_0$ (at 226.98 cm^{-1}). Generally, at the high temperatures of the upper ionosphere, the exponential terms are close to unity, and $Q(O)$ can be put equal to $\sum_J (2J + 1) = 9$. In these circumstances the equilibrium condition becomes

$$\frac{[H^+]}{[O^+]} = \frac{9}{8}\frac{[H]}{[O]} \tag{6.99}$$

Fehsenfeld and Ferguson[39] have noted that under solar minimum conditions, with a relatively low temperature in the thermosphere, the approximation that $E_J/kT \sim 0$ in eqn. 6.98 could lead to significant error. Values of K obtained without this approximation are as follows: $T = \infty$, $K = 0.89$; $T = 2000 \text{ K}$, $K = 0.84$; $T = 1000 \text{ K}$, $K = 0.78$; $T = 600 \text{ K}$, $K = 0.71$ and $T = 300 \text{ K}$, $K = 0.55$. From their observations of H^+ and O^+ in the F_2 region, Brinton et al.[98] concluded that chemical and thermal equilibrium prevail below $\sim 350 \text{ km}$; Hoffman et al.[99] found equilibrium below ~ 450 km. Above 350–450 km diffusive processes become increasingly important.

He^+ *ions.* Helium is introduced into the atmosphere at ground level as He^4, as a result of the radioactive decay of uranium and thorium, and into the upper atmosphere, as He^3, by solar flares. Its presence at high altitudes is a consequence of diffusive processes that eventually carry it into the exosphere from which it is lost. Atmospheric helium ions are formed by photoionization of neutral helium and removed by charge transfer to neutral constituents, especially O_2 and N_2:

$$He^+ + O_2 \rightarrow He + O + O^+ \tag{6.100}$$

$$k_{6.100} \sim 1 \times 10^{-9} \text{ cm}^3 \text{ molecule}^{-1} \text{ s}^{-1} \text{ at 300 K }[41]$$

and
$$He^+ + N_2 \rightarrow N^+ + N + He \tag{6.90a}$$
$$\rightarrow N_2^+ + He \tag{6.90b}$$

The overall rate coefficient, $k_{6.90} = 1.4 \times 10^{-9} \text{ cm}^3 \text{ molecule}^{-1} \text{ s}^{-1}$, is independent of energy over the temperature range of ionospheric interest and is insensitive to the N_2 vibrational temperature up to 6000 K.[95] The branching ratio $k_{6.90a}/k_{6.90b}$ is between 1 and 2. A spectroscopic study of the reaction has shown the initial step to be the near resonant charge transfer

$$He^+ + N_2(X^1\Sigma_g) \rightarrow N_2^+(C^2\Sigma_u^+, v = 3, 4) + He \tag{6.101}$$

producing excited molecular ions which may either radiate

$$N_2^+(C^2\Sigma_u^+) \rightarrow N_2^+(X^2\Sigma_g^+) + h\nu \tag{6.102}$$

or predissociate

$$N_2^+(C^2\Sigma_u^+) \rightarrow [N_2^+(^4\Pi_u)] \rightarrow N^+ + N \tag{6.103}$$

In the region where chemical reactions control the distribution of He^+ ions we have

$$[He^+] = \frac{J[He]}{k_{6.90}[N_2]} \tag{6.104}$$

where J is the rate coefficient for photoionization of helium. Deviations from helium ion densities predicted by eqn. 6.104 become apparent above ~ 400 km.[98,99]

The chemistry in the F_2 layer differs from that in the F_1 region in degree rather than kind. At altitudes near to and above the F_2 peak increasing discrepancies between observations and results based on photochemical models become evident as chemical and diffusive time constants become comparable. Above the F_2 peak the rate of diffusion of ions and electrons through the neutral gas (ambipolar diffusion) is similar to the rate of loss of ions by chemical processes, and transport mechanisms influence both ion and electron concentrations. At night the F_2 layer remains, despite the absence of primary photoionization processes, but the concentration of charged particles is greatly reduced. In the absence of some ion production mechanism processes as rapid as the principal F region reactions should rapidly deplete the ionosphere. Whether the persistence of the F_2 region at night is due to a genuine additional F_2 ionization source, to a continual supply of ions being available through downward diffusion, or to thermospheric winds which increase the height of the F_2 layer and thus decrease the effective loss rate for O^+, is a question which has not been answered satisfactorily at the time of writing.

6.9 The exosphere

As altitude increases, charged and neutral particles are able to travel larger and larger distances before they are removed by one of the available loss processes. The region where molecular collisions, and therefore also chemical reactions, become improbable because of the extremely low particle density is called the *exosphere*. The neutral species mean free path is greater than 10 km over the entire exosphere. At a sufficiently high altitude (above 1000–3000 km) the dominant atmospheric neutral constituent is atomic hydrogen, and the dominant ion, H^+. This region is called the *protonosphere*. At such great altitudes the only ions observed are the two lightest, H^+ and He^+, their relative concentrations appearing to be dependent to some extent on the level of solar activity. Helium presented a problem in connection with its mode of escape from the atmosphere. The simple picture for escape of particles from the earth's gravitational attraction is one based on particle velocities: if the vertical component of velocity of a particle, after its latest collision, exceeds the escape velocity it will be lost. The flux F of particles exceeding the escape velocity is given by[8]

$$F = \tfrac{1}{2}N(Z_c)\left(\frac{2kT}{\pi m}\right)^{1/2}\cdot\left[1 + \frac{GMm}{kTR_c}\right]\cdot\exp\left(-\frac{GMm}{kTR_c}\right) \qquad (6.105)$$

where $N(Z_c)$ is the number density of particles of mass m at a height Z_c, the critical altitude above which the particles will suffer no further collisions. R_c is the radius corresponding to the critical level Z_c, G is the gravitational constant, and M is the mass of the planet. For helium the rate of escape predicted by eqn. 6.105 is too low by at least an order of magnitude. Helium is being produced at a rate of 4×10^6 atoms $cm^{-2} s^{-1}$ over the surface of the earth,[102] and at this rate of helium production (and ignoring He produced from solar flares), the helium concentrations presently observed in the exosphere would be attained in one to two million years if thermal escape governed by 6.105 was the only method available. Therefore one or more non-thermal processes must be capable of imparting the necessary escape velocity, to either neutral He atoms or He^+ ions. Several processes, both chemical and physical, have been suggested, although in recent years some of them have been discarded for various reasons. One possible chemical process is the reaction between metastable helium and atomic oxygen:[103]

$$He(2^3S) + O \rightarrow He + O^+ + e + 6.2\,eV \qquad (6.106)$$

Reaction 6.106 is mainly responsible for the loss of triplet metastable helium atoms, which are produced in the earth's atmosphere at a rate $\sim 10^6$ atom $cm^{-2} s^{-1}$. The minimum energy required by a helium atom before it can escape from the earth is 2.4 eV, and, unless the O^+ were electronically excited, much more than this would be available to the helium atom from reaction 6.106. However it is most likely that helium is removed from the exosphere in its ionized form in the 'polar wind.'[104,105] The earth's magnetic field restricts the motion of ionized particles to a direction tangential to the field, so that ionized particles cannot escape except where the field lines open into outer space, which occurs mainly above the polar regions. The ions H^+ and He^+ receive rapid acceleration to supersonic speeds along these open magnetic field lines and the resulting outward flow is termed the *polar wind*.[105]

References

1   Hanson, W. B., in *'Satellite Environment Handbook'* Ed. Johnson, F. S., Stanford University Press, 1961.
2   Bourdeau, R. E., Aiken, A. C. and Donley, J. L., *J. geophys Res.*, **71**, 727 (1966).
3   Paulsen, D. E., Huffman, R. E. and Larrabee, J. C., *Radio Sci.* **7**, 51 (1972).

4   Reid, G. C., *J. geophys. Res.*, **75**, 2551 (1970).
5   Takayanagi, K. and Itikawa, Y., *Space Sci. Rev.*, **11**, 380 (1970).
6   Banks, P. M., *Annls Géophys.*, **22**, 577 (1966).
7   Mitra, S. K., '*The Upper Atmosphere*', 2nd edition, The Asiatic Society, Calcutta, 1952.
8   Whitten, R. C. and Poppoff, I. G., '*Fundamentals of Aeronomy*', John Wiley and Sons Inc., New York, 1971.
9   Mitra, A. P., *J. atmos. terr. Phys.*, **30**, 1065 (1968).
10 Biondi, M. A., *Can. J. Chem.*, **47**, 1711 (1969); Leu, M. J., Biondi, M. A. and Johnson, R. J., *Phys. Rev.* **A7**, 292 (1973).
11 Phelps, A. V., *Can. J. Chem.*, **47**, 1783 (1969).
12 Ferguson, E. E., *Acc. Chem. Res.*, **3**, 402 (1970).
13 Fehsenfeld, F. C., Albritton, D. L., Burt, J. A. and Schiff, H. I., *Can. J. Chem.*, **47**, 1793 (1969).
14a Parkes, D. A. and Sugden, T. M., *J. chem. Soc.*, *Faraday Transactions II*, **68**, 600 (1972).
14b Thomas, L., Gondhalekar, P. M. and Bowman, M. R., *J. atmos. terr. Phys.*, **35**, 397 (1973).
15 Schoen, R. I. *Can. J. Chem.*, **47**, 1879 (1969).
16 Branscomb, L. M., *Annls Géophys.*, **20**, 88 (1964).
17 Byerly, R. and Beaty, E. C. J., *J. geophys. Res.*, **76**, 4596 (1971).
18 Mechtly, E. A. and Smith, L. G., *J. atmos. terr. Phys.*, **30**, 1555 (1968).
19 Mechtly, E. A. and Smith, L. G., *J. atmos. terr. Phys.*, **30**, 363 (1968).
20 Hall, J. E. and Fooks, J., *Planet. Space Sci.*, **13**, 1013 (1965).
21 Thrane, E. V., *J. geophys. Res.*, **74**, 1311 (1969).
22 Hunten, D. M. and McElroy, M. B., *J. geophys. Res.*, **73**, 2421 (1968).
23 Reid, G. C., Proceedings of ESRIN/ESRO Symposium on Upper Atmospheric Models and Related Experiments, Ed. Fiocco, G., D. Reidel Publishing Co., Dordrecht, Netherlands, 1970. Strobel, D. F., *J. geophys. Res.* **77**, 1337 (1972).
24 Hinteregger, H. E., Hall, L. A. and Schmidtke, G., *Space Research*, **5**, 1175 (1965).
25 Hinteregger, H. E., *Annls Géophys.*, **26**, 547 (1970).
26 Timothy, A. F., Timothy, J. G., Willmore, A. P. and Wager, J. H., *J. atmos. terr. Phys.*, **34**, 969 (1972).
27 Donahue, T. M., *Planet. Space Sci.*, **14**, 33 (1966).
28 Ferguson, E. E., Fehsenfeld, F. C., Goldan, P. D. and Schmeltekopf, A. L., *J. geophys, Res.*, **70**, 4323 (1965).
29 Ogawa, T. and Tohmatsu, T., *Rep. Ionosph. Space Res. Japan*, **20**, 395 (1966).
30 Barth, C. A., *Planet. Space Sci.*, **14**, 623 (1966).
31 Whitten, R. C. and Poppoff, I. G., *Nature*, **224**, 1187 (1969).
32 Mitra, A. P. and Ramanamusty, Y. V., *Radio Sci.*, **7**, 67 (1972).
33 Tulinov, V. F., Shibaeva, L. V. and Jakovlev, S. G., *Space Research*, **9**, 231 (1969).
34 Young, J. M., Carruthers, S. G. R., Holmes, J. C., Johnson, C. Y. and Patterson, N. P., *Science*, *N.Y.* **160**, 990 (1968).
35 Prag, A. B., Morse, F. A. and McNeal, R. J., *J. geophys. Res.*, **71**, 3141 (1966).
36 *Handbook of Geophysics and Space Environments*, Air Force Cambridge Research Laboratories, McGraw-Hill Book Co. Inc., 1965.
37 Ferguson, E. E., *Rev. Geophys.*, **5**, 305 (1967).
38 Ferguson, E. E., *Rev. Geophys. and Space Phys.*, **9**, 997 (1971).
39 Fehsenfeld, F. C. and Ferguson, E. E., *J. chem. Phys.*, **56**, 3066 (1972).
40 Ferguson, E. E., *Atomic Data* (in press, 1973).
41 Fite, W. L., *Can. J. Chem.*, **47**, 1797 (1969).
42 Stebbings, R. F. and Rutherford, J. A., *J. geophys. Res.*, **73**, 1035 (1968).
43 Fehsenfeld, F. C. and Ferguson, E. E., *Radio Sci.*, **7**, 113 (1972).
44 Dunkin, D. B., McFarland, M., Fehsenfeld, F. C. and Ferguson, E. E., *J. geophys. Res.*, **76**, 3820 (1971).
45 Fehsenfeld, F. C., Dunkin, D. B. and Ferguson, E. E., *Planet. Space Sci.*, **18**, 1267 (1970).
46 Good, A., Durden, D. A. and Kebarle, P., *J. chem. Phys.*, **52**, 212 (1970).
47 Fehsenfeld, F. C., Schmeltekopf, A. L. and Ferguson, E. E., *J. chem. Phys.*, **46**, 2802 (1967).

48   Good, A., Durden, D. A. and Kebarle, P., *J. chem. Phys.*, **52**, 222 (1970).
49   Puckett, L. J. and Teague, M. W., *J. chem. Phys.*, **54**, 2564 (1971).
50   Puckett, L. J. and Teague, M. W., *J. chem. Phys.*, **54**, 4860 (1971).
51   French, M. A., Hills, L. P. and Kebarle, P., *Can. J. Chem.*, **51**, 456 (1973).
52   Niles, F. E., Heimerl, J. M., Keller, G. E. and Puckett, L. J., *Radio Sci.*, **7**, 117 (1972).
53   Johnson, C. Y., *J. geophys. Res.*, **71**, 330 (1966).
54   Narcisi, R. S. and Bailey, A. D., *J. geophys. Res.*, **70**, 3687 (1965).
55   Narcisi, R. S., *'Composition Studies of the Lower Ionosphere'*, Lectures presented at the International School of Atmospheric Physics, Erice, Sicily, 1970. To be published in *'Upper Atmosphere Physics'*, Gordon and Breach.
56   Krankowsky, D., Arnold, F., Wieder, H., Kissel, J. and Zahringer, J., *Radio Sci.*, **7**, 93 (1972).
57   Fehsenfeld, F. C. and Ferguson, E. E., *J. geophys. Res.*, **74**, 2217 (1969).
58   Ferguson, E. E. and Fehsenfeld, F. C., *J. geophys. Res.*, **74**, 5743 (1969).
59   Ferguson, E. E., Proceedings of the ESRIN/ESLAB Symposium, Frascatti, Ed. Fiocco, G., D. Reidel Publishing Co., Dordrecht, Netherlands, 1970.
60   Dunkin, D. B., Fehsenfeld, F. C., Schmeltekopf, A. L. and Ferguson, E. E., *J. chem. Phys.*, **54**, 3817 (1971).
61   Kebarle, P., Searles, S. K., Zolla, A., Scarborough, J. and Arshadi, M., *J. Am. chem. Soc.*, **89**, 6393 (1967).
62   Johnson, C. Y., Meadows, E. B. and Holmes, J. C., *Annls Géophys.*, **14**, 475 (1958).
63   Narcisi, R. S., Bailey, A. D., Della Lucca, L., Sherman, C. and Thomas, D. M., *J. atmos. terr. Phys.*, **33**, 1147 (1971).
64   Arnold, F., Kissel, J., Krankowsky, D., Wieder, H. and Zahringer, J., *J. atmos. terr. Phys.*, **33**, 1169 (1971).
65   Ferguson, E. E., Dunkin, D. B. and Fehsenfeld, F. C., *J. chem. Phys.*, **57**, 1459 (1972).
66   Ferguson, E. E., *Can. J. Chem.*, **47**, 1815 (1969).
67   Fehsenfeld, F. C., Schmeltekopf, A. L., Schiff, H. I. and Ferguson, E. E., *Planet. Space Sci.*, **15**, 373 (1967).
68   Fehsenfeld, F. C. and Ferguson, E. E., *Planet. Space Sci.*, **20**, 295 (1972).
69   Ferguson, E. E., *'D region, negative ion chemistry'*, COSPAR symposium, Urbana, July 1971.
70   Adams, N. G., Bohme, D. K., Dunkin, D. B., Fehsenfeld, F. C. and Ferguson, E. E., *J. chem. Phys.*, **52**, 3133 (1970).
71   Payzant, J. D. and Kebarle, P., *J. chem. Phys.*, **56**, 3482 (1972).
72   Fehsenfeld, F. C., Ferguson, E. E. and Bohme, D. K., *Planet. Space Sci.*, **17**, 1759 (1969).
73   Arnold, F. and Krankowsky, D., *J. atmos. terr. Phys.*, **33**, 1693 (1971).
74   Phelps, A. V., *'Rates of clustering of oxygen negative ions with water vapour'*, COSPAR symposium, Urbana, July 1971.
75   Payzant, J. D., Cunningham, A. J. and Kebarle, P., *Can. J. Chem.*, **50**, 2230 (1972).
76   Ferguson, E. E., *Annls Géophys.*, **26**, 589 (1970).
77   Kebarle, P., French, M. and Payzant, J. D., *'Reaction mechanisms and bond energies for ion hydrates of ionospheric interest'*, COSPAR symposium, Urbana, July 1971.
78   Stubbe, P., *J. atmos. terr. Phys.*, **34**, 519 (1972).
79   Johnson, C. Y., *Radio Sci.*, **7**, 99 (1972).
80   Keneshea, T. J., Narcisi, R. S. and Swider, W., *J. geophys. Res.*, **75**, 845 (1970).
81   Barth, C. A., *Annls Géophys.*, **22**, 198 (1966).
82   Istomin, V. G., *Space Research*, **3**, 209 (1963).
83   Young, J. M., Johnson, C. Y. and Holmes, J. C., *J. geophys. Res.*, **72**, 1473 (1967).
84   Narcisi, R. S., *Space Research*, **8**, 360 (1967).
85   Smith, L. G., *J. atmos. terr. Phys.*, **32**, 1247 (1970).
86   Layzer, D., *Radio Sci.*, **7**, 385 (1972).
87   Swider, W., *Annls Géophys.*, **26**, 595 (1970).
88   Rutherford, J. A., Mathis, R. F., Turner, B. R. and Vroom, D. A., *J. chem. Phys.*, **55**, 3785 (1971); **56**, 4654 (1972); **57**, 3087 (1972); Rutherford, J. A. and Vroom, D. A., *J. chem. Phys.*, **57**, 3091 (1972).
89   Ferguson, E. E., *Radio Sci.*, **7**, 397 (1972).

90   Spears, K. G. and Fehsenfeld, F. C., *J. chem. Phys.*, **56**, 5698 (1972).
91a Anderson, J. G. and Barth, C. A., *J. geophys. Res.*, **76**, 3723 (1971).
91b Gadsden, M., *J. geophys. Res.*, **77**, 1330 (1972).
92   Gadsden, M., *Annls Géophys.*, **76**, 141 (1970).
93   Pharo, M. W., Scott, L. R., Mayr, H. G., Brace, L. H. and Taylor, H. A., *Planet. Space Sci.*, **19**, 15 (1971).
94   Krankowsky, D., Kasprzak, W. T. and Nier, A. O., *J. geophys. Res.*, **73**, 7291 (1968).
95   Ferguson, E. E., *Annls Géophys.*, **25**, 819 (1969).
96   O'Malley, T. F., *J. chem. Phys.*, **52**, 3269 (1970).
97   Jacchia, L. G., Special Report 170, Smithsonian Institution Astrophysical Observatory, Cambridge, Mass., 1964.
98   Brinton, H. C., Pharo, M. W., Mayr, H. G. and Taylor, H. A., *J. geophys. Res.*, **74**, 2941 (1969).
99   Hoffman, J. H., Johnson, C. Y., Holmes, J. C. and Young, J. M., *J. geophys. Res.*, **74**, 6281 (1969).
100 Taylor, H. A., *Planet. Space Sci.*, **19**, 77 (1971).
101 Bailey, G. J. and Moffett, R. J., *Planet. Space Sci.*, **20**, 616 (1972).
102 McDonald, G. J. F., *Rev. Geophys.*, **1**, 305 (1963).
103 Ferguson, E. E., Fehsenfeld, F. C. and Schmeltekopf, A. L., *Planet. Space Sci.*, **13**, 925 (1965).
104 Axford, W. I., *J. geophys. Res.*, **73**, 6855 (1968).
105 Banks, P. M. and Holzer, T. E., *J. geophys. Res.*, **74**, 6317 (1969); **73**, 6846 (1968).

7
The Chemistry of a Polluted Atmosphere

7.1 Introduction

The label pollution can be attached to almost any deleterious material which is present in air in addition to the 'normal' atmospheric components, provided only that the material is present as a result of human activity. Thus it may apply to wind-borne soil and dust, as well as radioactive fallout, vehicle exhaust gases, and various kinds of smoke or smog. Our emphasis in this chapter, as in the rest of the book, is on the chemical reactions of gaseous materials in the atmosphere; consequently pollution by dust particles and aerosols will receive only brief mention. For a fuller discussion of the origin and properties of particulate matter in the atmosphere the reader is referred to books by Junge[1] and Cadle.[2] A large number of different chemical substances have been identified as contributing to air pollution, and there is a considerable body of literature describing experimental methods for measuring concentrations of individual pollutants.[3] Certain substances tend to be associated with particular industries; for example, soot and sulphur dioxide with coal fired power plants, and hydrocarbons with petroleum refineries. In such cases it is often possible to identify a single major source of the pollution at a given locality, and to take appropriate remedial action.

Pollution of the troposphere by gases commonly leads to the formation of a characteristic type of smog. Smog is a coined word, which was first used at least 70 years ago to describe the yellow mixture of smoke and fog which constituted a 'London pea-souper'. More recently it has been used to describe any smoky or hazy condition of the atmosphere associated with pollution, including Los Angeles type smog, which occurs under climatic conditions very far removed from those in which fog is likely to be encountered. Two extreme types of smog can be distinguished, namely that associated with pollution by vehicle exhaust gases containing oxides of nitrogen, and that associated with pollution by fumes or smoke which contain sulphur dioxide. For the first type of smog, photochemical processes are a necessary ingredient of the situation in which smog is formed; with the second variety, photochemical reactions may be involved but are not essential. These two basic forms of tropospheric pollution are discussed in the next two sections of this chapter.

During the last few years a new avenue for air pollution has appeared through the advent of supersonic transport aircraft, designed to travel long distances at high speed in the stratosphere. Concern about the likely effects of pollution of the stratosphere by aircraft exhaust gases arose initially in connection with the large amounts of water vapour which would be deposited in the stratosphere. More recently it has been pointed out that oxides of nitrogen in the exhaust gases can be expected to react with the ozone in the stratosphere, and in the process might destroy much of our protection from the sun's short wavelength ultraviolet radiation. Pollution of the stratosphere by high flying aircraft is discussed in the last section of this chapter.

7.2 NO$_2$ in the troposphere and photochemical smog

Much of what we say in this section will necessarily depend on the description of the nature and origin of photochemical smog which is given in the classic book by Leighton.[4] The main characteristics of photochemical smog, such as is experienced in the Los Angeles basin, are as follows:

(i)   it forms under conditions of bright sunshine and low humidity, with the peak concentration of irritant material occurring at or shortly after midday;

(ii)   it is chemically oxidizing, and causes cracking of rubber;

(iii)   it can be seen as a whitish haze; however, the loss of visibility is less important than its other effects, notably:

(iv)   it causes eye irritation in humans and leaf damage in plants;

(v)   the raw materials from which photochemical smog is formed consist of automobile exhaust gases at high dilution in air.

There is a large amount of evidence that oxides of nitrogen are an essential component of the gas mixture which is ultimately transformed into smog, and that the transformation is actually initiated by the photodissociation of NO$_2$ by near ultraviolet radiation. In particular, it has been shown[5] that the properties of Los Angeles type smog can be reproduced in the laboratory by the irradiation with near ultraviolet light of either dilute exhaust gases, or dilute mixtures of NO$_2$ with air and olefinic hydrocarbons. We shall therefore begin by considering the photochemistry of pure NO$_2$ at wavelengths greater than 290 nm, then discuss the photochemistry of NO$_2$ at high dilution in air, and finally consider the effects of sunlight on dilute mixtures of NO$_2$ with air in the presence of traces of organic materials.

7.2.1   PHOTOCHEMISTRY OF PURE NO$_2$

In Chapter 2 we noted that NO$_2$ absorbs throughout the visible and near ultraviolet, and that with light of wavelength less than about 395 nm the absorption process results in dissociation to NO and O. The final products of photolysis of NO$_2$ between 300 and 400 nm are NO and O$_2$, the basic mechanism being

$$NO_2 + h\nu \rightarrow [NO_2^*] \rightarrow NO + O \qquad (7.1)$$

$$O + NO_2 \rightarrow NO + O_2 \qquad (7.2)$$

In practice in the laboratory, even though reaction 7.2 is very fast, some oxygen atoms are lost by recombination on the walls of the vessel (forming O_2 directly rather than through 7.2) and possibly, at long photolysis times, by termolecular recombination with NO in the gas phase (reforming NO_2), so the quantum yield of O_2 is slightly less than unity. At wavelengths greater than 400 nm the intermediate excited molecule in process 7.1 normally disposes of its excess energy in the form of fluorescence, since it has insufficient energy to dissociate. The transition from fluorescence to dissociation is shown in Table 7.1 which summarizes the results of irradiation of NO_2

Table 7.1 Results of irradiation of NO_2 at room temperature with mercury radiation at various wavelengths λ.

λ (nm)	O_2 quantum yield	Fluorescence
435.8	<0.005	strong
404.7	0.36	weak
380.0	0.82	negligible
366.0	0.92	negligible
313.0	0.97	negligible

with a series of mercury lines.[6] The observation of a significant O_2 yield at 404.7 nm is at first sight surprising, since the theoretical dissociation limit lies below 400 nm. This difficulty disappears when it is noted that the yield at 404.7 nm is strongly temperature dependent. The results can be accounted for quantitatively by considering the absorption of light by molecules which already possess some vibrational energy as a result of normal thermal excitation.

7.2.2 PHOTOLYSIS OF NO_2-AIR MIXTURES

When NO_2 is photolysed in the presence of a large excess of air, reaction 7.2 has to compete with O atom removal by

$$O + O_2 + M \rightarrow O_3 + M \qquad (7.3)$$

where M is any third body (normally N_2 or O_2). Using $k_{7.2} = 5.0 \times 10^{-12}$ cm^3 molecule^{-1} s^{-1},[7] and $k_{7.3} = 7.5 \times 10^{-34}$ cm^6 molecules^{-2} s^{-1} for M = O_2,[8] and assuming a similar value of $k_{7.3}$ for M = N_2, we find that in air at 1 atm. and 300 K the rates of removal of O atoms by reactions 7.2 and 7.3 are equal at an NO_2 concentration of 1.8×10^{16} molecules cm^{-3}.

Thus when the concentration of NO$_2$ is less than this value, which corresponds to a partial pressure of 0.007 atm., reaction 7.3 will predominate over reaction 7.2, and NO$_2$ photolysis will lead to the formation of ozone. This is the normal situation in a polluted atmosphere. The fate of the NO molecule produced in reaction 7.1 is normally to reform NO$_2$ very slowly, either by

$$2NO + O_2 \rightarrow 2NO_2 \tag{7.4}$$

or

$$NO + O_3 \rightarrow NO_2 + O_2 \tag{7.5}$$

Listed values of the rate constants for these reactions near 300 K are $k_{7.4} = 4.3 \times 10^{-34}$ cm^6 molecule^{-2} s^{-1},[7] and $k_{7.5} = 4.0 \times 10^{-14}$ cm^3 molecule^{-1} s^{-1}.[9] For highly dilute ozone and nitric oxide in air at a pressure of one atmosphere the rates of removal of NO by reactions 7.4 and 7.5 become equal when the ratio of nitric oxide to ozone concentration is about 19; at higher relative ozone concentrations reaction 7.5 predominates. In what follows we shall generally ignore reaction 7.4.

Ozone in the troposphere is itself photolysed according to

$$O_3 + hv \rightarrow O_2 + O \tag{7.6}$$

as a result of absorption in the weak Chappuis bands in the visible region, and in the Huggins bands in the ultraviolet (see Section 2.3), at a rate which is about an order of magnitude slower than the rate of NO$_2$ photolysis. This does not have a significant effect on the ozone concentration, because of reaction 7.3, but does serve to augment the concentration of atomic oxygen. The alternative mode of removal of O atoms

$$O + O_3 \rightarrow 2O_2 \tag{7.7}$$

for which $k_{7.7} = 1.6 \times 10^{-14}$ cm^3 molecule^{-1} s^{-1},[9,10] is easily shown to be unimportant in comparison with reaction 7.3 in air at pressures near one atmosphere. A possible complicating factor is that, as a result of the absorption of light of wavelength between 290 and 310 nm, which represents the most energetic component of sunlight at sea level, it is possible for the oxygen atom in reaction 7.6 to be produced in the ^{1}D state. Atoms in this state are rapidly quenched to the ^{3}P level by collision with either O$_2$ or N$_2$,[11] yielding O$_2$($^1\Sigma_g^+$) or vibrationally excited O$_2$ or N$_2$, all of which can be expected to get rid of their excess energy by physical rather than chemical processes. In laboratory systems, with relatively high ozone concentrations, the quantum yield for O$_3$ decomposition is of the order of ten at wavelengths below 310 nm, and it has been postulated than an energy chain can result from the reaction of O(^{1}D) with O$_3$ to form O$_2$ in very high vibrational levels,[12] the vibrationally excited O$_2$ being able to dissociate further O$_3$ in energy transfer collisions. More recently[12a] it has been suggested that electronically rather than vibrationally excited O$_2$ is the chain carrier. At high dilution in the atmosphere, however, quenching of excited species by O$_2$ or N$_2$ predominates over reaction with O$_3$, so that the production of

$O(^1D)$ in reaction 7.6 can have very little influence on the overall course of events.

Ozone photolysis can also yield excited oxygen molecules in the $^1\Delta_g$ state ($\lambda < 590$ nm) and $^1\Sigma_g^+$ state ($\lambda < 460$ nm), by spin-forbidden processes, and $O_2(^1\Delta_g) + O(^1D)$ by a spin-allowed process for $\lambda < 310$ nm (cf. Table 2.2). The production of singlet oxygen molecules may trigger an important series of secondary reactions when olefins are present, but is probably not significant in the present context.

Ozone reacts slowly with NO_2 to yield the radical NO_3:

$$NO_2 + O_3 \rightarrow NO_3 + O_2 \tag{7.8}$$

where $k_{7.8} = 7.2 \times 10^{-17}$ cm^3 molecule^{-1} s^{-1}.[13] NO_3 radicals react rapidly with NO and NO_2 according to

$$NO_3 + NO \rightarrow 2NO_2 \tag{7.9}$$

$$NO_3 + NO_2 \rightarrow N_2O_5 \tag{7.10}$$

$$NO_3 + NO_2 \rightarrow NO + NO_2 + O_2 \tag{7.11}$$

with $k_{7.9} = 9 \times 10^{-12}$, $k_{7.10} = 3 \times 10^{-12}$ and $k_{7.11} = 4 \times 10^{-16}$ cm^3 molecule^{-1} s^{-1}.[14] The N_2O_5 molecule produced by reaction 7.10 is not stable, but undergoes unimolecular decomposition according to

$$N_2O_5 \rightarrow NO_2 + NO_3 \tag{7.12}$$

thereby regenerating NO_3, with a rate constant $k_{7.12} = 0.24$ s^{-1}.[13] Alternatively, N_2O_5 may react with water vapour to yield HNO_3. Reaction 7.8 does not predominate over reaction 7.5 as a mode of removal of ozone unless $[NO_2] > 550\,[NO]$. NO_3 has an absorption band between 500 and 700 nm, where it is photolysed according to

$$NO_3 + h\nu \rightarrow NO + O_2 \tag{7.13}$$

A value of $\sim 10^{-2}$ s^{-1} for the rate coefficient $J_{7.13}$ can be estimated from the extinction coefficient of NO_3.[14]

In the steady state, under conditions where the important processes involving oxygen atoms are 7.1, 7.3 and 7.6, the O atom concentration is given by

$$[O] = (J_{7.1}[NO_2] + J_{7.6}[O_3])/k_{7.3}[O_2][M] \tag{7.14}$$

where $J_{7.1}$ is the product of the absorption cross section of NO_2 and the intensity of sunlight at sea level in photons per second, integrated over the wavelength range from 290 to 400 nm, and $J_{7.6}$ is the similar quantity for O_3 integrated over the Chappuis and Huggins bands at wavelengths greater than 290 nm. At small solar zenith angles $J_{7.1} \sim 8 \times 10^{-3}$ s^{-1} and $J_{7.6} \sim 5 \times 10^{-4}$ s^{-1},[15] so that in air at a pressure of one atmosphere

$$[O] \sim 9 \times 10^{-8}[NO_2] + 6 \times 10^{-9}[O_3] \tag{7.14a}$$

If the production of NO_3 by reaction 7.12 is neglected we obtain, as a *lower limit* for the steady state NO_3 concentration,

$$[NO_3] \sim k_{7.8}[NO_2][O_3]/(k_{7.9}[NO] + k_{7.10}[NO_2] + J_{7.13}) \quad (7.15)$$

The concentrations of O atoms and of NO_3 are both very small in comparison with those of the stable species NO_2, NO and O_3, so that in the NO_2/air system we must have, as a consequence of processes 7.1, 7.3 and 7.5,

$$[NO] \sim [O_3] \quad (7.16)$$

In the NO_2/air system at high dilution the NO and O_3 concentrations are governed by 7.1 and 7.5, and we therefore have in the steady state

$$[NO][O_3] \sim (J_{7.1}[NO_2]/k_{7.5}) = 2.0 \times 10^{10}[NO_2] \quad (7.17)$$

A typical concentration of NO_2 in Los Angeles type smog would be 5 pphm (5 parts per hundred million), or 1.2×10^{12} molecule cm^{-3}. Inserting this figure in eqn. 7.17 and using 7.16, we obtain $[NO] = [O_3] = 5 \times 10^{11} = 2$ pphm, which in turn gives, from eqns. 7.14 and 7.15, $[O] = 5 \times 10^{-7}$ pphm, and $[NO_3] = 2 \times 10^{-5}$ pphm. The calculated quantity of ozone is of similar magnitude to the amount which is normally present in clean air as a result of downward transport from the stratosphere.

7.2.3 PHOTOLYSIS OF NO_2 PLUS AIR AND HYDROCARBONS

In practice reactions involving the olefinic hydrocarbons in Los Angeles type smog destroy the equality 7.16, the effect being to increase $[O_3]$ and decrease $[NO]$. Representative observational data are shown in Fig. 7.1. The main additional processes which must be considered in dilute automobile exhaust gases, or in dilute mixtures of NO_2 with air and hydrocarbons, are the reactions of ozone and atomic oxygen with olefins, and the photochemical reaction of nitrogen oxides with olefins. The role of NO_3 radicals has not been fully established.

Ozone reacts with olefins to yield, as major products, water, aldehydes, carbon monoxide and carbon dioxide. The rate constants for these reactions are typically 10^{-16}–10^{-17} cm^3 molecule^{-1} s^{-1} near 300 K. This is two to three orders of magnitude less than the rate of reaction with NO (reaction 7.5), and since the NO concentration is usually within an order of magnitude of the hydrocarbon concentration, the olefin reactions should not have a significant effect on the steady state ozone concentration. However, the rate of production of aldehydes by such reactions in Los Angeles smog is estimated to be several parts per hundred million per hour,[4] which is significant. Atomic oxygen reacts with olefins, with rate constants typically 2×10^{-12} cm^3 molecule^{-1} s^{-1} at 300 K. The final products include, in addition to those already noted for ozone, organic acids, ketones, alkene oxides and paraffins. Because of the low steady state O atom concentration it would

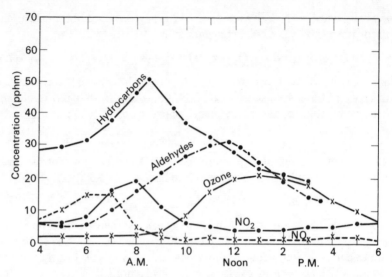

Fig. 7.1   Average concentrations of pollutants during days of eye irritation in downtown Los Angeles. Data for NO and NO$_2$ are for 1958, other data for 1953–4. (After P. A. Leighton, *The Photochemistry of Air Pollution*, Academic Press, New York, 1961.)

appear that these reactions are probably not important in comparison with the rather similar reactions of O$_3$. It has been suggested[16] that free radicals, formed by the reaction of atomic oxygen with hydrocarbons may add to O$_2$ to yield organic peroxy radicals RO$_2$. Such peroxy radicals could then react further with O$_2$ to form ozone, according to

$$RO_2 + O_2 \rightarrow RO + O_3 \tag{7.18}$$

which would help account for the increase in the steady state ozone concentration when hydrocarbons are present. An interesting alternative involves the reaction of singlet molecular oxygen[17] with olefins to form hydroperoxides

$$O_2^* + R\!-\!CH = CH\!-\!CH_2R' \rightarrow R\!-\!\underset{\underset{\displaystyle OOH}{|}}{CH}\!-\!CH\!\!=\!\!CHR' \tag{7.19}$$

followed by breakdown of the hydroperoxide to radicals such as RCO.[18] Subsequent reaction might then include

$$RCO + O_2 \rightarrow RCO_3 \tag{7.20}$$

$$RCO_3 + NO \rightarrow RCO_2 + NO_2 \tag{7.21}$$

and $\qquad\qquad RCO_3 + O_2 \rightarrow RCO_2 + O_3 \tag{7.22}$

These reactions are similar to those usually postulated to account for the conversion of NO to NO$_2$ in the presence of hydrocarbons.[19] Molecular oxygen in the $^1\Delta_g$ state can be formed in the atmosphere by photolysis of O_3, by energy transfer from electronically excited NO_2,[20] and probably also by the reaction of NO with O_3.[17]

Possibly the most interesting feature of the photochemical reaction of NO_2 with olefins is that it leads to the production of the material peroxyacetyl nitrate, usually referred to as PAN. The structure of PAN has been established[21] as

$$CH_3—\underset{\underset{O}{\|}}{C}—O—O—NO_2$$

and it can be regarded as a mixed anhydride of peroxyacetic and nitric acids. Other peroxyacyl nitrates, with a variable group R replacing the methyl radical, can also be formed. The infrared, ultraviolet and mass spectra of PAN have been described; the infrared spectrum has been used to monitor its concentration during kinetic studies.[22] PAN is virtually a concentrated essence of Los Angeles type smog, in that it is a powerful oxidizer, with both phytotoxic† and eye irritating properties. It is photolysed only very slowly by sunlight, so that its concentration can readily build up into the pphm range. The detailed mechanism by which PAN is produced has not been established. It is likely that the final step is the reaction of RCO_3 radicals (eqn. 7.20) with NO_2; alternatively one might postulate a reaction of NO_3 (or N_2O_5) with RCO_2.

The results of some laboratory studies of the irradiation of dilute automobile exhaust gases in air, and of a dilute mixture of NO_2 with trans but-2-ene in air, are shown in Figs. 7.2 and 7.3a, respectively.[4] The initial conversion

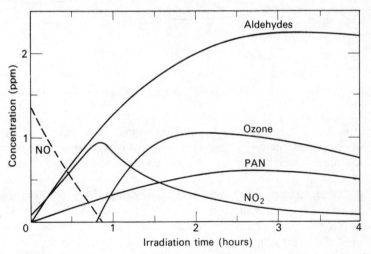

Fig. 7.2 Results of irradiation of dilute automobile exhaust gases in air (after Leighton[4]).

† Greek phyton = a plant.

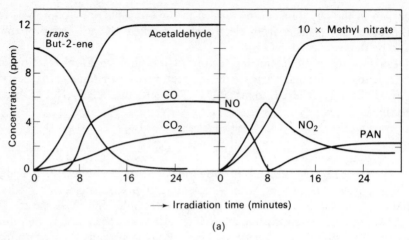

(a)

Fig. 7.3a   Results of irradiation of a dilute mixture of trans but-2-ene and NO in air.

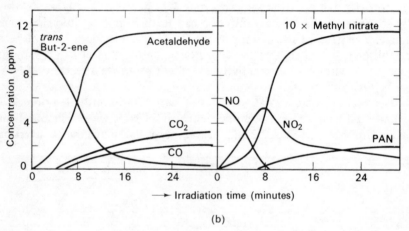

(b)

Fig. 7.3b   Concentration profiles obtained by computer simulation of the reaction system of Fig. 7.3a.[23]

of NO to NO_2, shown in Fig. 7.2, and subsequent build-up of the O_3 concentration, is a characteristic feature of this type of experiment. The detailed reaction mechanisms which are required to account for observations such as those in Fig. 7.3a are necessarily very complex. Nevertheless, as shown in Fig. 7.3b, a striking degree of success has been achieved in the computer simulation of such systems, on the basis of known and estimated rate constants for the two hundred or more reactions involved.[23] When allowance is made for the varying rate of absorption of sunlight and the continuing input of un-photolysed material during the day, the similarity of the results in Figs. 7.2 and 7.3 to the observational data in Fig. 7.1 shows that the main characteristics of photochemical smog, listed at the beginning of this section,

Table 7.2 Analysis of particulate matter from Los Angeles smog, collected with an electrostatic precipitator.[2]

		Elements (by emission spectrography)
	$>10\%$	Si, Al, Fe
	$1-9\%$	Ti, Ca
	1%	Mg, Ba, Na, K
	$0.1-0.9\%$	Pb, Zn
	0.1%	V, Mn, Ni
	trace	Sn, Cu, Zr, Sr, B, Cr, Bi, Co
Minerals and inorganic substances, about 60%; (threequarters of this, insoluble in water.)		*Substances identified chemically*
	14.3%	SiO_2
	7.8%	Fe, Al
	5.2%	Ca
	4.8%	NO_3^- (as HNO_3)
	4.6%	Na (as NaCl)
	2.5%	SO_4^{--} (as H_2SO_4)
	0.7%	NH_3
	0.26%	Cl^- (as NaCl)

Individual organic compounds about 15%; fibres, pollen, carbon, highly polymerized organic material, about 15%; water, volatile organic substances (by difference), about 15%.	Mainly hydrocarbons, plus 0.27% organic acids, plus aldehydes, and including 0.04% peroxides (calculated as H_2O_2)

can be accounted for satisfactorily. The only point not covered so far is particle or aerosol formation, with its attendant loss of visibility. Laboratory experiments have shown that particulate matter does form, presumably as the result of free radical polymerization reactions, during the irradiation of dilute automobile exhaust gases in air. These particles consist of organic material to the extent of 70% or more. In the laboratory studies, the production of aerosol particles has been shown to be greatly enhanced by the presence of a trace of sulphur dioxide, and we may note that some SO_2 is invariably present in the atmosphere on account of the normal sulphur content ($>0.1\%$) of hydrocarbon fuels. The enhancement presumably results from formation of a sulphuric acid aerosol, as described in the next section. Chemical analysis of particulate matter from Los Angeles smog (Table 7.2) has shown it to contain a rather high proportion of inorganic material. Thus there may be little or no direct causative relationship between the aerosol particles in the smog and the photochemical processes we have been considering.

7.3 Reactions of sulphur dioxide in the troposphere

Pollution by SO_2 is generally associated with the burning of coal or fuel oil, for either industrial or domestic heating.* The total amount of sulphur

* For a survey of SO_2 pollution in Britain see S. Rose and L. Pearse, *New Scientist*, **53**, 376 (1972).

dioxide entering the atmosphere each year is measured in millions of tons in the USA alone.[1] Fortunately the build-up of the concentration of SO_2 is limited by the scrubbing effect of rainfall, and by the absorption of SO_2 (and other pollutants) in soil.[24] In areas where soils are deficient in sulphur, pollution by SO_2 may actually be beneficial to the flora, if not to the fauna. In the atmosphere sulphur dioxide is gradually oxidized to sulphuric acid, the oxidation being able to occur by both photochemical and non-photochemical mechanisms. Sulphur dioxide reacts photochemically with hydrocarbons, with the production of sulphinic acids, and, as noted in the last section, certain reactions of SO_2 readily lead to the formation of aerosols. In this section we shall first consider the photochemistry of pure SO_2, and then discuss photochemical reactions of SO_2 at high dilution in air, both alone and in the presence of other pollutants such as hydrocarbons. We shall also describe some important non-photochemical processes which can occur in air polluted by SO_2.

7.3.1 PHOTOCHEMISTRY AND PHOTO-OXIDATION OF SO_2

As described in Chapter 2, the absorption spectrum of SO_2 in the region of interest consists of a very weak band system between 340 and 390 nm, arising from the transition $\tilde{a}^3B_1 \leftarrow \tilde{X}^1A_1$, and a fairly intense system due to the transition $\tilde{A}^1B_1 \leftarrow \tilde{X}^1A_1$ below 320 nm, with maximum intensity near 290 nm. Absorption in the intense band system lying just below 230 nm does not occur naturally in the troposphere, but is of some interest because it leads to fluorescence from the excited SO_2. Measurement of the intensity of this fluorescence has been proposed as a sensitive means of estimating the SO_2 concentration in the atmosphere.[25] From the viewpoint of the photochemistry of SO_2 in the troposphere the most important absorption occurs in the band at 290 nm, but absorption in the weak band above 340 nm may also be significant when the absorption path is measured in kilometres. The observation that photochemical reactions do occur as a result of excitation in the 290 nm band, even though photodissociation is energetically impossible at wavelengths greater than 218 nm, clearly indicates that reactions of excited SO_2 molecules are important.

In pure SO_2 irradiated at 313 nm Hall[26] found SO_3 and sulphur to be formed with a quantum yield of the order of 10^{-2}; in mixtures of SO_2 and oxygen, SO_3 alone was formed with a similar quantum yield. Studies of the fluorescence of SO_2 excited in the 290 nm band[27] show that as the pressure of SO_2 or other quencher is increased, the quantum efficiency of fluorescence due to the $\tilde{A} \rightarrow \tilde{X}$ transition decreases, and phosphorescence, arising from the $\tilde{a} \rightarrow \tilde{X}$ transition, begins to appear. This phosphorescence is attributed to population of the $\tilde{a}$ state by collision induced intersystem crossing; with SO_2 embedded in solid inert gases at low temperatures only phosphorescence, and no fluorescence, is observed.[28] The mean lifetime of the $\tilde{a}$ state has been measured as 2.5×10^{-5} s at an SO_2 pressure of 1.55 torr, extrapolating to about 1 ms at zero pressure. The lifetime estimated from the area under the $\tilde{a} \leftarrow \tilde{X}$ absorption bands is between 10^{-2} and 2×10^{-3} s;[6] a value of 17.5 ms was found for $SO_2(\tilde{a})$ at high dilution in a neon matrix at 4 K.[28] The rate constant for quenching by $SO_2(\tilde{X})$ is 6.5×10^{-13} cm^3 molecule^{-1} s^{-1} at 298 K.[29] In view of the high probability of collisional quenching, it is reasonable to identify $SO_2(\tilde{a}^3B_1)$ as the excited species responsible for both the photo-decomposition of pure SO_2 and the photo-oxidation of SO_2 by O_2 during irradiation in the 290 nm band, and this is supported by the observation that addition of biacetyl, which is known to quench $SO_2(\tilde{a})$ by energy transfer, reduces the quantum yield of SO_3.[29,30]

Several groups of workers have followed Hall in measuring quantum yields for SO_2 photo-oxidation, and have obtained rather widely varying results. Cox[30] has argued that the scatter of the results is due to a surface reaction, which has caused most of the values obtained to be too high. His own data give an SO_3 quantum yield of $(4 \pm 1) \times 10^{-3}$ from pure SO_2, and about 1×10^{-3} from SO_2 at low concentrations in oxygen or air. The extrapolated yield from SO_2 at concentrations in the parts per million range, with irradiation by sunlight, is 3×10^{-4}. The actual oxidation steps are thought to be [4,30]

$$SO_2(\tilde{a}) + SO_2 \rightarrow SO_3 + SO \qquad (7.23)$$

in pure SO_2, and also

$$SO_2(\tilde{a}) + O_2 \rightarrow SO_4 \qquad (7.24)$$

followed by

$$SO_4 + SO_2 \rightarrow 2SO_3 \qquad (7.25)$$

in SO_2/O_2 mixtures. In pure SO_2 the SO radicals formed by reaction 7.23 react with one another to form sulphur and oxygen; when O_2 is present they react, probably in the presence of a third body, to form SO_3.

7.3.2 PHOTOLYSIS OF SO_2–HYDROCARBON MIXTURES

Dainton and Ivin[31] showed that the products of the photochemical reaction of SO_2 with hydrocarbons consist largely of sulphinic acids (RSO_2H), the stoichiometry of the reaction being approximately

$$RH + SO_2^* \rightarrow RSO_2H \qquad (7.26)$$

They measured quantum yields of reactant consumption which ranged from 0.006 for methane to 0.26 for pentane, with values for olefins typically in the range 0.02–0.05. Somewhat lower quantum yields were obtained by Timmons.[32] Subsequent workers have confirmed the suggestion of Dainton and Ivin that triplet SO_2 molecules are involved in the reaction. However, since the measured intersystem crossing yield during quenching of the excited singlet state by hydrocarbons is only 0.09 ± 0.01,[29] it appears that the quantum yields which Dainton and Ivin obtained with propane and higher paraffins must be in error. Calvert and co-workers[29] find good agreement between the product yield data of Timmons and the theoretical estimates based on their own measured quenching rates and intersystem crossing yields.

In Section 7.2 we noted that a trace of SO_2 increased the rate of aerosol formation in NO_2–hydrocarbon systems. Dainton and Ivin found the sulphinic acids resulting from photochemical reactions of SO_2 with hydrocarbons to be produced in the form of a mist, which settled into droplets of a colourless or pale yellow involatile oil with a disagreeable odour. This, then, might be a possible source of an atmospheric aerosol. However, Schuck and Doyle[33a] found that olefins partially suppress aerosol formation during the photo-oxidation of SO_2 at high dilution in air. Also, at the low hydrocarbon concentrations which exist in moderately polluted air, the major reaction of $SO_2(\tilde{a})$ will be with O_2 rather than with RH. Therefore the observed effect of SO_2 on aerosol formation in the olefin–NO_2–air photochemical system must involve a reaction of *ground state* SO_2 with relatively long-lived intermediates, such as organic free radicals, formed in the olefin–NO_2 reaction. More recently Cox and Penkett[33b] have reported that at low concentrations in air SO_2 is readily oxidized by a mixture of ozone and hydrocarbons, to form a sulphuric acid aerosol. The reactive intermediate in this case is thought to be a peroxide diradical.

The aerosol produced with air and SO_2 alone consists of droplets of sulphuric acid formed from SO_3 and water vapour. If water droplets are already present, as in smoke or fog, the rate of formation of sulphuric acid is considerably enhanced, because sulphurous acid is much easier to oxidize than gaseous SO_2.[34] The oxidation of aqueous SO_2 is a dark, i.e. non-photochemical, reaction. Oxidation of gaseous SO_2 to SO_3 in the dark is presumably able to be catalysed by NO_2, as in the lead chamber process for making sulphuric acid. Surprisingly, the oxidation of SO_2 by ozone in air is stated to be negligibly slow, even at high relative humidity.[2]

When sulphur dioxide dissolves and becomes oxidized in drops that contain dissolved sodium chloride, evaporation of the drop must lead to loss of HCl, with the formation, ultimately, of solid particles of mixed sodium sulphate and sodium chloride. Particles of this composition have often been found in the atmosphere. In coastal regions where there is a large amount of atmospheric pollution by NO_2, a similar process leads to the formation of particles of mixed sodium chloride and sodium nitrate. Particles of ammonium sulphate have been found in regions where combustion of the local fuel produces both ammonia and sulphur dioxide. Ammonium sulphate is also thought to be a major constituent of an aerosol layer in the stratosphere.[1,35]

Possibly the most significant statement which can be made about pollution by SO_2 is that several major cities have shown that the problem can be alleviated by controlling the type of fuel which is burnt and the kind of smoke which is emitted from domestic and industrial heating plants. An outstanding example of the effectiveness of such control is London, where a gloomy mixture of smoke and fog, redolent of Sherlock Holmes and Jack the Ripper, is no longer the most characteristic feature of a winter's evening.

7.4 Pollution of the stratosphere

As noted in the introduction of this chapter, initial concern about pollution of the stratosphere by exhaust gases from jet aircraft was connected with possible effects of the introduction of large amounts of water vapour. Theoretical studies soon showed that the effect of added water vapour on the amount of ozone in the stratosphere, and on climate by way of cloud formation, should be negligible. Since the effects of other exhaust gas constituents such as carbon monoxide and nitrogen oxides had been discounted, this led to the conclusion that the operation of a fleet of supersonic transport aircraft (SST's) in the stratosphere would produce no ill effects on the earth below. The view that nitrogen oxides could safely be ignored was challenged by Johnston,[36] who showed that on the basis of published figures for the projected number of SST flights by 1985, currently accepted values for the rate constants of the major reactions involved, and a fairly conservative estimate of the nitrogen oxide concentration resulting from the introduction of SST exhaust gases, a reduction in the amount of stratospheric ozone by a factor of the order of two could be predicted. Results which differ in their detailed conclusions from those of Johnston have subsequently been obtained by other workers, who used different combinations of rate constants and atmospheric parameters in their calculations. Nevertheless the main point of Johnston's *Science* article, that reactions of nitrogen oxides must not be discounted, has not been seriously contested. At the time of writing a completely definitive calculation of the effects to be expected is still lacking; in the remainder of this section we shall describe the basic features of the chemistry of the ozone layer, and consider the general nature of the effects which can be expected to result from the introduction of photo-chemical smog into the stratosphere.*

A basic reaction scheme for the stratosphere, in the presence of water and 'odd nitrogen' oxides (NO and NO_2), is as follows:[37]

$$NO_2 + h\nu(\lambda < 395 \text{ nm}) \rightarrow NO + O \tag{7.1}$$

$$O_2 + h\nu(\lambda < 242 \text{ nm}) \rightarrow O + O \tag{7.27}$$

$$O_3 + h\nu(\lambda > 310 \text{ nm}) \rightarrow O_2 + O \tag{7.6}$$

$$O_3 + h\nu(\lambda < 310 \text{ nm}) \rightarrow O_2 + O(^1D) \tag{7.28}$$

* Another pollutant which could have a significant effect on O_3 in the stratosphere is chlorine. Chlorine compounds might originate from the present industrial use of freon gases, and from the exhaust gases of the American space shuttles to fly in the 1980's. See articles by S. C. Wofsy and M. B. McElroy; R. S. Stolarski and R. J. Cicerone, *Can. J. Chem.*, **52**, 1582, 1610 (1974).

$$O(^1D) + M \rightarrow O + M \qquad (7.29)$$

$$O(^1D) + H_2O \rightarrow OH + OH \qquad (7.30)$$

$$O + NO_2 \rightarrow NO + O_2 \qquad (7.2)$$

$$O + O_2 + M \rightarrow O_3 + M \qquad (7.3)$$

$$NO + O_3 \rightarrow NO_2 + O_2 \qquad (7.5)$$

$$O + O_3 \rightarrow O_2 + O_2 \qquad (7.7)$$

$$O + OH \rightarrow O_2 + H \qquad (7.31)$$

$$H + O_2 + M \rightarrow HO_2 + M \qquad (7.32)$$

$$HO_2 + O \rightarrow OH + O_2 \qquad (7.33)$$

$$OH + O_3 \rightarrow HO_2 + O_2 \qquad (7.34)$$

$$OH + OH \rightarrow H_2O + O \qquad (7.35)$$

$$OH + HO_2 \rightarrow H_2O + O_2 \qquad (7.36)$$

$$HO_2 + HO_2 \rightarrow H_2O_2 + O_2 \qquad (7.37)$$

$$H_2O_2 + OH \rightarrow HO_2 + H_2O \qquad (7.38)$$

$$H_2O_2 + h\nu(\lambda < 565 \text{ nm}) \rightarrow OH + OH \qquad (7.39)$$

$$OH + NO_2 + M \rightarrow HNO_3 + M \qquad (7.40)$$

$$HO_2 + NO + M \rightarrow HNO_3 + M \qquad (7.41)$$

$$HNO_3 + h\nu(\lambda < 546 \text{ nm}) \rightarrow NO_2 + OH \qquad (7.42)$$

The following four points may be noted in connection with this scheme:

(i) the concentrations of the odd nitrogen oxides are related to one another by

$$[NO]/[NO_2] = (J_{7.1} + k_{7.2}[O])/k_{7.5}[O_3] \qquad (7.43)$$

and the concentrations of the 'odd oxygen' species O and O_3 are related by

$$[O]/[O_3] = (J_{7.6} + J_{7.28})/k_{7.3}[O_2][M] \qquad (7.44)$$

These relationships will hold in the atmosphere provided photochemical equilibrium is not upset by vertical transport, which appears to be a justifiable assumption. The stratosphere itself constitutes a stable temperature inversion, with prevailing mean vertical wind velocity less than 0.5 cm s^{-1},[38] and eddy diffusivity $K \sim 10^3$ cm^2 s^{-1}.[39]

(ii) Typically [O] is several orders of magnitude less than [O_3] in the stratosphere, and the steady state concentrations of other minor oxygen containing constituents such as OH and HO_2 are also very small, so that the production or destruction of odd oxygen species is equivalent to the production or destruction of ozone. A close inspection of the reaction scheme now reveals that processes such as 7.6, 7.29 or 7.3 convert one odd oxygen species into another, but bring about no net production or removal, while process 7.5 appears to remove atomic oxygen, but is ineffectual in the daytime because of reaction 7.1. The only net production of odd oxygen species is by reaction 7.27, and the only significant removal steps are 7.2, 7.7, 7.31, 7.33 and 7.34. Thus for the total rate of production of odd oxygen species, P, we have

$$P = 2J_{7.27}[O_2] \tag{7.45}$$

and for the rate of destruction, D, we have

$$D = D_H + D_N + D_O \tag{7.46}$$

where

$$D_N = 2k_{7.2}[O][NO_2] \tag{7.47}$$

is the rate of O_2 formation, i.e. the rate of odd oxygen removal, by reactions involving odd nitrogen oxides,

$$D_H = k_{7.31}[O][OH] + k_{7.34}[O_3][OH]$$
$$+ k_{7.33}[O][HO_2] \tag{7.48}$$

is the rate of formation of O_2 by reactions of O and O_3 with species derived from H_2O, and

$$D_O = 2k_{7.7}[O][O_3] \tag{7.49}$$

is the rate of loss of odd oxygen species by mutual self-destruction.

(iii) The reaction scheme does not include a process for producing odd nitrogen oxides. In his calculations based on this model Crutzen[37] used concentrations of odd nitrogen oxides which had been calculated by Bates and Hays[40] on the assumption that photolysis of N_2O in the absorption band at 180 nm gives the products NO and N with 20% probability, the other 80% of the primary processes giving $N_2 + O(^1D)$.[41] Later work has shown that the yield of NO in the primary process is actually less than 1%,[42,43] so that the odd nitrogen oxide concentrations given by Bates and Hays are definitely too large. McElroy and McConnell[44] have calculated the rate of NO production by the process

$$N_2O + O(^1D) \rightarrow NO + NO \tag{7.50}$$
$$\rightarrow N_2 + O_2 \tag{7.50a}$$

where 60% of the reactions follow the path 7.50,[43] and the overall rate constant for the removal of $O(^1D)$ has the very large value of 1.8×10^{-10} cm^3 molecule^{-1} s^{-1}.[45] Their results, which are shown in Fig. 7.4b, were dependent on the value chosen for the eddy diffusivity K. As Fig. 7.4a shows, a K value of 10^3 cm^2 s^{-1}, which is consistent with studies of the diffusion of radioactive debris,[39] leads to good agreement with experimental data for the mixing ratio of N_2O as a function of altitude. With this value

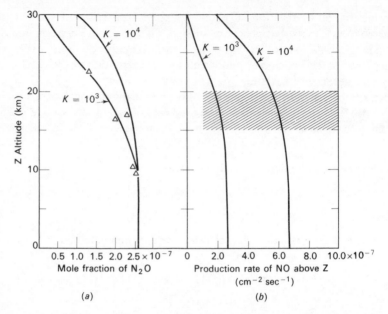

Fig. 7.4 Calculated rate of NO production from N_2O by reaction 7.45.[44] (a) Variation of N_2O concentration with altitude for different values of the eddy diffusion coefficient K. The calculations were constrained to give a mixing ratio (= mole fraction) of 2.6×10^{-7} in the troposphere. The triangles are experimental values. (b) Calculated rate of NO production above a particular altitude, as a function of altitude. The shaded area shows the range covered by different estimates of the NO production which would result from the projected number of SST flights for 1985.

of K, the NO production rates of McElroy and McConnell are less than those of Bates and Hays by about a factor of three.

(iv) Nitric acid has been observed in the stratosphere by ground based spectroscopy;[46] the deduced mixing ratio[47] of $\sim 2 \times 10^{-9}$ would make the concentration of HNO_3 comparable with that of odd nitrogen oxides in the lower stratosphere, below 15 or 20 km. This large concentration of HNO_3 is difficult to understand if the reaction of HNO_3 with O atoms is believed to be as fast as has been deduced from some laboratory photolysis studies.[48] However, direct kinetic studies by Morris and Niki[49] have shown that HNO_3 reacts only very slowly with O and H atoms, upper limits to the rate constants being 2×10^{-14} and 10^{-13} cm³ molecule^{-1} s^{-1} for O and H, respectively. Mulcahy and Smith[50] have observed a rapid reaction between OH and NO_2 in a flow system, corresponding to a value of the order of 10^{-29} cm⁶ molecule^{-2} s^{-1} for $k_{7.40}$ but it is very likely that a heterogeneous reaction was involved, in which case the result has no bearing on processes in the stratosphere. Nitrogen oxides, once introduced, are not easily lost from the stratosphere. The importance of nitric acid is that it is unreactive towards ozone and ground state oxygen atoms, and so represents a sink into which odd nitrogen oxides from SST exhaust gases might ultimately be consigned.

Crutzen[51] has made elaborate numerical calculations for the stratosphere on the basis of a model which incorporates fifty six different reaction processes, including 7.50. He also drew attention to the danger of ignoring the nitrogen oxides in SST exhaust gases. In these calculations he used ozone concentrations obtained from the compilation by Dutsch,[52] together with mixing ratios for N_2O, CH_4 and H_2O equal to 2.5×10^{-7}, 1.5×10^{-6} and 3×10^{-6}, respectively, at the tropopause. Table 7.3 shows the resulting

Table 7.3 **Calculated concentrations of odd nitrogen species in the stratosphere** $(3.6(8) = 3.6 \times 10^8$ molecules $cm^{-3})$.

Altitude (km)	[Odd N]	$[NO]_{day}$	$[NO_2]_{day}$	$([NO] + [NO_2])$ (see text)
55	3.6(8)	—	—	—
50	7.8(8)	7.5(8)	2.7(7)	9(8)
45	1.5(9)	1.2(9)	2.2(8)	—
40	3.0(9)	1.8(9)	1.1(9)	—
35	6.8(9)	2.4(9)	3.6(9)	4.5(9)
30	1.2(10)	3.3(9)	7.6(9)	6(9)
25	2.1(10)	3.8(9)	1.2(10)	6(9)
20	2.1(10)	4.2(9)	1.1(10)	3(9)
15	1.3(10)	5.1(9)	5.2(9)	3(9)

values of the total odd nitrogen concentration,

$$[Odd\ N] = [NO] + [NO_2] + [NO_3] + 2[N_2O_5]$$
$$+ [HNO_2] + [HNO_3] \tag{7.51}$$

and the daytime concentrations of odd nitrogen oxides as a function of altitude, using an average value of 4×10^{-11} cm^3 molecule^{-1} s^{-1} for $k_{7.30}$, and a value of 10^3 cm^2 s^{-1} for the eddy diffusivity. (For $M = O_2$ reference 44 gives $k_{7.30} = 6 \times 10^{-11}$, and for $M = N_2$, $k_{7.30} = 4.2 \times 10^{-11}$.) The last column of the table contains values of $([NO] + [NO_2])$ as obtained by Crutzen[37] from the paper by Bates and Hays,[40] but reduced by a factor of three, in accordance with the findings of McElroy and McConnell. The agreement between the last column and the rest of the data is satisfactory in view of the oversimplification that is involved in our having divided the data of Bates and Hays by a constant factor of three.

Table 7.4 shows the results of Crutzen's calculation of the daily amounts of production and destruction of ozone.[51] The first three columns of the table contain P, the total odd oxygen production by reaction 7.27, D_O, the amount of destruction by reaction 7.7, and D_H, the amount of destruction by species derived from water. The last column shows the values of odd nitrogen oxide concentration which would be required to balance the production rate P, if reaction with odd N oxides were the only mode of

Table 7.4   Calculated production and destruction rates of odd oxygen species $(2.5(11) = 2.5 \times 10^{11}$ molecules cm^{-3} day$^{-1})$.

Altitude (km)	P	D_O	D_H	$([NO] + [NO_2])_{max}$
55	2.5(11)	3.1(10)	4.5(11)	1.5(10)
50	4.0(11)	2.0(11)	5.0(11)	4.0(9)
45	5.0(11)	1.4(11)	1.9(11)	3.0(9)
40	5.0(11)	6.3(10)	5.2(10)	5.2(9)
35	3.8(11)	1.9(10)	1.7(10)	1.3(10)
30	1.5(11)	5.6(9)	5.5(9)	1.9(10)
25	2.9(10)	1.6(9)	1.5(9)	1.5(10)
20	1.4(9)	1.8(8)	1.4(8)	6.0(9)

destruction of odd oxygen species. Thus these concentrations must represent firm upper limits for the natural abundance of odd nitrogen oxides.

Comparing the last column of Table 7.4 with the entries in Table 7.3, we find that the calculated nitrogen oxide concentrations in Table 7.3 are too high at 20 and 25 km. The discrepancy could be due to the assumed mixing ratio of N_2O at the tropopause, 2.5×10^{-7}, being too great. As Crutzen points out, a lower value, 1.5×10^{-7}, has been reported.[53] Thus the concentrations in the first three columns of Table 7.3 are possibly almost a factor of two too large. This does not affect the most important feature of Table 7.4 from the viewpoint of stratospheric pollution, which is that the sum of D_O and D_H is considerably less than P at all altitudes below 45 km. It follows that the destruction of ozone in the altitude range between 20 and 45 km must occur mainly by reaction with odd nitrogen oxides. In the photochemical steady state we should have

$$P = D_N + D_H + D_O \tag{7.52}$$

However, it is noticeable that the calculated values of D_H in Table 7.4 exceed P at 50 km and 55 km. One explanation of this result might be that the assumed mixing ratio of H_2O is too large. Although the calculated values of D_O and D_H depend on a number of parameters whose values can in some cases be estimated only very roughly, this does not invalidate the general conclusion that reaction with odd nitrogen oxides is normally the most important mode of destruction of odd oxygen species in the region of the stratosphere which contains the bulk of the earth's protective ozone layer.

On the basis of the foregoing discussion we may conclude that the addition of between 10^9 and 10^{10} molecule cm^{-3} of odd nitrogen oxides to the stratosphere near an altitude of 20 km would approximately double the rate of destruction of odd oxygen species in the region of maximum ozone concentration, and thereby almost halve the amount of ozone in the protective layer. Such a reduction would be likely to have adverse effects on plant and animal life on the surface.[36,54] Unfortunately it is quite difficult to estimate the number of SST flights which would be needed to bring about a specified increase in nitrogen oxide concentration. The figure given initially for the quantity of nitric oxide in SST exhaust gases appears to have been about an order of magnitude too high.[55] Johnston[36] multiplied this figure by a factor of 0.35, which may still be too large. McElroy and McConnell[44] conclude simply, on the basis of the data in Fig. 7.4b, that if the projected number of flights for 1985 did eventuate, the natural and artificial inputs of nitric oxide would be of similar magnitude.

An additional factor which must be considered is the effectiveness of horizontal transport above the tropopause. The normal circulation of ozone in the stratosphere, with winds of velocity typically 50 m s^{-1} blowing from the equator to the poles, should ensure fairly rapid mixing over a complete hemisphere of the NO and NO_2 introduced at any given altitude. However, the local concentration of odd nitrogen oxides along a busy air corridor would still be expected to exceed the average value. Fluctuations of the order of 10% in the amount of ozone above any particular observation point occur naturally. Thus the onset of the effect of SST flights on the ozone layer above any one place might be difficult to establish. If a large worldwide reduction in the ozone layer were to occur it should be readily detected by the routine monitoring of ozone in the upper atmosphere which is already carried out at many localities.* Whether, at that point, action would be taken to reduce the frequency of SST flights, is another question.

References

1　Junge, C. E., *'Air Chemistry and Radioactivity'*, Academic Press, New York, 1963.
2　Cadle, R. D., *'Particles in the Atmosphere and Space'*, Reinhold, New York, 1966.
3　See, for example: Altshuller, A. P., *Analyt. Chem.*, **39**(5), 10R (1967).
4　Leighton, P. A., *'The Photochemistry of Air Pollution'*, Academic Press, New York, 1961.
5　Haagen-Smit, A. J., Bradley, C. E. and Fox, M. M., *Ind. Eng. Chem.*, **45**, 2086 (1953).
6　Calvert, J. G. and Pitts, J. N. Jnr., *'Photochemistry'*, John Wiley and Sons Inc., New York, 1966.
7　Baulch, D. L., Drysdale, D. D. and Horne, D. G., *'High Temperature Reaction Rate Data'*, No. 5, The University, Leeds, 1970.
8　Kaufman, F. and Kelso, J. R., *Discuss. Faraday Soc.*, **37**, 26 (1964).
9　Phillips, L. F. and Schiff, H. I., *J. chem. Phys.*, **36**, 1509 (1962); **37**, 924 (1962).
10　Mathias, A. and Schiff, H. I., *Discuss. Faraday Soc.*, **37**, 38 (1964).
11　Zipf, E. C., *Can. J. Chem.*, **47**, 1863 (1969).
12　Norrish, R. G. W., *Proc. chem. Soc.*, 1958, 247; McCullough, D. W. and McGrath, W. D., *Chem. Phys. Lett.*, **12**, 98 (1971).
12a　Wayne, R. P., *Discuss. Faraday Soc.*, **53**, 172 (1972).

* Johnston, Whitten and Birks[56] have recently demonstrated, by the analysis of worldwide data, that a statistically significant decrease in the total ozone occurred during the years 1960–62, and was followed by a somewhat larger increase in total ozone during the years 1963–70. These changes, which are of the order of 5%, are correlated with an intensive period of atmospheric testing of nuclear bombs from 1960 to 1962, and the cessation of such testing from 1963 onwards. The observations can be accounted for quantitatively in terms of the effect of nitric oxide introduced into the stratosphere by the explosions.

13 Johnston, H. S. and Jost, D. M., *J. chem. Phys.*, **17**, 386 (1949).
14 Schott, G. and Davidson, N., *J. Am. chem. Soc.*, **80**, 1841 (1958).
15 For convenient tables see reference 4, p. 19, 49 and 59.
16 Cadle, R. D. and Johnston, H. S., '*Proceedings of the Second National Air Pollution Symposium*', Stanford Research Institute, California, 1952.
17 Wayne, R. P., *Adv. Photochem.*, Eds. Noyes, W. A., Jnr., Hammond, G. S. and Pitts, J. N., Jnr., Volume 7, p. 311, Wiley–interscience, New York, 1969; Pitts, J. N., Jnr., *Adv. Environ. Sci. and Technol.*, Eds. Pitts, J. N., Jnr. and Metcalf, R. L., Volume 1, p. 289, Wiley–Interscience, New York, 1969.
18 Pitts, J. N., Jnr., Khan, A. V., Smith, E. B. and Wayne, R. P., *Environ. Sci. and Technol.*, **3**, 241 (1969); Altshuller, A. P. and Bufalini, J. J., *Environ. Sci. and Technol.*, **5**, 39 (1971).
19 Altshuller, A. P. and Bufalini, J. J., *Photochem. and Photobiol.*, **4**, 97 (1965).
20 Jones, I. T. N. and Bayes, K. D., *Chem. Phys. Lett.*, **11**, 163 (1971).
21 Nicksie, S. W., Harkins, J. and Mueller, P. K., *Atmos. Environ.*, **1**, 11 (1967); Stevens, E. R., *Atmos. Environ.*, **1**, 19 (1967).
22 Stevens, E. R., *Adv. Environ. Sci.*, **1**, 119 (1969); Fild, I. and Lovell, D. J., *J. opt. Soc. Am.*, **60**, 1315 (1970).
23 Kerr, J. A., Calvert, J. G. and Demerjian, K. L., *Chem. Brit.*, **8**, 252 (1972).
24 Forrance, L. E. and Leather, G. R., *Science, N.Y.*, **173**, 914 (1971).
25 Okabe, H., *J. Am. chem. Soc.*, **93**, 7093 (1971).
26 Hall, T. C., Ph.D. Thesis, University of California at Los Angeles, 1953.
27 Strickler, S. J. and Howell, D. B., *J. chem. Phys.*, **49**, 1947 (1968); Mettee, H. D., *J. phys. Chem.*, **73**, 1071 (1969).
28 Phillips, L. F., Smith, J. J. and Meyer, B., *J. molec. Spectrosc.*, **29**, 230 (1969); Meyer, B., Phillips, L. F. and Smith, J. J., *Proc. nat. Acad. Sci.* (U.S.A.), **61**, 7 (1968).
29 Sidebottom, H. W., Badcock, C. C., Calvert, J. G., Rabe, B. R. and Damon, E. K., *J. Am. chem. Soc.*, **93**, 3121 (1971), and references therein.
30 Cox, R. A., *J. phys. Chem.*, **76**, 814 (1972).
31 Dainton, F. S. and Ivin, K. J., *Trans. Faraday Soc.*, **46**, 374, 382 (1950).
32 Timmons, R. B., *Photochem. and Photobiol.*, **12**, 219 (1970).
33a Schuck, E. A. and Doyle, G. J., Report no. 29, Air Pollution Foundation, San Marino, California, 1959.
33b Cox, R. A. and Penkett, S. A., *J. chem. Soc., Faraday Transactions I*, **68**, 1735 (1972).
34 Johnstone, H. F. and Coughanowr, D. R., *Ind. Eng. Chem.*, **50**, 1169 (1958).
35 Lamb, D., *J. Atmos. Sci.*, **28**, 1082 (1971).
36 Johnstone, H. S., *Science, N.Y.*, **173**, 517 (1971); Lawrence Radiation Laboratory Report UCRL-20568, University of California, Berkeley, 1971; *Search* (*Journal of the Australian and New Zealand Association for the Advancement of Science*), **3**, 276 (1972).
37 Crutzen, P. J., *Q. J. R. met. Soc.*, **96**, 320 (1970).
38 Teweles, S., Technical Note 70, Secretariat of the World Meteorological Organization, Geneva, 1965.
39 Gudikson, P. H., Fairhall, A. W. and Reed, R. J., *J. geophys. Res.*, **73**, 4461 (1968).
40 Bates, D. R. and Hays, P. B., *Planet. Space Sci.*, **15**, 189 (1967).
41 Doering, J. P. and Mahan, B. H., *J. chem. Phys.*, **36**, 1682 (1962).
42 Preston, K. F. and Barr, R. F., *J. chem. Phys.*, **52**, 3347 (1971).
43 Greenberg, P. I. and Heicklen, J., *Int. J. Chem. Kinet.*, **2**, 185 (1970).
44 McElroy, M. B. and McConnell, J. C., *J. Atmos. Sci.*, **28**, 1095 (1971).
45 Young, R. A., Black, G. and Slanger, T. G., *J. chem. Phys.*, **49**, 4758 (1968).
46 Murcray, D. G., Kyle, T. G., Murcray, F. H. and Williams, W. J., *Nature*, **218**, 78 (1968).
47 Rhine, P. E., Tubbs, L. D. and Williams, D., *Appl. Opt.*, **8**, 1500 (1969).
48 Berces, T. and Forgeteg, S., *Trans. Faraday Soc.*, **66**, 640 (1970).
49 Morris, E. D., Jnr. and Niki, H., *J. phys. Chem.*, **75**, 3193 (1971).
50 Mulcahy, M. F. R. and Smith, R. H., *J. chem. Phys.*, **54**, 5215 (1971).
51 Crutzen, P. J., *J. geophys. Res.*, **76**, 7311 (1971).
52 Dutsch, H. U., in '*Climate of the Free Atmosphere*', Ed. Rex, D. F., Elsevier, New York, 1969.

53 Goldman, A., Murcray, D. G., Murcray, F. H., Williams, W. J., Kyle, T. G. and Brooks, J. N., *J. opt. Soc. Am.*, **60,** 1466 (1970).
54 Tranquilli, W., *Ann. Rev. Pl. Physiol.*, **15,** 359 (1964).
55 Carter, L. J., *Science, N.Y.*, **169,** 660 (1970).
56 Johnston, H. S., Whitten, G. and Birks, J., Lawrence Berkeley Laboratory Report UCLBL-1421, 1972.

8
The Atmospheres of Other Planets

8.1 Introduction

Once a chemist has begun to appreciate the possibilities of the earth–sun combination as a photochemical system it is natural, even inevitable, that his attention will be drawn to the chemistry of planetary atmospheres in general. This is a large field, with ramifications in most areas of laboratory photochemistry; there are also strong links with observational astronomy and with the exciting practical business of space exploration. Clearly it also provides a topic which will allow the present book to end on a suitably expansive note.

Table 8.1 lists the known planets of the solar system, together with values of planetary mass, distance from the sun, mean radius, acceleration due to gravity at the surface, albedo (the percentage of diffuse reflection of sunlight), and observed and calculated equilibrium temperature. The distance from the sun, R, is given in astronomical units ($1AU$ = the mean earth–sun distance of 149.6×10^6 km) and the mass is expressed as a multiple of the earth's mass (6.00×10^{24} kg). The equilibrium temperature T_{eq} is calculated on the assumption that the planet is in thermal balance with the incoming solar radiation, i.e.

$$p\pi r^2 \sigma T^4 = \pi r^2 (1 - A)F/R^2 \qquad (8.1)$$

where r is the radius of the planet in metres, A the albedo, F the solar constant (1400 J m^{-2}, equal to the solar flux at a distance of 1 AU from the sun), and σ is the Stefan-Boltzmann constant (5.67×10^{-8} J m^{-2} s^{-1} K^{-4}). The factor p is equal to four if the dark and light hemispheres of the planet have almost the same temperature, as will be the case for a planet with a dense atmosphere and rapid rotation. If the dark side becomes very cold so that it does not radiate significantly, as with Mercury and Mars, p is equal to two. The calculated temperature for the earth is the effective temperature of the radiating level, which is probably in the vicinity of the tropopause.[1] The surface temperature is higher because the lower atmosphere is relatively transparent in the visible and opaque in the infrared; this is known as the 'greenhouse effect'. For Venus and Jupiter the temperatures shown refer to a region near the top of the reflecting cloud layer. A comparison of the observed and calculated temperatures for Jupiter and Saturn leads to the interesting conclusion that these planets radiate more than twice as much energy as they receive from the sun.[2]

Table 8.1   Physical characteristics of the planets[3]

Planet	Mass	R(AU)	Radius (km)	$g(m\ s^{-2})$	Albedo	$T_{obs}(K)$	$T_{eq}(K)$
Mercury	0.556	0.3871	2435	3.73	0.058	557	527
Venus	0.8161	0.7233	6055	8.85	0.75	207	243
Earth	1.0123*	1.0000	6370	9.80	0.36	270	250
Mars	0.1076	1.5237	3385	3.77	0.171	250	257
Jupiter	317.89	5.203	71 600	22.88	0.45	134	105
Saturn	95.18	9.539	60 000	9.05	0.61	97	71
Uranus	14.6	19.18	25 400	8.30	0.35	55	57
Neptune	17.2	30.06	24 750	11.00	0.35	53	45
Pluto	0.11	39.54	3200?	4.30	0.14	—	42

*   Total mass of earth–moon system

Because of their low surface gravity Mercury and Mars possess only very tenuous atmospheres; the high surface temperature is an additional obstacle to the retention of an atmosphere in the case of Mercury. The oxygen–nitrogen atmosphere of earth, which has a biological origin, is unique in the solar system. Venus and Mars have atmospheres based on carbon dioxide, the former with a surface pressure approximately one hundred times that on earth. These atmospheres are believed to be derived from out-gassing of the minerals of which the planets are composed. The 'Jovian' planets, i.e. Jupiter, Saturn, and probably also Uranus and Neptune, have atmospheres based on hydrogen, methane, and ammonia, the ammonia being frozen out at low levels. Pluto is an enigma; if the information shown in Table 8.1 is correct the planet should have an atmosphere, presumably with everything but hydrogen and helium frozen out. However, Pluto was first detected by its perturbing effect on the orbit of Neptune, and a planet as small as the data in the table would indicate could not produce the effect observed! The measured perturbations might be erroneously large. Alternatively, it has been suggested that the size estimate for Pluto could be in error because the apparent visible disc which G. P. Kuiper measured with the 200-inch telescope might correspond to an area of relatively high reflectivity on a larger surface.

At present a great deal of information is available about the atmospheres of Venus and Mars, as a result of both a continuing series of earth based spectroscopic observations and a number of recent spectacular experiments using space probes. For Jupiter there is also a fairly considerable body of information available, at present virtually all derived from earth based spectroscopy, although this has begun to be supplemented by information gained from space probes. For Saturn and the outer Jovian planets, and also for Mercury, the data are relatively sparse, while for Pluto we depend almost entirely on speculation. In the next two sections of this chapter we discuss current experimental data and theoretical models for the two nearest neighbour planets, Venus and Mars. In the last section we consider the available data for Jupiter, and give brief attention to Saturn, Uranus, and Neptune, and the satellites Io, Ganymede, and Titan.

8.2   The atmosphere of Venus

8.2.1   THE VISIBLE DISC

Probably the most characteristic feature of Venus is its dense cloud cover. Temperature contours for the disc, based on the intensity of infrared emission between 8 and 14 μm, are shown in Fig. 8.1.[4] The temperatures shown refer to a region near the top of the cloud layer which is inferred to be complete, since any break in the layer would appear as a hot spot in the infrared. The clouds are penetrated by radio frequency radiation, so that the radius of the planet can be determined quite precisely from radar observations, and some surface features and differences in elevation have been picked out.[5] The brightness of the planet, as measured at a wavelength of 10 cm, corresponds to a surface temperature of 750 K. An interesting feature of Fig. 8.1 is that the terminator, which marks the edge of the visible crescent and which would correspond to the day–night line on a planet with a transparent atmosphere, does not mark any obvious transition in the temperature of the cloud layer. This implies that at a given altitude below the cloud tops the temperature is maintained essentially constant by large scale circulation of the dense

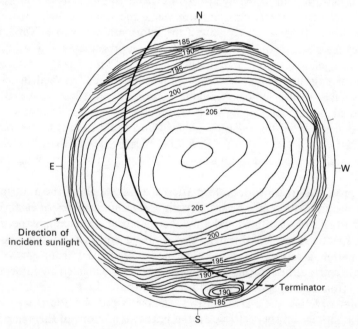

Fig. 8.1   Infrared brightness temperatures (Kelvin) of Venus on December 15, 1962. From Murray *et al.*[4]

strength of the atmospheric circulation, we may note that radar observations of the surface lead to a value of 243 days for the period of rotation of the planet, whereas visual observations of surface markings or shadings suggest that the outer atmosphere rotates with a period of only four days.[6]

The chemical composition of the Venus clouds has been the subject of much speculation; suggested possibilities include ice, carbon suboxide, sulphuric acid, hydrocarbons, mercurous chloride, ammonium chloride, and hydrated ferrous chloride. The likelihood is that there are multiple cloud layers with different compositions. Data obtained by Hansen and Arking[7] from studies of the polarization of sunlight reflected from the clouds lead to the conclusion that most of the particles in the outer layer are spherical, with mean radius near 1 μm, and that the refractive index of the material at 550 nm is 1.45 $\pm$ 0.02 (1.44 $\pm$ 0.01, Hunten[1]). The pressure in the region from which the scattering is observed is estimated to be 50 millibars,* which corresponds to an altitude of about 85 km. Hansen and Arking state that their results eliminate the possibility that the clouds are composed of pure water (refractive index 1.33) or ice (1.31), and that solid particles of materials such as SiO_2, NaCl, NH_4Cl, $FeCl_2$, and Hg_2Cl_2 can also be ruled out, at least as far as the visible layer of clouds is concerned. Lewis[8] has predicted, on the basis of the measured abundances of water and HCl, that cloud particles consisting of 25% aqueous hydrochloric acid should be present at the level probed by the polarization measurements, if the temperature in that region is around 205 K, i.e. similar to the maximum infrared brightness temperature of Fig. 8.1. Such droplets would have a refractive index of 1.39, which is still low, but other solutes might also be present. The most serious objection to this theory is that rotational energy level distributions found for CO_2 bands present in spectra of the reflected light correspond to rotational temperatures near 240 K,[9] which would appear to require an unacceptably large amount of water vapour to be present if droplets were to exist, unless the droplets consisted of an extremely concentrated solution. Hansen and Arking point out that although liquid carbon suboxide has a refractive index of 1.454 at 273 K, there is spectroscopic evidence to indicate that too little C_3O_2 is present for condensation to occur. Velikovsky's contention that the clouds should consist of hydrocarbons has been disproved.[10] Hunten[1] summarized the situation in 1971 as follows: 'Least violence to all the data is done by a model containing a layer of C_3O_2 particles, at a pressure level of 10–50 mbar, and of optical thickness about unity. At 100–200 mbar is the discrete surface of a layer of ice cloud, extending down to 230–300 mbar. Alternatively, one of the exotic clouds proposed by Lewis might be present. Clearly the Venus clouds are still a prime subject for investigation.' In 1973 Sill, Young and Young, and Pollack, independently pointed out that the outer layer's refractive index requirements would be met by a cloud of droplets of 75% sulphuric

---

* 1 bar = 10^5 N m^{-2} = 0.9869 atm.

acid, Prinn proposed a formation mechanism for H_2SO_4 based on photo-oxidation of carbonyl sulphide, and Young showed that the infrared emission spectrum of the clouds supported this identification.[11]

8.2.2   TEMPERATURE AND COMPOSITION OF THE ATMOSPHERE

The surface temperature of 750 K obtained from radar measurements has been confirmed by Mariner and Venera space probes. Mariners 2 and 5 viewed the planet from a distance; information about the atmosphere was obtained from the observed refraction of microwave signals from the probes as they passed out of sight behind the planet, i.e. as they were occulted, and again as they returned into view. The procedure by which atmospheric information is extracted from occultation data is described by Hunten.[1] Veneras 4–7 parachuted through the atmosphere, and Venera 7 definitely reached the surface, from whence it transmitted a temperature reading of 747 ± 20 K.[11] The surface pressure derived from the Venera 7 experiment is 88 bars, with an uncertainty of ± 15 bars. These results were confirmed by Venera 8. In view of the high pressure, corrosive clouds, high temperature, and violent circulation, it is not too surprising that Veneras 4, 5 and 6 did not survive the trip through the atmosphere, or that Veneras 7 and 8 stopped transmitting after only a short time (twenty-three and fifty-five minutes, respectively) on the surface. Pressure and temperature profiles obtained by combining data from Mariner 5 and Veneras 4–6 are given by Avduevsky et al.[12] A complete analysis of the Mariner occultation data[13] leads to a temperature profile with more detailed structure, especially above 60 km. A decrease of roughly a factor of two in the temperature lapse rate at an altitude of 60 km, the 'tropopause', implies the presence of a region corresponding to a stratosphere. A 'summary view' of the atmospheric structure up to an altitude of 250 km is given in Fig. 8.2.[14] The temperature of the Venus exosphere is approximately 650 K on the bright side, falling to below 300 K on the dark side.[15]

The bulk composition of the Venus atmosphere, as determined by Veneras 4, 5 and 6, is given in Table 8.2.[16] The results in the table represent a refinement of the data from Venera 4, on the basis of data from similar gas analysers carried by the later probes.[12,17]

The Venera 4 analyses are summarized in Table 8.3. Eleven gas analysers were used, five being designed to operate at a pressure of 550 torr and the rest at 1500 torr. Both 'amplitude' and 'threshold' detectors were used; the first type indicated the actual amount of a particular substance present, the second indicated whether the amount of a substance exceeded a preset value. A basic form of amplitude detector consisted of a tube divided into two parts by a thin membrane. A chemical absorber was placed in one part of the tube, and the whole tube was sealed and evacuated. In operation the tube was opened to the Venus atmosphere, then immediately closed, and the

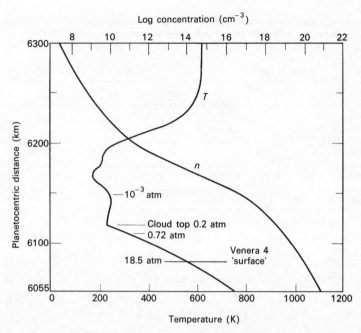

Fig. 8.2 Summary view of the atmosphere of Venus. The curves given by Johnson[14] have been modified to give a surface pressure of 88 bars and temperature of 750 K at a planetocentric distance of 6055 km.

pressure difference resulting from the absorption of gas in one compartment was measured. Thus, for example, in three of the four carbon dioxide determinations this kind of tube was employed with KOH as the absorber. The amount of nitrogen and inert gases was found by difference after CO_2 and O_2 had been removed. An upper limit of 2.5% N_2 was set by observing the pressure reduction due to reaction of N_2 with hot, powdered zirconium. Lower and upper limits for H_2O were determined from the electrical conductivity of moist P_2O_5, and by absorption in anhydrous $CaCl_2$, respectively. Oxygen was determined using two threshold sensors, the first being a tungsten filament which was adjusted to burn out at an oxygen pressure of about

Table 8.2 Chemical composition of the atmosphere of Venus as measured by Veneras 4, 5 and 6

Component	Relative Abundance, %	Comments
CO_2	97(+3, −4)	Altered from earlier figure of 95 ± 2.
H_2O	0.4 to 1	6–11 mg dm^{-3} at 0.6–2 atm total pressure.
N_2	2	3.5 ± 1.5 total for nitrogen, argon, and other inert gases.
O_2	0.1	Original finding of ca. 1% O_2 by Venera 4 was incorrect.

Table 8.3   Chemical analyses performed by Venera 4 (altitudes corrected from the original paper in the light of Venera 7 and radar results)

Analysis Conditions	Substance Determined	Type of Detector	Method	Range (%)	Result (%)
Group 1,	CO_2	Threshold	Heat conductivity	1	>1
altitude =	CO_2	Amplitude	KOH absorption	7–100	90 ± 10
52 ± 3 km,	N_2	Amplitude	CO_2, O_2 removal	7–100	<7
$P \sim 550$ torr,	O_2	Threshold	Tungsten filament	0.4	>0.4*
$T = 25 \pm 10°$ C	H_2O	Threshold	Conductivity P_2O_5	0.1	>0.1
Group 2,	CO_2	Amplitude	KOH absorption	2–30	>30
altitude =	CO_2	Threshold	KOH absorption	1	>1
45 ± 3 km,	N_2	Amplitude	CO_2, O_2 removal	2.5–50	<2.5
$P \sim 1500$ torr,	$O_2(+H_2O)$	Threshold	P vaporizing	1.6	<1.6
	H_2O	Amplitude	$CaCl_2$ absorption	Above 0.7	<0.7
$T = 90 \pm 10°$ C	H_2O	Threshold	Conductivity P_2O_5	0.05	>0.05

(* = see text)

3 torr, and the second being based on the selective absorption of O_2 by sublimed phosphorus, which in the presence of H_2O would lead to the production of moist P_2O_5. This second detector was set to have a threshold at 1.6% total $O_2 + H_2O$. The only case of detector malfunction on the Venera 4 flight involved the tungsten filament which apparently burnt out, implying that more than 3 torr or 0.4%, of O_2 was present. The large proportion of O_2 which this would indicate is inconsistent with spectroscopic data,[18] and the breaking of the filament can reasonably be attributed to the unexpectedly violent conditions encountered.

Earth based spectroscopic studies have yielded a great deal of information about the composition of the Venus atmosphere. CO_2 has long been known as a major constituent, but prior to the direct analyses by the Russian space probes the relative abundance of CO_2 could be given only as 60 ± 40%. Now that a more precise figure is available for the CO_2 abundance it is possible to determine the abundances of other constituents by comparing their absorption lines with nearby CO_2 lines of similar intensity. Connes and co-workers have made a series of elegant investigations by means of Fourier spectroscopy with a Michelson interferometer, and have published an important compilation of the infrared spectra of Venus, Mars, Jupiter and Saturn.[19] Some unexpected findings have resulted from these measurements, notably the conclusion that quite large amounts of HCl and HF are present in the upper atmosphere of Venus. The amount of HCl is such[8] that if liquid water is present it must be approximately 25% HCl by weight.* Part of the spectrum of HCl, as obtained by Connes *et al.*,[20] is shown in Fig. 8.3. The

---

* Lewis [21] has pointed out that the determination of $[H_2O]$ by electrical conductivity measurements, as in the Venera probes, may be in error because of the large effect of dissolved HCl on the conductivity of water.

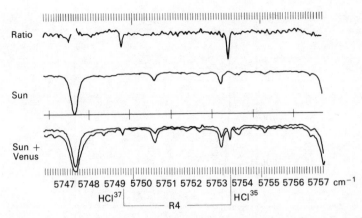

Fig. 8.3   Part of the spectrum of HCl in the atmosphere of Venus. (From Connes *et al.*[20])

lower part of the figure shows two traces of the planetary spectrum, the middle part is a solar spectrum, and the upper trace gives the ratio of the planetary and solar spectra, with Fraunhofer and telluric absorptions thereby largely cancelled out. (Telluric lines are those produced by light absorption in the earth's atmosphere.) The upper trace shows R-branch lines arising from the $J'' = 4$ level for HCl^{37} and HCl^{35}. The abundances relative to CO_2 of minor constituents of the Venus atmosphere, as determined by absorption spectroscopy, are summarized in Table 8.4. Jenkins and co-workers[26] give upper limits for the total amounts of a number of substances on the basis of the absence of characteristic ultraviolet bands from spectra obtained with a rocket-borne spectrograph.

The structure of the ionosphere of Venus was investigated by radio occultation experiments during the flight of Mariner 5.[27] Electron densities obtained for both daytime and night-time hemispheres are shown in Fig. 8.4. Theoretical models for the two cases have been presented by McElroy[28] and Stewart[29] for daytime, and by McElroy and Strobel[30] for night-time conditions. The models are successful in accounting for the broad features of the

Table 8.4   Spectroscopic abundances, relative to CO_2, of minor constituents of the atmosphere of Venus

Constituent	Mixing Ratio	Reference
CO	4.6×10^{-5}	22
HCl	6×10^{-7}	20
HF	1.5×10^{-9}	20
O_2	8×10^{-5}	23
H_2O, above clouds	10^{-4} (variable)	1
H_2O, lower atmosphere	$5–10 \times 10^{-3}$	24
C_3O_2	10^{-5}	25, 26
O_3	10^{-8}	26

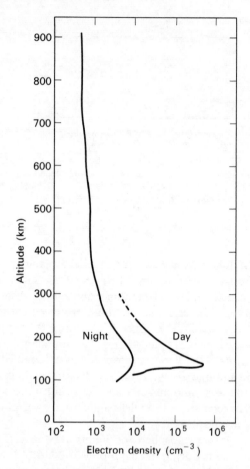

Fig. 8.4   Mariner 5 electron density profiles for the atmosphere of Venus.

observed electron density profiles, although it appears that the observed concentrations of electrons are considerably larger than would be predicted on the basis of the measured solar photon flux.[29] Best agreement is obtained on the basis of an atmosphere of essentially pure CO_2; however, the large value of the scale height at night-time above about 250 km altitude implies that ions of low mass are present. The preferred model of McElroy and Strobel involves a reactive light ion, such as He^+ or H_2^+, which could be supplied by lateral transport from the dayside. McElroy and Strobel also discuss the possibility that ion–molecule reactions might contribute to a detectable airglow, such as the weak Lyman-α airglow which was detected by Mariner 5,[31] or the emission lines of atomic oxygen at 130.4 nm and 135.6 nm which have been measured with a rocket-borne spectrograph.[32] The Venus ionosphere appears to be strongly coupled to the interplanetary medium.

8.2.3 PHOTOCHEMISTRY OF THE VENUS ATMOSPHERE

The atmosphere of Venus, consisting as it does of almost pure CO_2, presents a thorny photochemical problem by its mere existence. The same is necessarily true of the atmosphere of Mars, which we will be considering in the next section of this chapter. The results of numerous laboratory investigations indicate that CO_2 is photolysed by sunlight to yield CO and O at an average rate of about 2.5×10^{12} molecules $cm^{-2} s^{-1}$ on Mars,[33] which corresponds to a figure of 1.1×10^{13} on Venus when allowance is made for the relative distances of the two planets from the sun. According to McElroy and McConnell[33] this rate is sufficient to photochemically modify the whole Martian atmosphere in as little as 2000 years, and the total amount of CO that is observed to be present could be produced in only two years. The problem appears less serious on Venus because there the atmosphere as a whole is more than 10^4 times as massive; nevertheless the very small extent of dissociation to CO and O_2, as shown in Table 8.4, is somewhat difficult to understand. In what follows we shall first discuss laboratory studies of the photolysis of CO_2 in relation to conditions on Venus and Mars, and then consider the factors which might be responsible for the photochemical stability of a pure CO_2 atmosphere.

The absorption spectrum of CO_2 is described in Chapter 2. Table 8.5 shows wavelength thresholds for the production of various electronic states of CO and O in the primary photolysis process. The photo-ionization limit of CO_2 occurs at 98.6 nm, and at the five shortest wavelengths in the table the preferred primary process is actually one which yields $CO_2^+ + e^-$ rather than neutral fragments. Processes which produce ground state CO are spin-forbidden unless the O atom is also in a singlet state; nevertheless photolysis will occur in the wavelength interval between 165 and 224 nm if there is a sufficiently long absorption path, as is the case in a planetary atmosphere. Processes yielding triplet CO and singlet oxygen are also spin-forbidden, and will be unimportant in comparison with spin-allowed processes in the same wavelength region. Young and co-workers[34] have observed emission of the 557.7 nm ($^1S \rightarrow {}^1D$) line of atomic oxygen during the photolysis of CO_2 at 116.5 and 123.6 nm, which demonstrates that a significant amount of $O(^1S)$ is produced by irradiation in the 112 nm band of CO_2. The onset of

Table 8.5 Wavelength thresholds for production of excited states of O and CO during CO_2 photolysis. (Upper wavelength limits in nm)

CO state	$O(^3P)$	$O(^1D)$	$O(^1S)$
$X\,^1\Sigma^+$	224.0	165.0	127.3
$a\,^3\Pi$	107.0	91.7	78.7
$A\,^1\Pi$	91.3	79.8	67.7

intense CO_2 absorption at 165 nm, the theoretical threshold for $O(^1D)$ production, is a clear indication that $O(^1D)$ results from photolysis in the absorption region between 125 and 165 nm.

Most studies of CO_2 photolysis which have been reported have involved absorption in the region between 107 and 165 nm, where singlet oxygen atoms are produced. There is fairly general agreement among the different workers that the quantum yield of CO in this region is unity, but until recently it was not established that the yield of O_2 has the expected value of 0.5. Mahan[35] found the products of photolysis in a glass system to be deficient in O_2, and this observation was confirmed by later studies.[36] Katakis and Taube[37] observed the exchange reaction

$$OCO^{18} + O(^1D) \rightarrow CO_2 + O^{18} \qquad (8.2)$$

to occur when ozone was photolysed with 253.7 nm radiation in the presence of labelled CO_2, and proposed that exchange occurred via an intermediate CO_3 radical. Moll, Clutter and Thompson[38] observed an infrared spectrum which could be attributed to CO_3 to be produced by the photolysis of ozone in a matrix of CO_2; similar spectra were obtained by Taube and co-workers[39] who photolysed solid CO_2, and by Arvis[40] who trapped the products of 147 nm photolysis of gaseous CO_2. Thus it seemed plausible to assume that a stable CO_3 species could be formed from $O(^1D)$ and CO_2 in the gas phase. This would account for the deficiency of O_2 in CO_2 photolysis products, and would also provide a means of regenerating CO_2 on Venus and Mars, in the form of reaction 8.3.[41]

$$CO_3 + CO \rightarrow 2CO_2 \qquad (8.3)$$

Unfortunately for this theory, subsequent work has shown that the O_2 deficiency in the CO_2 photolysis products can be understood in terms of a wall reaction of atomic oxygen,[42] the yield being restored to the theoretical value of 0.5 by suitable treatment of the walls of the photolysis vessel,[43] that CO_3 formation is not significant in the gas phase except at very high pressures,[44] and that in any case the CO_3 mechanism does not account satisfactorily for the observed data relating to the atmospheres of Mars and Venus.[33] The quenching of $O(^1D)$ by CO_2 is extremely fast ($k_{8.4} = (2 \pm 1) \times 10^{-10}$ cm^3 molecule^{-1} s^{-1} [45]) so that it is in fact very unlikely that reactions of $O(^1D)$, other than

$$CO_2 + O(^1D) \rightarrow CO_2 + O(^3P) \qquad (8.4)$$

(direct quenching) can be significant in CO_2 photolysis systems. Quenching of $O(^1S)$ is relatively inefficient ($k_{8.5} = 3.6 \times 10^{-13}$ cm^3 molecule^{-1} s^{-1} [46]) but $O(^1S)$ is noticeably less reactive than $O(^1D)$,

$$CO_2 + O(^1S) \rightarrow CO_2 + O(^3P) \qquad (8.5)$$

and quenching to the ground state is still the most probable fate of $O(^1S)$

in a CO_2 atmosphere. The radiative lifetime of $O(^1S)$ is 0.74 s, so there is an equal probability for deactivation by emission of the 557.7 nm line and for quenching by reaction 8.5 when the CO_2 concentration is 2×10^{12} molecules cm^{-3}, which from Fig. 8.2, corresponds to an altitude of about 140 km on Venus. At wavelengths below 107 nm there is the additional possibility that the primary step in CO_2 photolysis will yield $O(^3P)$ and $CO(a^2\Pi)$; however, flow system studies have shown that the a-state of CO is also quenched very rapidly by CO_2 ($k_{8.6} \sim 2 \times 10^{-11}$ cm^3 molecule^{-1} s^{-1} [47]). Thus the immediate products of photolysis of CO_2 at wavelengths between the ionization limit and 165 nm, namely CO and O in their electronic ground states, are the same as the primary products of photolysis between 165 and 224 nm.

$$CO(a^3\Pi) + CO_2 \rightarrow CO(X^1\Sigma) + CO_2 \qquad (8.6)$$

Studies of the photolysis of CO_2 at wavelengths greater than 175 nm[48] show that the primary process gives ground state CO and O with a quantum yield of unity. Because the solar flux increases very sharply with wavelength above 165 nm this is the most important primary process for CO_2 in the atmospheres of Mars and Venus. Calculations by McElroy and McConnell[33] indicate that photolysis to $CO(^1\Sigma) + O(^3P)$ dissociates CO_2 ten times as rapidly as the next most important primary process, which is that yielding CO and $O(^1D)$. The peak rate of CO_2 photodissociation occurs between the surface and an altitude of about 15 km on Mars, which corresponds, in terms of particle density, to an altitude of about 110 km on Venus.

The simplest possible mechanism for dissociation and reformation of CO_2 on Venus involves the steps

$$CO_2 + hv \rightarrow CO + O \qquad (8.7)$$

$$CO + O + CO_2 \rightarrow 2CO_2 \qquad (8.8)$$

$$O + O + CO_2 \rightarrow O_2 + CO_2 \qquad (8.9)$$

$$O_2 + hv' \rightarrow O + O \qquad (8.10)$$

At a level near the peak of CO_2 dissociation on Venus the photodissociation rate $J_{8.7}[CO_2]$ is approximately 2×10^7 molecule cm^{-3} s^{-1},[49] while for $k_{8.8}$ and $k_{8.9}$ we can use values of 2×10^{-35} and 2.7×10^{-33} cm^6 molecule^{-2} s^{-1}, respectively.[50,41] Putting $[CO_2] = 10^{16}$ molecule cm^{-3}, and assuming that the O_2 is largely dissociated to O atoms, we find a steady state CO concentration of 1.2×10^{13} molecule cm^{-3}, or a CO mixing ratio of 1.2×10^{-4}. If the O_2 were not assumed to be largely dissociated, the steady state value of $[O]$ would be lower, and that of $[CO]$ correspondingly higher. This is one aspect of the problem, i.e. the observed CO mixing ratio (Table 8.4) is significantly lower than the calculated lower limit for the photochemical steady state. However, a more serious aspect appears when the consequences of vertical transport of CO and O_2 are considered. The constant arrival of fresh CO_2 in the upper atmosphere, and the relative resistance of CO to photodissociation (in this it resembles N_2), should long ago have resulted in conversion of the atmosphere to one rich in CO and O_2. According to Donahue,[51] in the course of 4.5×10^9 years irradiation about 10^{28} molecules of CO_2 have been dissociated per square centimetre of surface, whereas the total CO_2 content of the atmosphere amounts to less than 5×10^{27} molecules cm^{-2}. Therefore we require an additional mechanism for oxidizing CO to CO_2, which may be located in either the upper or lower atmosphere.

As discussed earlier, the possibility that CO is oxidized by CO_3 at a significant rate can now be discounted. Another suggested mechanism[52] involves oxidation of CO by HO_2 and OH, the important processes being

$$H + O_2 + CO_2 \rightarrow HO_2 + CO_2 \tag{8.11}$$

$$HO_2 + CO \rightarrow OH + CO_2 \tag{8.12}$$

and $\qquad OH + CO \rightarrow CO_2 + H \tag{8.13}$

This reaction sequence is evidently too slow to be effective on Venus,[49] but a mechanism of this type is more successful on Mars where water is relatively more abundant in comparison with CO_2. An interesting possibility, unfortunately not applicable to the atmosphere of Mars, has been raised by Prinn,[49] who has considered the photochemistry of HCl and other minor constituents of the Venus atmosphere. Prinn concludes, on the basis of a detailed reaction scheme, that the principal source of H_2 on Venus is likely to be the photolysis of HCl at low levels in the atmosphere, and that the low abundance of water is probably the result of the conversion of H_2O to HCl by interaction with hot minerals on the surface, with subsequent loss of the hydrogen through photodissociation. The stability of the CO_2 atmosphere is suggested to be the result of the catalytic cycle

$$Cl + O_2 + CO_2 \rightarrow ClOO + CO_2 \tag{8.14}$$

$$ClOO + CO \rightarrow ClO + CO_2 \tag{8.15}$$

$$ClO + CO \rightarrow Cl + CO_2 \tag{8.16}$$

In evaluating the consequences of this model Prinn assigned values of 10^{-12} cm^3 molecule^{-1} s^{-1} to both $k_{8.15}$ and $k_{8.16}$. In view of the generally low reactivity of CO these rates seem very large, especially the second one, but the ability of chlorine to catalyse the photo-oxidation of CO is documented.[53]

Further laboratory studies appear to be required to establish whether a purely photochemical solution to the problem of the stability of the Venus atmosphere can be obtained. Failing a photochemical solution, it becomes almost unavoidable to postulate that CO and O_2 are transported into a region where CO_2 is able to be regenerated efficiently by a heterogeneous process. Such a region might exist in the cloud layer, with solution chemistry making an unexpected appearance, or in the dust laden[54] furnace of the lower atmosphere.

8.3 The atmosphere of Mars

8.3.1 THE VISIBLE DISC

In contrast to Venus, Mars is almost free of obscuring clouds, and a considerable amount of surface detail is visible from the earth through quite a small telescope. Several Mariner space probes have returned excellent high resolution photographs of the surface,[55] and have shown the presence of numerous craters, strikingly similar in appearance to those on the lunar surface. Detailed inspection of the photographs reveals that Mars differs significantly from the moon, in that there is evidence of some weathering of the surface features and of continuing geological activity. The Martian surface, as viewed through a telescope, undergoes seasonal colour changes, with which are associated changes in the size of the white polar caps. The combination of orbital eccentricity and inclination of the axis accentuates seasonal differences in the southern hemisphere and moderates them in the northern hemisphere. The minimum temperature of the southern polar cap has been measured as 148 K.[56] At 148 K the vapour pressure of solid CO_2 is 6.4 mbar; when combined with the measured CO_2 pressure of about 6 mbar in the atmosphere this provides strong evidence that the caps are actually composed of solid CO_2,[57] a conclusion which has been confirmed by means of an earth based spectroscopy. Traces of other gases, notably ozone and water are believed to be trapped in the solid layer, with probably a discrete layer of water-ice which persists through the Martian year.[58] The 'polar hood', long observed by astronomers, is believed to be a cloud layer, composed of water ice crystals, above the polar region in the autumn hemisphere. The rapidity with which the polar caps grow and shrink indicates that they are probably only a few centimetres thick over much of their area. At times the surface detail is obscured by dust clouds which can last for several weeks. There is also an intermittent bluish haze, which may be a condition of the surface, and a white cloudiness which is associated with condensation of H_2O and CO_2 at high levels in the atmosphere.[60] Nevertheless, in normal circumstances the surface of the planet is open to the sun's radiation at all wavelengths between about 195 nm and the near infrared.

The characteristic reddish colour of Mars has often been attributed to the presence of iron bearing minerals such as limonite, but such materials do not give matching reflection spectra. One alternative possibility, suggested by Plummer and Carson,[61] is that the colouration is in fact due to the presence of polymerized carbon suboxide. They obtained reflection spectra similar to that of Mars from polymerized C_3O_2 in the laboratory. The build-up of surface layers by photopolymerization is a known laboratory phenomenon,[62] which occurs with a very wide variety of monomers. Plummer and Carson also suggest that the dark areas on the planet reflect differing degrees of polymerization in the surface layer, and that seasonal colour changes could be due to changes in the amount of water in the hygroscopic polymer. It is to be expected that this and other theories will be tested by landing instruments on Mars in the resonably near future.

8.3.2   TEMPERATURE AND COMPOSITION OF THE ATMOSPHERE

Our information about the temperature and composition of the Martian atmosphere comes from ground based spectroscopic studies and from the radio occultation and spectroscopic experiments performed by Mariner space probes. Mariners 4, 6 and 7 each provided occultation data at two locations on the planet, first as they were occulted and then as they came back into view. Mariner 9 orbited Mars for an extended period and in the process gave data from a large number of occultations.

Vertical temperature distributions derived by Rasool and co-workers[63] from the four occultation measurements of Mariners 6 and 7 are shown in Fig. 8.5. Surface pressures and temperatures are given in Table 8.6 on the basis of data from Mariners 4, 6 and 7. The results were calculated for a pure CO_2 atmosphere; with 20% of N_2 or Ar the calculated surface temperatures and pressures would be about 10% higher. (Later work has indicated that the N_2 abundance is of the order of 1%). The different temperature profiles of Fig. 8.5 refer to differing latitudes, seasons, and day/night conditions, as summarized in Table 8.6. In particular, the very low temperatures of the

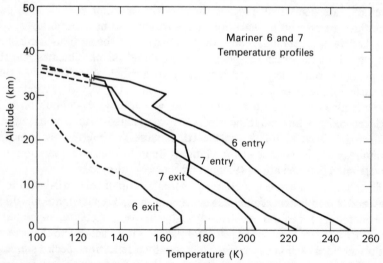

Fig. 8.5   Temperature profiles for Mars from occultation data of Mariners 6 and 7.[63] Near the surface the estimated error is ± 5 K; at 30 km it is ± 20 K.

Table 8.6 **Surface pressures and temperatures on Mars from occultation data of Mariners 4, 6 and 7. ('Entry' refers to observations made as the probe moved behind the planet, 'exit' to observations made as the probe re-emerged.** χ **is the solar zenith angle; values of** χ **greater than** $90°$ **indicate that the sun is below the horizon)**

Occultation	Latitude	Longitude	Season	χ	P (mbar)	T (K)
4, entry	50.5°S	177.0°E	Winter	67°	4.5	160
4, exit	60.0°N	34.0°W	Summer	104°	8.0	210
6, entry	3.7°N	4.3°W	Autumn	57°	6.0	250
6, exit	79.3°N	87.1°E	Winter	107°	7.6	164
7, entry	58.2°S	30.3°E	Spring	56°	4.9	224
7, exit	38.1°N	148.3°W	Autumn	130°	7.5	205

Mariner 6 'exit' profile refer to the night side at a latitude of 79° North. The differences between the surface pressures in Table 8.6 are real, and correspond to differences in altitude at the surface, i.e. to varying topography. The low surface temperature at Mariner 4 entry is appropriate for winter at a high altitude. As noted in the last section, the vapour pressure of CO_2 at 148 K, the measured temperature of the coolest part of the south polar cap, is 6.4 mbar, which agrees very closely with the observed pressure of about 7 mbar at a mean surface altitude. In the earth's atmosphere this pressure would correspond to an altitude of about 33 km.

When Mariner 9 went into orbit around Mars in mid-November 1971 the whole planet was in the grip of the most intense dust storm ever observed, one which at its peak, three weeks earlier, had obscured even the south polar cap from telescopic observation. By the time Mariner 9 arrived the polar cap had reappeared, together with some high spots on the surface, and after about three further weeks the atmosphere had cleared sufficiently for satisfactory photographs of the surface to be obtained. Both infrared radiometry measurements[64] and occultation data[65] show that the presence of the dust clouds had a considerable effect on the temperature of the atmosphere. A preliminary analysis of fifteen of the occultation measurements[65] showed that the effect of the dust storm was to cause the lower atmosphere to have an approximately constant temperature up to a distance of about 3405 km from the centre of the planet, corresponding to a mean altitude of about 20 km. Some typical profiles are shown in Fig. 8.6. The effect is attributed to absorption of solar radiation by fine dust particles suspended in the atmosphere. A similar effect is noticeable in one of the Mariner 4 temperature profiles.[66]

Stable gaseous species which have been detected by spectroscopic methods on Mars include CO_2, CO, O_2, H_2O and O_3. The upper limit for N_2 abundance is about 5 %, on the basis of the absence of spectra derived from nitrogen in the Martian airglow.[67] Other inert gases, notably argon, might also be present but undetected. The abundance of CO_2 has been estimated from the intensity of absorption of the 123–000 and 203–000 combination tone vibration–rotation bands at 1050 and 1038 nm in the near infrared.[1] The

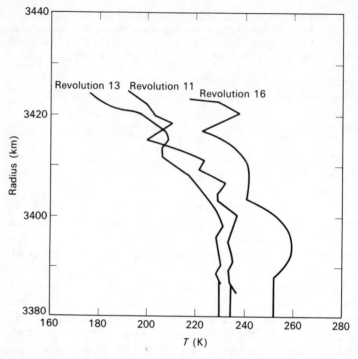

Fig. 8.6 Temperature profiles from Mariner 9 occultation data, showing the near isothermal lower atmospheres which resulted from the presence of entrained dust particles. (From the preliminary analysis by Kliore et al.[65])

total abundance thus found corresponds to an absorption path of 78 ± 11 metre atmospheres (at NTP), or a surface pressure of 5.5 ± 0.8 mbar. Most of the uncertainty of this result is stated to be due to the difficulty of measuring the band strength in the laboratory. The total gas pressure at the surface has been estimated as 5.3 mbar from the extent of Lorentz broadening of a strong absorption line of CO.[68] Taking into account possible errors in both of these results, Hunten[1] gives the fractional abundance of CO_2 as 90 ± 10 %. The CO abundance estimated from its weak 3–0 infrared band is 5.6 cm atm,[68] corresponding to a mixing ratio of 8×10^{-4}. O_2 can be detected from the absorption of the atmospheric band near 762 nm, provided the spectrum is observed at a time of large relative velocity in the earth–Mars system so that use can be made of the Doppler shift to separate the Martian and telluric absorptions.[1] The results of Belton and Hunten[23] give a value of 20 cm atm for the abundance of O_2, a figure which is probably best regarded as an upper limit. The corresponding mixing ratio with CO_2 is 2.6×10^{-3}. The high abundance of O_2 relative to CO is accounted for by assuming a continuous process of photodissociation of water, with escape of hydrogen from the upper atmosphere. H_2O has been determined in a similar manner to O_2, using the 211–000 band at 820 nm.[69] A typical mixing ratio would be of the order of 10^{-4}. The abundance of H_2O shows marked seasonal variations,[70] and there are indications that the detectable water vapour may at times be confined to one hemisphere. Ozone has been detected over the polar caps of Mars from ultraviolet reflectance spectra obtained by Mariner 7.[71] The observed absorption corresponds to about 10^{-3} cm atm of O_3, i.e. to a mixing ratio of about 1.5×10^{-7} if the O_3 were evenly distributed. The absence of detectable O_3 absorption in reflectance spectra of desert areas can be explained by postulating either that most of the ozone is below the hazy region from which light is scattered above the deserts, or that the ozone is physically trapped in the polar caps.[58] Mariner 9 observations have revealed a strong negative correlation between the concentration of ozone and that of water vapour in the Martian atmosphere.

Atomic oxygen, atomic hydrogen, and CO were detected in the ultraviolet dayglow from the upper atmosphere of Mars by spectrometers carried on Mariners 6, 7 and 9. At an altitude of 135 km, corresponding to the peak electron density in the Martian ionosphere, the mixing ratios of O and CO have been deduced to be about 3×10^{-2} and 3×10^{-3}, respectively.[72] For the atomic hydrogen density there is a figure of $(3 \pm 1) \times 10^4$ cm^{-3} at 250 km.[73] The other minor constituent which has been demonstrated to be present is the $CO_2{}^+$ ion,[71] which mainly originates from photoionization of CO_2.[74] The ion is detected by means of the near ultraviolet emission bands which arise from excited states formed when CO_2 is photoionized at wavelengths below 70 nm. A recent model of the upper atmosphere,[75] in terms of the major neutral constituents, is shown in Fig. 8.7a.

The first information yielded by spacecraft occultation data is an electron density profile for the planet's ionosphere; in fact the effect of the ionosphere must always be subtracted before dayside temperature profiles, such as those in Fig. 8.5, can be derived. Electron density profiles for Mars obtained by Mariners 4, 6 and 9 are shown in Fig. 8.8. (The Mariner 7 curve is almost identical with that obtained from Mariner 6.) The low maximum density found by Mariner 4 is correlated with a period of low solar activity. The greater altitude of the whole ionosphere found by Mariner 9 is attributed to the higher temperature of the lower atmosphere due to the dust storm that prevailed at the time.[65] Results of a model calculation for the Martian ionosphere are shown in Fig. 8.7b. Although the major neutral species is CO_2, and the most important primary process produces $CO_2{}^+$, the fast reactions[76]

$$O + CO_2{}^+ \rightarrow CO + O_2{}^+ \tag{8.17}$$

and
$$O + CO_2{}^+ \rightarrow CO_2 + O^+ \tag{8.18}$$

$(k_{8.17} + k_{8.18} = 2.6 \times 10^{-10}$ cm^3 molecule^{-1} s^{-1}, $k_{8.17}/k_{8.18} \sim 1.7)$ with

$$O^+ + CO_2 \rightarrow CO + O_2{}^+ \tag{8.19}$$

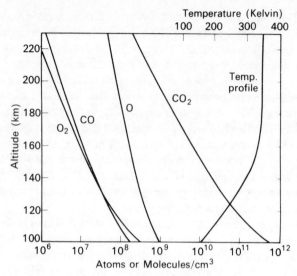

Fig. 8.7a   Mars model atmosphere: neutral constituents.[75]

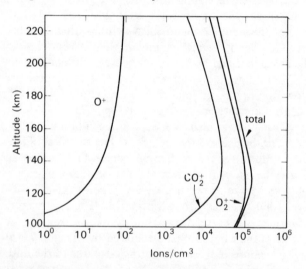

Fig. 8.7b   Mars model ionosphere.[75]

($k_{8.19} = 1.1 \times 10^{-9}$ cm^3 molecule^{-1} s^{-1}) ensure that the most abundant ion is O_2^+.

An unusual feature of the ionosphere of Mars is that it appears to play an important role in determining the overall composition of the atmosphere. McElroy and Hunten[77] point out that the fast ion–molecule reaction

$$CO_2^+ + H_2 \rightarrow HCO_2^+ + H \qquad (8.20)$$

($k_{8.20} = 1.4 \times 10^{-9}$ cm^3 molecule^{-1} s^{-1})

followed by the dissociative recombination process

$$HCO_2^+ + e^- \rightarrow CO_2 + H \tag{8.21}$$

can convert H_2 efficiently into H atoms, which must then escape the planet's gravitational field. Molecular hydrogen, at a mixing ratio of the order of 10^{-6}, is believed to result from the photodecomposition of H_2O near the surface. This process depends on the planet continuing to outgas, and McElroy[78] has concluded that in fact the atmosphere of Mars is still evolving, with fast atoms of oxygen, nitrogen and carbon also escaping from the upper atmosphere. The escape velocity for Mars is only 4.9 km s^{-1} (compared with 11 and 10 km s^{-1} for earth and Venus, respectively) so that the vertical component of kinetic energy required for escape amounts to only 192 kJ mol^{-1} for O, 168 for N, 144 for C, and 336 for CO. Assuming ground state

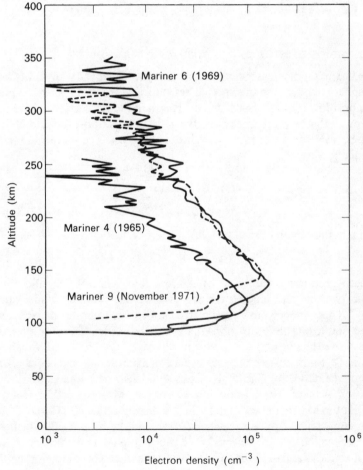

Fig. 8.8 Daytime electron density profiles for Mars, from Mariners 4, 6 and 9.

species except where otherwise indicated, the following dissociative recombination processes are capable of forming products with sufficient translational energy to escape from Mars:

$$CO_2^+ + e^- \rightarrow CO \,(290 \text{ kJ}) + O \,(510 \text{ kJ}) \tag{8.22}$$

$$O_2^+ + e^- \rightarrow O(^1D) + O \,(\text{both } 240 \text{ kJ}) \tag{8.23}$$

$$N_2^+ + e^- \rightarrow N + N \,(\text{both } 280 \text{ kJ}) \tag{8.24}$$

$$CO^+ + e^- \rightarrow C \,(160 \text{ kJ}) + O \,(121 \text{ kJ}) \tag{8.25}$$

The calculated loss rate for O is 6×10^7 atom $cm^{-2} s^{-1}$; the observed loss rate for H is just twice as large as this, i.e. H_2O is being lost and the ratio of CO to O_2 in the atmosphere is maintained at 2:1. Further it would appear that if this balance should be upset, giving an excess of either CO or O_2, secondary processes in the photolysis of H_2O would be affected, and the relative rates of loss of H_2O dissociation products would alter in such a way as to restore the balance.

8.3.3 PHOTOCHEMISTRY OF THE MARTIAN ATMOSPHERE

The main problem associated with the photochemistry of the Martian atmosphere, namely the unexpected resistance of the atmosphere to photodecomposition, was discussed at some length in the last section. There it was mentioned that a mechanism involving decomposition products of water might be applicable to Mars, though not to Venus. The following reaction sequence

$$H + O_2 + CO_2 \rightarrow HO_2 + CO_2 \tag{8.11}$$

$$HO_2 + O \rightarrow O_2 + OH \tag{8.26}$$

$$OH + CO \rightarrow CO_2 + H \tag{8.13}$$

which amounts to an overall process

$$CO + O \rightarrow CO_2 \tag{8.27}$$

has been proposed to account for the low abundance of CO, and appears to be capable of reproducing the observed mixing ratios in model calculations.[79] These models also predict the negative correlation between water vapour and ozone concentrations that we noted earlier. The only alternative to the above scheme appears to be a mechanism based on rapid transport of CO and O_2 to the surface of Mars, where an efficient heterogeneous reaction converts them to CO_2. Clark[45] has considered heterogeneous catalysis of CO oxidation by metal oxides at the surface of Mars. McElroy and McConnell[33] find that the Mariner 6 and 7 data for the mixing ratios of CO and O in the upper atmosphere can be explained if the eddy diffusion coefficient has the very large value of 5×10^8 cm^2 s^{-1}, which implies that there is indeed a large flux of CO and O to the surface. Their calculated production rates for O and CO are given as a function of altitude in Fig. 8.9. It is interesting to note

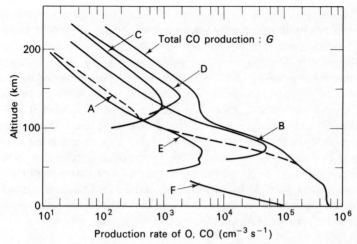

Fig. 8.9 Production rates of atomic oxygen and carbon monoxide on Mars, calculated by McElroy and McConnell for a solar zenith angle of 60°, with the flux reduced by a factor of 2 for day–night averaging.[33] Curves refer to processes as follows:

A: $O(^3P) + CO(^1\Sigma)$ from electron impact + photodissociation with $\lambda > 167$ nm.

B: $O(^1D) + CO(^1\Sigma)$, λ 108–167 nm, with $O(^1S)$ radiating to $O(^1D)$.

C: $O(^3P) + CO^*$ (triplet states) by electron impact and photodissociation with $\lambda < 108$ nm.

D: $O(^1D) + CO$ from ion–molecule reactions.

E: $O(^3P) + O(^1D)$ from O_2 photolysis, $\lambda < 175$ nm.

F: $O(^3P)$ from O_2 photolysis, λ 175–242 nm.

G: Total CO production.

that the peak rate of production of O occurs right at the surface, and that the peak rate of CO production also occurs very low in the atmosphere.

An important feature of the photochemistry of the Mars atmosphere is the dayglow emission which was observed by Mariners 6, 7 and 9. The emission features observed are the CO_2^+ A $\to$ X and B $\to$ X bands, the CO a $\to$ X and A $\to$ X bands, the CO^+ B $\to$ X bands, the 156.1 and 165.7 nm lines of atomic carbon, the 130.4, 135.6, and 297.2 nm lines of atomic oxygen, and the Lyman-α line (121.6 nm) of atomic hydrogen. Mariners 6 and 7 each viewed the atmosphere twice, the spectroscopic observations being made tangentially to the bright disc as the spacecraft flew by the planet. Mariner 9 obtained intensity measurements from numerous passes across the limb of the planet (i.e. across the edge of the disc) during orbits up to at least number 174, and provided a very large amount of information about the altitude variation of the emission of the CO Cameron and fourth positive bands, and the O(I) and H(I) lines at 130.4 and 121.6 nm.* In addition, photometric and spectral mapping was carried out by all three probes by pointing the spectrometer directly at the disc. Representative spectra of the Martian dayglow from

---

* This information is obtainable, in the form of *Data Reports for the Mariner 9 Ultraviolet Spectrometer Experiment*, from Dr. C. A. Barth of the Laboratory for Atmospheric and Space Physics, University of Colorado, Boulder. See also reference 81.

Mariners 6 and 7 are shown in Fig. 8.10.[80] The CO_2^+ spectra appear to result mainly from excitation during photoionization (equations 8.28 and 8.29) of CO_2, either directly or by impact of energetic photoelectrons, with an additional contribution from resonant scattering by ground state CO_2^+. The measured intensity of CO Cameron (a $\rightarrow$ X) bands varied directly with the 10.7 cm solar flux (correlation coefficient = 0.80), while the intensities of the CO_2^+(B $\rightarrow$ X) and CO(a $\rightarrow$ X) bands appeared to be proportional to one another.[81] Laboratory studies have shown that the cross sections for excitation by photoionization and electron impact are large enough to account for the observations.[82] The vibrational distribution in the CO Cameron bands (a $\rightarrow$ X) suggests that they also originate from electron or photon impact on CO_2 (equations 8.30 and 8.31). The same applies to the CO fourth positive bands (A $\rightarrow$ X) except that some of the bands are reduced in intensity because of absorption by CO in the atmosphere.

$$CO_2 + h\nu \rightarrow CO_2^{+*} + e \qquad (8.28)$$

$$CO_2 + e \rightarrow CO_2^{+*} + 2e \qquad (8.29)$$

$$CO_2 + h\nu \rightarrow CO^* + O \qquad (8.30)$$

$$CO_2 + e \rightarrow CO^* + O + e \qquad (8.31)$$

The atomic lines, with the exception of the O(I) resonance triplet at 130.4 nm and the H(I) line at 121.6 nm, are similarly believed to result from dissociative excitation of CO_2, since they can be produced in this way in the laboratory, and they have a similar height distribution to the CO_2^+ and CO bands, with a scale height appropriate for a cold CO_2 atmosphere. Altitude distributions from Mariner 9 are shown in Fig. 8.11. The O(I) line at 130.4 nm extends to much higher altitudes than the emissions derived from CO_2, and it is this which demonstrates the presence of atomic oxygen in the atmosphere. Similarly the observation of Lyman-α emission at altitudes of up to several planetary radii establishes the presence of atomic hydrogen in the exosphere, and incidentally shows the need for a continuous source of hydrogen atoms in the lower atmosphere.

A theoretical analysis of the photochemical behaviour of atomic carbon in the atmospheres of Mars and Venus has been presented by McElroy and McConnell.[83] The major removal process for carbon atoms is considered to be

$$C + O_2 \rightarrow CO + O \qquad (8.32)$$

In situations where the resonance scattering of sunlight provides a major contribution to the intensity of the carbon atom emission lines, measurements of this intensity should therefore provide information about the abundance of O_2.

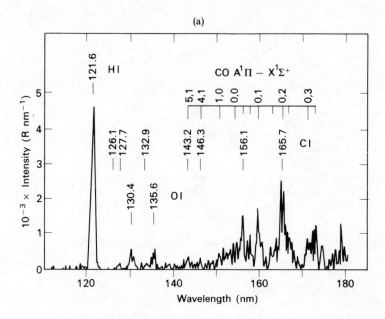

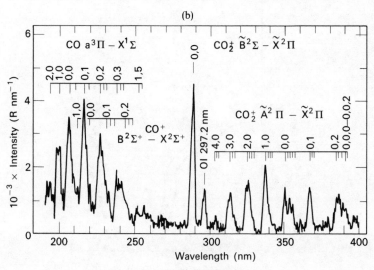

Fig. 8.10 Averaged ultraviolet spectra of the Martian dayglow, from Barth *et al.*[80] (a) λ 110–180 nm at 1 nm resolution; (b) λ 190–400 nm at 2 nm resolution.

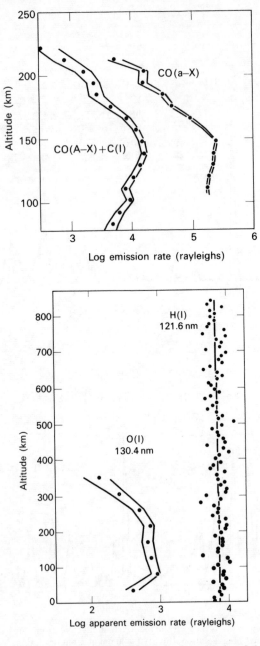

Fig. 8.11 Altitude profiles for dayglow emission features from Mariner 9 data. (a) For the CO Cameron bands and for the sum of C(I) lines and CO fourth positive bands; (b) for the O(I) and H(I) resonance lines. The line through the H(I) points is a least squares fit; for the other data the envelopes indicate the error limits corresponding to plus or minus two standard deviations from the mean intensity value at a given altitude.

8.4   Jupiter and the outer planets

8.4.1   JUPITER: STRUCTURE AND COMPOSITION

Jupiter is a fascinating object to view through a telescope.* Like Venus it is covered by clouds, but the clouds are marked with distinct coloured bands or belts, in pastel shades, running parallel to the equator, and there are also various light and dark spots, including the famous Red Spot, which has been known to astronomers since the seventeenth century. The Red Spot is quite variable in size and colour, sometimes seeming to disappear below the surface. Its latitude and longtitude also vary remarkably. Various suggestions have been made as to the nature of the Red Spot; television pictures sent back by Pioneer 10 suggest that it is in fact the Jovian equivalent of a hurricane, with a diameter greater than that of the earth. The planet as a whole rotates with a period of 9 hours 51 minutes, measured at the equator, but the different belts have rotation periods up to five minutes longer than this. Some infrared hot spots have been observed, mostly in the dark north equatorial belt,[84] which shows that the top layer of cloud cover is not as complete as on Venus. The volume of Jupiter is 1312 times that of earth, but the average density is only 1.3 g cm^{-3}, giving a total mass 318 times that of earth. There is no indication of any solid surface below the clouds, and the planet is aptly described as a gas giant. The observation that Jupiter apparently radiates twice as much heat as it absorbs from the sun implies that effective convection is present to great depths. The source of the additional energy is presumed to be gravitational collapse; in order to conserve angular momentum the velocity of rotation must increase steadily as the radius of the planet decreases. The present rate of rotation is already sufficient to produce a marked flattening of the globe from north to south. The rate of radial collapse required to account for the excess radiation is about 1 mm per year.

Particular interest is attached to the composition of Jupiter, since it appears that its composition, unlike that of the inner planets, may correspond quite closely to that of primitive stellar material.[85] In the absence of precise analytical data such as is now available for Mars and Venus it has been customary to assume, for the purpose of model calculations, that the elements are present with solar abundances. Thus, for example, helium was always assumed to be a major constituent of the atmosphere, even though there was, until the recent Pioneer 10 fly-by, no direct spectroscopic evidence for its presence.

Measurements of the relative abundances of the substances which *have* been found spectroscopically are complicated by the lack of a satisfactory model for the region which scatters sunlight back to the telescope. The simplest model, that of a clear absorbing atmosphere overlying a reflecting

---

* For some excellent photographs see *The Atlas of the Universe* by Patrick Moore, Hamlyn, London, 1970 and the cover of *Science*, N.Y., for 25 January 1974.

cloud deck, leads to the prediction that absorption lines should increase in intensity from the centre to the limb of the disc, because of the increasing absorption path, whereas exactly the opposite behaviour is often found. McElroy[86] has discussed different models for the scattering process, and has concluded that the observations can be accounted for by a model of absorption in a homogeneous, isotropically scattering medium. The Pioneer television pictures show the presence of towering cloud banks, with gaps between, so the depth of penetration of sunlight must depend very markedly on the angle of incidence. However, virtually all of the available abundance data has been derived on the basis of the simple reflecting model, and quoted values of absolute abundance must possess considerable uncertainty. Ratios of the abundances of different substances are likely to be more reliable provided the data for the substances compared were obtained in the same spectral region, i.e. are similarly affected by light scattering.

Molecular hydrogen at low pressures cannot absorb or emit dipole radiation during rotational or vibrational transitions, but vibration–rotation transitions involving quadrupole radiation are allowed, though exceedingly weak. Herzberg[87] predicted that such transitions should be observable in planetary atmospheres, and subsequently measured their wavelengths in the laboratory. Once it was known where to look, lines belonging to several quadrupole bands in the red and near infrared were detected in the atmosphere of Jupiter (Fig. 8.12). On the basis of the measured

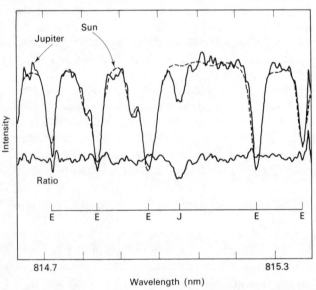

Fig. 8.12 Spectra of the sun and of Jupiter near 815 nm, with a ratio spectrum showing the presence (at J) of the 3–0 S(1) line of H_2 in the Jupiter spectrum. Lines marked with an E result from absorption in the earth's atmosphere. (From Fink and Belton.[88])

intensities and a reflecting layer model, Fink and Belton[88] obtained an abundance of 67 $\pm$ 17 km atm for H_2; Owen[85] gives 85 $\pm$ 15 km atm. Fink and Belton also obtained an effective temperature of 145 $\pm$ 20 K for the hydrogen above the cloud deck. It is apparent that different wavelengths probe different depths in the atmosphere, since the brightness temperature measured at 5 μm is 230 K.[89] At 145 K the calculated pressure at the bottom of a clear, 75 km atm column of H_2 is only 1.8 bar, or even less in a highly scattering atmosphere.

Methane overtone and combination tone bands in the red and near infrared are very prominent in the absorption spectrum of Jupiter. From a study of the $3\nu_3$ overtone at 1.1057 μm Belton[90] has obtained a figure of 30 m atm for the methane abundance, at a temperature of 163 K. Margolis and Fox,[91] using the same experimental data (which had been obtained in the first place by Walker and Hayes[92]) calculated an abundance of about 45 m atm, with two different photographs of the spectrum giving rotational temperatures of 156 K and 200 K.

Ammonia is believed to condense on Jupiter, forming the visible clouds. The intensity of the near infrared bands corresponds to an abundance of 12 m atm.[93] An abundance of only 2×10^{-3} cm atm, has been deduced from low resolution observations near 200 nm in the ultra-violet, but this figure would be expected to err on the low side for at least two reasons. The first reason is the small extent of penetration of a scattering medium by ultraviolet radiation (the probability of Rayleigh scattering is inversely proportional to the fourth power of the wavelength), and the second is the probable absence of a true continuum of NH_3 near 200 nm, so that a low resolution spectrum cannot give a correct indication of the NH_3 abundance.[94]

The presence of helium was at first inferred on the basis of indirect evidence. Baum and Code[95] observed the occultation by Jupiter of the star σ-Arietis, and concluded that the variation of light intensity as the star moved behind the planet corresponded to a scale height of 8.3 km ($+4.2$, -2.1). Assuming the region probed to be approximately isothermal, the mean molecular weight is given by the formula

$$M = T/(3.1H) \tag{8.33}$$

where H is the scale height in kilometres and T is the temperature. A reasonable value of T for the outer atmosphere is about 100 K,[85] so that M is 3.9 $\pm$ 1.3. The solar abundance ratio of hydrogen atoms to helium is 11($+7$, -5),[96] which corresponds to $M = 2.3(+0.2, -0.1)$. This just fails to overlap with the range given by Baum and Code's data; however, McElroy[83] has argued that the signal-to-noise ratio in the data of Baum and Code is such that it can accommodate a scale height as great as 16 km, which is the value corresponding to $M = 2$. This may be an extreme view, but clearly the data of Baum and Code do not rule out an H/He ratio corresponding to the solar abundance ratio. Following Spinrad and Trafton,[97] Owen and

Table 8.7   Abundances of elements relative to hydrogen

Atom Ratio	Jupiter	Sun
H/He	9	11
H/C	3.0×10^3	2.9×10^3
H/N	1.6×10^4	1.2×10^4

Mason[98] estimated the total pressure of gas which would be necessary to produce the observed broadening of weak lines in the methane spectrum, and estimated the helium abundance on the assumption that the only gases present were hydrogen and helium. The calculation gave a lower limit of 9 for the H to He ratio. This is consistent with model calculations of the interior structure of the planet; Peebles,[99] for example, gives H/He $\sim$ 8. A similar helium abundance derives from Pioneer 10 observations of the dayglow at 58.4 nm.

In addition to the major constituents, ethane, acetylene, and phosphine have now been detected on Jupiter. Water has not been detected, which is not surprising in view of the low temperature of the region accessible to spectroscopy. Other substances which are expected to be present in small amounts include H_2S (and NH_4HS + $(NH_4)_2S$), and HCN.

With the aid of present figures for the abundances of hydrogen, helium, methane and ammonia it is possible to compare the relative abundances of the elements on Jupiter with the corresponding abundances deduced for the sun. The results, given in Table 8.7, are consistent with the idea that Jupiter represents the composition of the primitive material from which the solar system was formed.

Model calculations for the structure and composition of cloud layers on Jupiter have been reported by Lewis.[100] A representative set of results for a model with solar composition are shown in Fig. 8.13. Lewis and Prinn[101] have discussed the development of an orange-brown colouration in the North Equatorial Belt, and the occurrence of hot spots with temperatures up to 310 K in the belt,[83] which otherwise has a uniform infrared brightness temperature of 225 K. They suggest that the cloud at 225 K is composed mainly of NH_4HS and $(NH_4)_2S$ in equilibrium with NH_3 and H_2S, in accordance with the model in Fig. 8.13, and that the colouring matter in the cloud layer consists of a mixture of hydrogen polysulphides, ammonium polysulphides, and sulphur produced by photo-decomposition of H_2S. They point out that the photolysis of H_2S provides a fast mechanism for producing colour changes, such as are observed to occur on Jupiter in times of the order of weeks or months, whereas the build-up of significant concentrations of complex organic molecules, which have also been proposed as a source of the observed colours, would appear to require a time of the order of thousands of years.[102] McNesby[103] has discussed the photochemistry of

Jupiter at wavelengths above 100 nm in relation to laboratory studies of NH_3 and CH_4 photolysis, with a model atmosphere consisting of hydrogen, methane and ammonia. He concludes that in the presence of a large excess of H_2 the photolysis of CH_4 and NH_3 is not likely to produce complex organic molecules, because the free radicals resulting from the primary process should simply react with H_2 to reform the parent species. Complex molecules are produced only if the concentration of radicals is very large, or if molecular ions are present, as in an electrical discharge through the gas mixture[104] or during photolysis of CH_4 in the extreme ultra-violet.[105] On the other hand Strobel[106] concludes that CH_4 is irreversibly converted by photolysis into more complex hydrocarbons, which are transported into hotter regions of the atmosphere and there decompose, regenerating methane. Clearly further laboratory studies are needed.

The aeronomy of Jupiter's upper atmosphere has been reviewed by Hunten.[107] A general survey of the outer planets is given by Newburn and Gulkis.[3] The Lyman-α albedo of the planet has been measured,[32] and provides a datum which can be used to calculate an effective eddy diffusion coefficient from a model of the H_2 photochemistry. A temperature profile adopted for purposes of discussion in Hunten's review is shown in Fig. 8.14. In the absence of any discernible solid surface the zero of the altitude scale is placed at the tropopause, which is assigned a temperature of 95.5 K and a pressure of 25 mbar on the basis of theoretical models for the lower atmosphere. Even in the upper atmosphere, which in principle should be able to be understood completely in terms of processes which can be studied under ordinary laboratory conditions, there are very large uncertainties relating to the heat balance and to the photochemistry of methane and ammonia. Other interesting problems are associated with the satellites Io and Ganymede: Brown[108] has observed intense sodium D-line emission from Io; Pioneer 10 found Io to possess a surprisingly dense ionosphere; Ganymede

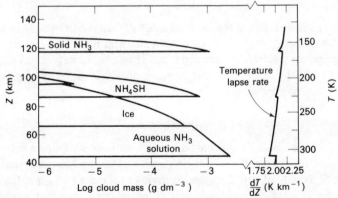

Fig. 8.13 Cloud composition, temperature, and temperature gradient on Jupiter as a function of altitude. Note the prominence of NH_4SH clouds. (From one of several model calculations by Lewis.[100])

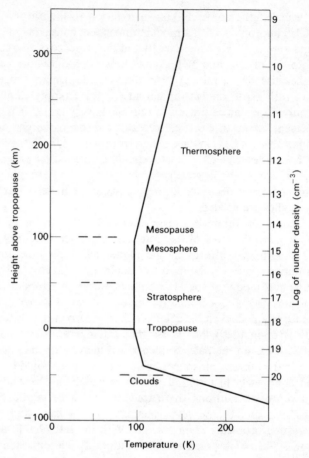

Fig. 8.14 Tentative temperature profile for the upper atmosphere of Jupiter. (From Hunten.[107])

probably resembles Titan, the large satellite of Saturn, in having an atmosphere based on methane. (Io is similar in size and mass to our own moon, which has no atmosphere; Ganymede and Titan are about twice as massive.) Pioneer 10, like the early Mars and Venus probes, has raised at least as many questions as it has answered.

8.4.2 SATURN, URANUS AND NEPTUNE

Apart from its uniquely spectacular ring system, Saturn bears a very strong resemblance to Jupiter. It is a gas giant, with a volume 763 times that of the earth, but a density of only 0.7 g cm^{-3}, giving a total mass 95 times that of the earth. There is a dense cloud layer, with coloured equatorial bands similar to, but less prominent than, those of Jupiter. The intensities of hydrogen quadrupole lines have been measured, the resulting abundance being

190 ± 40 km atm, and the rotational temperature 90 K.[109] Infrared brightness temperatures between 5 and 100 μm are in the range 90–120 K. The spectrum of methane in the near infrared is quite strong; however, the figure of 350 m atm for the abundance of methane given by Kuiper[110] in 1952 is considered by McElroy[85] to be probably too large. Because of the low temperature ammonia is expected to freeze out at a low level in the atmosphere, which would explain why its absorption spectrum has not been detected with certainty at the time of writing. Ethane has been detected as a trace constituent, and both methane and ethane are known to be present on the large satellite Titan.

Uranus and Neptune are smaller and also somewhat denser than Jupiter, but there is still a marked family resemblance. Uranus is described as pale green in colour, Neptune as being slightly bluish. Hunten[1] suggests that the colours may result from the H_2 quadrupole absorption. Banded surface markings have been reported for Uranus. Belton, McElroy and Price[111] have estimated an abundance of 1450 km atm for H_2 on Uranus from the quadrupole spectrum, with a rotational temperature of 118 ± 40 K, well above the equilibrium temperature of Table 8.1. The hydrogen quadrupole bands have not been observed for Neptune, but the presence of a large amount of hydrogen is demonstrated by the occurrence of bands arising from pressure induced dipole transitions. Methane is observed with an abundance of about 3.5 km atm on Uranus and 6 km atm on Neptune.[112] The greater abundance found for Neptune is thought to be due to deeper penetration of the radiation into the colder atmosphere. Helium is assumed to be present on all three planets but has not been demonstrated experimentally.

This concludes our account of the Jovian planets, and ends our discussion of the chemistry of the atmospheres of other planets. Many of the references in this last section are to papers presented at the Third Arizona Conference on Planetary Atmospheres; a useful resumé of this conference has been given by Goody.[113] Some of the results discussed here were presented, by various authors, at a meeting in Liége, Belgium, in August 1974,[114] and at the time of writing have still to appear in the literature. By this time it should be obvious to the reader that the investigation of planetary atmospheres constitutes a vast and intriguing field for future laboratory, theoretical, and observational studies.

References

1 Hunten, D. M., *Space Sci. Rev.*, **12**, 539 (1971).
2 Aumann, H. H., Gillespie, C. M. and Low, F. J., *Astrophys. J.*, **157**, L69 (1969).
3 Kaula, W. M., '*An Introduction to Planetary Physics*', John Wiley and Sons Inc., New York, 1968; Newburn, R. L., Jnr., and Gulkis, S., Space Sci. Rev., **3**, 179 (1973); Wilkins, G. A. and Sinclair A. T., *Proc. Roy. Soc.* (*London*) A**336**, 85 (1974).
4 Murray, B. C., Wildey, R. L. and Westphal, J. A., *J. geophys. Res.*, **68**, 4813 (1963).
5 Goldstein, R. M. and Rumsey, H., Jnr., *Science, N.Y.*, **169**, 974 (1970); Smith, W. B., *Science, N.Y.*, **169**, 1001 (1970).

6   Hunten, D. M., *Comments Astrophys. and Space Phys.*, **3**, 94 (1971); Snyder, C. W., *Icarus*, **15**, 555 (1971).

7   Hansen, J. E. and Arking, A., *Science, N.Y.*, **171**, 669 (1971).

8   Lewis, J. S., *Icarus*, **11**, 367 (1969); *Astrophys. J.*, **152**, L79 (1968).

9   Young, L. D. G., Schorn, R. A., Barker, E. S. and McFarlane, M., *Icarus*, **11**, 390 (1969).

10   Plummer, W. T., *Science, N.Y.*, **163**, 1191 (1969).

11   Sill, G. T., *Bull. Am. Astron. Soc.*, **5**, 299 (1973); Young, A. T. and Young, L. D. G., *Astrophys. J.*, **179**, L39 (1973); Pollack, J. B., *Bull. Am. Astron. Soc.*, **5**, 299 (1973); Prinn, R. G., *Science, N.Y.*, **182**, 1132 (1973); Young, A. T., *Science, N.Y.*, **183**, 407 (1974).

12   Avduevsky, V. S., Marov, M. Ya., Rozhdestvensky, M. K., Borodin, N. F., and Kerzhanovitch, V. V., *J. Atmos. Sci.*, **28**, 263 (1971); Avduevsky, V. S., Marov, M. Ya., and Rozhdestvensky, M. K., *Radio Sci.*, **5**, 333 (1970).

13   Fjeldbo, G., Kliore, A. J. and Eshleman, V. R., *Astr. J.*, **76**, 123 (1971).

14   Johnson, F. S., *J. Atmos. Sci.*, **25**, 661 (1968).

15   Dickinson, R. E., *J. Atmos. Sci.*, **28**, 885 (1971).

16   Kuzmin, A. D., *Radio Sci.*, **5**, 339 (1970); Vinogradov, A. P., Surkov, V. A. and Florensky, C. P., *J. Atmos. Sci.*, **25**, 535 (1968); Vinogradov, A. P., Surkov, V. A., Andreichikov, B. M., Kalinkina, O. M. and Grechischeva, I. M., International Astronomical Union Symposium No. 40, '*Planetary Atmospheres*', Eds. Sagan, C., Owen, T. and Smith, H. J., p. 3, D. Reidel Publishing Co., Dordrecht, Holland 1971.

17   Avduevsky, V. S., Marov, M. Ya and Rozhdestvensky, M. K., *J. Atmos. Sci.*, **27**, 561 (1970).

18   Belton, M. J. S., Broadfoot, A. L. and Hunten, D. M., *J. Atmos. Sci.*, **25**, 582 (1968).

19   Connes, P., Connes, J. and Maillard, J. P., '*Atlas des spectres infrarouges de Venus, Mars, Jupiter et Saturne*', Editions du Centre National de la Recherche Scientifique, Paris, 1969.

20   Connes, P., Connes, J., Benedict, W. S. and Kaplan, L. D., *Astrophys. J.*, **147**, 1230 (1967).

21   Lewis, J. S., *J. Atmos. Sci.*, **27**, 333 (1970).

22   Connes, P., Connes, J., Kaplan, L. D. and Benedict, W. S., *Astrophys. J.*, **152**, 731 (1968).

23   Belton, M. J. S. and Hunten, D. M., *Astrophys. J.*, **153**, 963 (1968) (Erratum: **156**, 797 (1969)).

24   Pollack, J. B. and Morrison, D., *Icarus*, **12**, 376 (1970).

25   Owen, T., *J. Atmos. Sci.*, **25**, 583 (1968).

26   Jenkins, E. B., Morton, D. C. and Sweigert, A. V., *Astrophys. J.*, **157**, 913 (1969).

27   Kliore, A., Levy, A. S., Cain, D. L., Fjeldbo, G. and Rasool, S. I., *Science, N.Y.*, **158**, 1683 (1967).

28   McElroy, M. B., *J. Atmos. Sci.*, **25**, 574 (1968).

29   Stewart, R. W., *J. Atmos. Sci.*, **25**, 578 (1968); **28**, 1069 (1971).

30   McElroy, M. B. and Strobel, D. F., *J. geophys. Res.*, **74**, 1118 (1969).

31   Barth, C. A., Wallace, L. and Pearce, J. B., *J. geophys. Res.*, **73**, 2541 (1968).

32   Moos, H. W., Fastie, W. G. and Bottema, M., *Astrophys. J.*, **155**, 887 (1969).

33   McElroy, M. B. and McConnell, J. C., *J. Atmos. Sci.*, **28**, 879 (1971).

34   Young, R. A., Black, G. and Slanger, T. G., *J. chem. Phys.*, **48**, 2067 (1968).

35   Mahan, B. H., *J. chem. Phys.*, **33**, 959 (1960).

36   Warneck, P., *Discuss. Faraday Soc.*, **37**, 57 (1964).

37   Katakis, D. and Taube, H., *J. chem. Phys.*, **36**, 416 (1962).

38   Moll, N. G., Clutter, D. R. and Thompson, W. E., *J. chem. Phys.*, **45**, 4469 (1966).

39   Weissberger, E., Breckenridge, W. H. and Taube, H., *J. chem. Phys.*, **47**, 1764 (1967).

40   Arvis, M., *J. chim. phys.*, **66**, 517 (1969).

41   McElroy, M. B. and Hunten, D. M., *J. geophys. Res.*, **75**, 1188 (1970).

42   Sehthi, D. H. and Taylor, H. A., *J. chem. Phys.*, **49**, 3669 (1968); Slanger, T. G. and Black, G., *J. chem. Phys.*, **54**, 1889 (1971).

43   Slanger, T. G., *J. chem. Phys.*, **49**, 3669 (1968); Ung, A. Y-M. and Schiff, H. I., *Can. J. Chem.*, **44**, 1981 (1966).

44 DeMore, W. B. and Dede, C., *J. chem. Phys.*, **74**, 2621 (1970).

45 Clark, I. D., *J. Atmos. Sci.*, **28**, 847 (1971).

46 Filseth, S. V., Stuhl, F. and Welge, K. H., *J. chem. Phys.*, **52**, 239 (1970).

47 Taylor, G. W. and Setser, D. W., *J. Am. chem. Soc.*, **93**, 4930 (1971); Wauchop T. S. and Broida, H. P., *J. chem. Phys.*, **56**, 330 (1972).

48 Inn, E. C. Y. and Heimerl, J. M., *J. Atmos. Sci.*, **28**, 839 (1971); DeMore, W. B. and Mosesman, M., *ibid*, **28**, 842 (1971).

49 Prinn, R. G., *J. Atmos. Sci.*, **28**, 1058 (1971).

50 Slanger, T. G. and Black, G., *J. chem. Phys.*, **53**, 3722 (1970).

51 Donahue, T. M., *J. Atmos. Sci.*, **25**, 568 (1968).

52 Reeves, R. R., Harteck, P., Thompson, B. A. and Waldron, R. W., *J. phys. Chem.*, **70**, 1637 (1966).

53 Porter, G. and Wright, F. J., *Discuss. Faraday Soc.*, **14**, 23 (1953).

54 Anderson, A. D., *Science, N.Y.*, **163**, 275 (1969).

55 Leighton, R. B., Horowitz, N. H., Murray, B. C., Sharp, R. P., Herriman, A. G., Young, A. T., Smith, B. A., Davies, M. R. and Leovy, C. B., *Science, N.Y.*, **165**, 684, 788 (1969); Masursky, H. *et al.*, *Science, N.Y.*, **175**, 294 (1972) (Mariner 9 issue).

56 Neugebauer, G., Munch, G., Chase, S. C., Jnr., Hatzenbeler, H. M., Miner, E. and Schofield, D., *Science, N.Y.*, **166**, 98 (1969).

57 Larson, H. and Fink, U., *Astrophys. J.*, **171**, L91 (1972).

58 Broida, H. P., Lundell, O. R., Schiff, H. I. and Ketcheson, R. D., *Science, N.Y.*, **170**, 1402 (1970); Barth, C. A. and Hord, C. W., *Science, N.Y.*, **173**, 197 (1971); Murray B. C. *et al.*, *Icarus*, **17**, 328 (1972); Miller, S. L. and Smythe, W. D., *Science, N.Y.*, **170**, 531 (1970).

59 Baum, W. A. and Martin, L. J., *Bull. Am. Astron. Soc.*, **5**, 296 (1973).

60 Herr, K. C. and Pimentel, G. C., *Science, N.Y.*, **167**, 47 (1970); Leovy, C. B., Smith, B. A., Young, A. T. and Leighton, R. B., *J. geophys. Res.*, **76**, 297 (1969); Leovy, C. B. *et al.*, *Icarus*, **17**, 373 (1972).

61 Plummer, W. T. and Carson, R. K., *Science, N.Y.*, **166**, 1141 (1969).

62 Wright, A. N., *Nature*, **215**, 953 (1967).

63 Kliore, A., Fjeldbo, G., Seidel, B. L. and Rasool, S. I., *Science, N.Y.*, **166**, 1393 (1969); Rasool, S. I., Horgan, J. S., Stewart, R. W. and Russell, L. H., *J. Atmos. Sci.*, **27**, 841 (1970); Rasool, S. I. and Stewart, R. W., *J. Atmos. Sci.*, **28**, 869 (1971).

64 Chase, S. C., Jnr., Hatzenbeler, H., Kieffer, H. H., Miner, E., Munch, G. and Neugebauer, G., *Science, N.Y.*, **175**, 308 (1972).

65 Kliore, A. J., Cain, D. L., Fjeldbo, G., Seidel, B. L. and Rasool, S. I., *Science, N.Y.*, **175**, 313 (1972).

66 Fjeldbo, G. and Eshleman, V. R., *Planet. Space Sci.*, **16**, 1035 (1968).

67 Dalgarno, A. and McElroy, M. B., *Science, N.Y.*, **170**, 167 (1970).

68 Kaplan, L. D., Connes, J. and Connes, P., *Astrophys. J.*, **157**, L187 (1969).

69 Owen, T. and Mason, H. P., *Science, N.Y.*, **165**, 893 (1969); Barker, E. S., Schorn, R. A., Woszczyk, A., Tull, R. G. and Little, S. J., *Science, N.Y.*, **170**, 1308 (1970).

70 Schorn, R. A., International Astronomical Union Symposium No. 40, '*Planetary Atmospheres*', Eds. Sagan, C., Owen, T. and Smith, H. J., p. 223, D. Reidel Publishing Co., Dordrecht, Holland, 1971.

71 Barth, C. A., Fastie, W. G., Hord, C. W., Peake, J. B., Kelly, K. K., Stewart, A. I., Thomas, G. E., Anderson, G. P., and Raper, O. F., *Science, N.Y.*, **165**, 1004 (1969).

72 Thomas, G. E., *J. Atmos. Sci.*, **28**, 859 (1971).

73 Anderson, D. E. and Hord, C. W., *J. geophys. Res.*, **76**, 6666 (1971).

74 Dalgarno, A., Degges, T. C. and Stewart, A. I., *Science, N.Y.*, **167**, 1490 (1970).

75 Barth, C. A., Stewart, A. I., Hord, C. W. and Lane, A. L., *Icarus*, **17**, 457 (1972).

76 Fehsenfeld, F. C., Dunkin, D. B. and Ferguson, E. E., *Planet. Space Sci.*, **18**, 1267 (1970); Fehsenfeld, F. C., Ferguson, E. E. and Schmeltekopf, A. L., *J. chem. Phys.*, **44**, 3022 (1966); Paulson, J. F., Mosher, R. L. and Dale, F., *J. chem. Phys.*, **44**, 3025 (1966).

77 McElroy, M. B. and Hunten, D. M., *J. geophys. Res.*, **74**, 5807 (1969); Hunten, D. M. and McElroy, M. B., *J. geophys. Res.* **75**, 5989 (1970).

78 McElroy, M. B., *Science, N.Y.*, **175**, 443 (1972).

79   McElroy, M. B. and Donahue, T. M., *Science, N.Y.*, **177**, 986 (1972); Parkinson, T. D. and Hunten, D. M., *J. Atmos. Sci.*, **29**, 1380 (1972).
80   Barth, C. A., Hord, C. W., Pearce, J. B., Kelly, K. K., Anderson, G. P. and Stewart, A. I., *J. geophys. Res.*, **76**, 2213 (1971).
81   Barth, C. A., *Ann. Rev. Earth and Planetary Sci.*, April 1974.
82   Wauchop, T. S. and Broida, H. P., *J. geophys. Res.*, **76**, 21 (1971); Ajello, J. M., *J. chem. Phys.*, **55**, 3169 (1971).
83   McElroy, M. B. and McConnell, J. C., *J. geophys. Res.*, **76**, 6674 (1971).
84   Westphal, J. A., *Astrophys. J.*, **157**, L63 (1969).
85   Owen, T., *Science, N.Y.*, **167**, 1675 (1970).
86   McElroy, M. B., *J. Atmos. Sci.*, **26**, 798 (1969).
87   Herzberg, G., *Astrophys. J.*, **87**, 428 (1938); '*The Atmospheres of the Earth and Planets*', Ed. Kuiper, G. P., chapter 13, University of Chicago Press, 1952.
88   Fink, U. and Belton, M. J. S., *J. Atmos. Sci.*, **26**, 952 (1969).
89   Gillett, F. C., Low, F. J. and Stein, W. A., *Astrophys. J.*, **157**, 925 (1969).
90   Belton, M. J. S., *Astrophys. J.*, **157**, 469 (1969).
91   Margolis, J. S. and Fox, K., *J. Atmos. Sci.*, **26**, 862 (1969).
92   Walker, M. F. and Hayes, S., *Publs. astr. Soc. Pacif.*, **79**, 464 (1967).
93   Owen, T. and Mason, H. P., *Astrophys. J.*, **154**, 317 (1968).
94   Anderson, R. C., Pipes, J. G., Broadfoot, A. L. and Wallace, L., *J. Atmos. Sci.*, **26**, 874 (1969); Hudson, R. D., *Rev. Geophys. and Space Phys.*, **9**, 305 (1971).
95   Baum, W. A. and Code, A. D., *Astr. J.*, **58**, 108 (1953).
96   Biswas, S. and Fichtel, C. E., *Space Sci. Rev.*, **4**, 709 (1965).
97   Spinrad, H. and Trafton, L., *Icarus*, **2**, 19 (1963).
98   Owen, T and Mason, H. P., *Astrophys. J.*, **154**, 317 (1968).
99   Peebles, P. J. E., *Astrophys. J.*, **140**, 328 (1964).
100   Lewis, J. S., *Icarus*, **10**, 365 (1969).
101   Lewis, J. S. and Prinn, R. G., *Science, N.Y.*, **169**, 472 (1970).
102   Sagan, C., *Astr. J.*, **65**, 499 (1960).
103   McNesby, J. R., *J. Atmos. Sci.*, **26**, 594 (1969).
104   Sagan, C. E., Lippincott, E. R., Dayhoff, M. O. and Eck, R. V., *Nature*, **213**, 273 (1967); Miller, S. L., *J. Am. chem. Soc.*, **77**, 2351 (1955).
105   Rebbert, R. E. and Ausloos, P., *J. Am. chem. Soc.*, **90**, 7370 (1968); Jensen, C. A. and Libby, W. F., *J. chem. Phys.*, **49**, 2831 (1968).
106   Strobel, D. F., *J. Atmos. Sci.*, **26**, 906 (1969).
107   Hunten, D. M., *J. Atmos. Sci.*, **26**, 826 (1969).
108   Brown, R., quoted by Metz, W. D., *Science, N.Y.*, **183**, 293 (1974). See also ref. 114.
109   Owen, T., *Icarus*, **10**, 355 (1969).
110   Kuiper, G. P., '*The Atmospheres of the Earth and Planets*', p. 306, University of Chicago Press, 1952.
111   Belton, M. J. S., McElroy, M. B. and Price, M. J., *Astrophys. J.*, **164**, 191 (1971).
112   Owen, T., *Icarus*, **6**, 108 (1967).
113   Goody, R., *J. Atmos. Sci.*, **26**, 997 (1969).
114   '*Physics and Chemistry of Atmospheres*', Ed. McCormac, B. M., Summer Advanced Study Institute, Liége, Belgium, 1974.

Appendix

Reaction rates of neutral species of atmospheric importance as recommended by the Climatic Impact Assessment Program of the U.S. Department of Transportation.[a]

In this table the units of the rate coefficient k are s^{-1} for first order, cm^3 molecule^{-1} s^{-1} for second order, and cm^6 molecule^{-2} s^{-1} for third order. The altitude, temperature and number density model adopted is

altitude (km)	temp (K)	$log[M]$ (molecule cm^{-3})
15	220	18.60
20	217	18.27
25	222	17.93
30	227	17.58
35	235	17.26
40	250	16.92
45	260	16.60

tions of neutral species in their ground electronic states

	Temp Range (K)	Rate Constant	Comments
$+ M \rightarrow O_2 + M$	$1000 < T < 8000$	$3.8 \times 10^{-30} T^{-1} \exp(-170/T)$	$M = O_2$, 4.8×10^{-33} at 300 K
$+ O_2 + M \rightarrow O_3 + M$	200–346	$6.6 \times 10^{-35} \exp(510/T)$	$M = Ar$, $Ar(1.0)$, $N_2(1.6)$, $O_2(1.7)$
$O_2 + M \rightarrow O_3^* + M$	300	5.4×10^{-34}	$M = O_2(O_2(1.0)N_2O(2.4))$
$+ O_3 \rightarrow O_2 + O_2$	220–1000	$1.9 \times 10^{-11} \exp(-2300/T)$	
$O \rightarrow NO_2 + h\nu$	300	4.2×10^{-18}	
$_2 \rightarrow OH + H$	400–2000	$3.0 \times 10^{-14} (T) \exp(-4480/T)$	
$NO + NO \rightarrow NO_2 + NO_2$	273–660	$3.3 \times 10^{-39} \exp(530/T)$	
$M \rightarrow O + O_2 + M$	200–1000	$1.65 \times 10^{-9} \exp(-11\,400/T)$	$O_3(1.0)$, $O_2(0.44)$, $N_2(0.39)$
$h\nu \rightarrow O(^3P) + O_2(^3\Sigma_g^-)$		$\varphi = 1, 450 < \lambda < 750$ nm $\varphi = 0, 250 < \lambda < 350$ nm	Chappuis bands
$h\nu \rightarrow O(^3P) + O_2(^1\Delta$ or $^1\Sigma)$		$\varphi = 1, 310 < \lambda < 350$ nm $\varphi = 0, \lambda < 310$ nm	Huggins bands
$h\nu \rightarrow O(^1D) + O_2(^1\Delta)$		$\varphi = 1, 250 < \lambda < 310$ nm $\varphi = 0, \lambda > 310$ nm	Hartley bands
$h\nu \rightarrow O(^1D) + O_2(^1\Sigma_g^+)$		$\varphi = 0, 250 < \lambda < 350$ nm	
$O(^3P) \rightarrow O_2 + O_2$	220–1000	$1.9 \times 10^{-11} \exp(-2300/T)$	
$_2 + M \rightarrow HO_2 + M$	203–404	$6.7 \times 10^{-33} \exp(290/T)$	$M = Ar$ or He, Relative M efficiencies $Ar(1.0)$, $He(1.0)$, $N_2(3.1)$, $O_2(3.1)$, $H_2O(25)$
$H + M \rightarrow H_2 + M$	300	8.3×10^{-33}	
$_3 \rightarrow OH + O_2$	300	2.6×10^{-11}	

ta have been taken from Chemical Kinetics Data Survey IV (NBSIR 73-203), Data Survey BSIR 73-206) and Data Survey VII (NBSIR 74-430) Interim Reports, Ed. D. Garvin, as anuary 1974, prepared for Climatic Impact Assessment Program.

Reactions of neutral species in their ground electronic states (*continued*)

	Temp Range (K)	Rate Constant	Comments
$OH + OH \rightarrow H + HO_2$	290–800	$2.0 \times 10^{-11} \exp(-20\,200/T)$	
$OH + OH \rightarrow H_2O + O$	300–2000	$1.0 \times 10^{-11} \exp(-550/T)$	
$OH + O \rightarrow H + O_2$	300–2000	$3.8(\pm 1.7) \times 10^{-11}$	
$OH + O_3 \rightarrow HO_2 + O_2$	220–450	$1.6 \times 10^{-12} \exp(-1000/T)$	
$OH + H \rightarrow H_2 + O$	400–2000	$1.4 \times 10^{-14} T \exp(-3500/T)$	
$OH + H_2 \rightarrow H_2O + H$	300–2500	$3.6 \times 10^{-11} \exp(-2590/T)$	
$OH + H + M \rightarrow H_2O + M$	1000–3000	$6.1 \times 10^{-26} T^{-2}$	$M = N_2$
$OH + OH + M \rightarrow H_2O_2 + M$	700–1500	$2.5 \times 10^{-33} \exp(2250/T)$	
$HO_2 + O_3 \rightarrow OH + 2O_2$	225–298	$1 \times 10^{-13} \exp(-1250/T)$	
$HO_2 + M \rightarrow H + O_2 + M$	300–2000	$3.5 \times 10^{-9} \exp(-23\,000/T)$	$M = Ar$
$HO_2 + H \rightarrow H_2 + O_2$	290–800	$4.2 \times 10^{-11} \exp(-350/T)$	
$HO_2 + H \rightarrow OH + OH$	290–800	$4.2 \times 10^{-10} \exp(-950/T)$	
$HO_2 + H_2 \rightarrow H_2O_2 + H$	300–800	$1.2 \times 10^{-12} \exp(-9400/T)$	
$HO_2 + OH \rightarrow H_2O + O_2$	220–300	$2 \times 10^{-11} < k < 2 \times 10^{-10}$	
$HO_2 + HO_2 \rightarrow H_2O_2 + O_2$	300–1000	$3 \times 10^{-11} \exp(-500/T)$	
$H_2O + H \rightarrow H_2 + OH$	300–2500	$1.5 \times 10^{-10} \exp(-10\,250/T)$	
$H_2O + O \rightarrow OH + OH$	300–2000	$1.1 \times 10^{-10} \exp(-9240/T)$	
$H_2O + HO_2 \rightarrow H_2O_2 + OH$	300–800	$4.7 \times 10^{-11} \exp(-16\,500/T)$	
$H_2O_2 + h\nu \rightarrow OH + OH$		*Absorption cross section*	

λ (nm)	σ (cm^2 molecule^{-1}), base e
190	80×10^{-20}
195	60×10^{-20}
200	54×10^{-20}
205	46×10^{-20}
210	40×10^{-20}
215	34×10^{-20}
220	28×10^{-20}
225	24×10^{-20}
254	7.4×10^{-20}

Quantum Yield

$$\varphi(-H_2O_2) = 1.0, \lambda > 200 \text{ nm}$$

	Temp Range (K)	Rate Constant	Comments
$H_2O_2 + O(^3P) \rightarrow HO_2 + OH$ (a)			
$ \rightarrow H_2O + O_2$ (b)	283–373	$k_{(a+b)} = 2.75 \times 10^{-12} \exp(-2125/T)$	
$H_2O_2 + H \rightarrow H_2 + HO_2$	300–800	$2.8 \times 10^{-12} \exp(-1900/T)$	
$H_2O_2 + H \rightarrow H_2O + OH$		no recommendation	
$H_2O_2 + OH \rightarrow H_2O + HO_2$	300–800	$1.7 \times 10^{-11} \exp(-910/T)$	
$N + O + M \rightarrow NO + M$	200–400	$1.8 \times 10^{-31} T^{-0.5}$	$M = N_2$
$N + O_2 \rightarrow NO + O$	300–3000	$1.1 \times 10^{-14} T \exp(-3150/T)$	
$N + O_3 \rightarrow NO + O_2$	300	5.7×10^{-13}	
$N + OH \rightarrow NO + H$	300	5.3×10^{-11}	
$N + N + M \rightarrow N_2 + M$	100–600	$8.3 \times 10^{-34} \exp(+500/T)$	$M = N_2$
			$\dfrac{-d[N]}{dt} = 2k[N]^2[M]$
$N_2 + M \rightarrow N + N + M$	6000–15\,000	$6.1 \times 10^{-3} T^{-1.6} \exp(-113\,200/T)$	$M = N_2$
$N_2 + O \rightarrow N + NO$	2000–5000	$1.3 \times 10^{-10} \exp(-38\,000/T)$	
$N_2 + O_2 \rightarrow N_2O + O$	1200–2000	$1.0 \times 10^{-10} \exp(-52\,200/T)$	
$N_2 + OH \rightarrow N_2O + H$	700–2500	$2.5 \times 10^{-12} \exp(-40\,400/T)$	
$NO + M \rightarrow N + O + M$	4200–6700	$6.6 \times 10^{-4} T^{-1.5} \exp(-75\,500/T)$	$M = Ar, O_2, N_2$ Insufficient data for reliable recommendatio… (use with caution)
$NO + O \rightarrow N + O_2$	1000–3000	$2.5 \times 10^{-15} T \exp(-19\,500/T)$	
$NO + O + M \rightarrow NO_2 + M$	200–500	$3.0 \times 10^{-33} \exp(940/T)$	$M = O_2$ Relative M efficienci… $O_2(1.0), Ar(1.0), N_2(1.4\ldots$
$NO + O + M \rightarrow NO_2 + M + h\nu$	300	7×10^{-32}	
$NO + O_3 \rightarrow NO_2 + O_2$	198–330	$9 \times 10^{-13} \exp(-1200/T)$	

.ctions of neutral species in their ground electronic states (*continued*)

	Temp Range (K)	Rate Constant	Comments
+H+M → HNO+M	200–400	$2.1 \times 10^{-32} \exp(300/T)$	M = H_2
+H_2 → HNO+H	2000	4×10^{-18}	Value based on reverse rate
+OH → NO_2+H	298–633	$2.8 \times 10^{-12} \exp(-15\,100/T)$	Value based on reverse rate
+HO_2 → NO_2+OH	300	2×10^{-13}	
+H_2O → HNO+OH	2000	3×10^{-18}	Value based on reverse rate
+H_2O_2 → OH+HNO_2	300	$<5 \times 10^{-20}$	
+N → N_2+O	300–5000	2.7×10^{-11}	
+NO → N+NO_2			Endothermic—unimportant compared to NO+NO → N_2O+O
+NO+O_2 → NO_2+NO_2	270–660	$3.3 \times 10^{-39} \exp(526/T)$	
+NO → N_2O+O	1200–2000	$2.2 \times 10^{-12} \exp(-32\,100/T)$	$-\dfrac{d[NO]}{dt} = 2k[NO]^2$
+NH_2 → N_2+H_2O*	300	8×10^{-12}	
+OH(+M) → HNO_2(+M)	300	2×10^{-12}	2nd order high pressure limit
+hv → NO($X^2\Pi$)+O(^3P)	λ (nm)	$\varphi(O(^3P))$	

λ (nm)	$\varphi(O(^3P))$
295–398	1.0
400	0.70
405	0.29
410	0.12
420	0.02
435	0.002
440	0.001
>440	0

	Temp Range (K)	Rate Constant	Comments
+M → NO+O+M	1400–2400	$1.8 \times 10^{-8} \exp(-33\,000/T)$	M = Ar
+O(^3P) → NO+O_2	230–550	9.1×10^{-12}	$k_r = k_f/K_{eq} = 2.8 \times 10^{-12} \times \exp(-23\,400/T)$
+O+M → NO_3+M	298	1.0×10^{-31}	M = N_2
+O_2 → NO+O_3	200–350	$2.8 \times 10^{-12} \exp(-25\,400/T)$	Value based on reverse rate
+O_3 → NO_3+O_2	220–340	$1.1 \times 10^{-13} \exp(-2450/T)$	
+N → all channels	300	1.85×10^{-11}	No reliable estimate can be made for the relative importance of → NO+NO, → N_2O+O, → N_2+O_2, → N_2+O+O.*
+NO_2 → NO+NO+O_2	600–2000	$3.3 \times 10^{-12} \exp(-13\,540/T)$	
+NO+O_2 → NO_2+NO_3	300–500	$8 \times 10^{-41} \exp(400/T)$	Value based on reverse rate
+H → OH+NO	300	4.8×10^{-11}	
+HO_2 → HNO_2+O_2	300	$\sim 3 \times 10^{-14}$	
+OH(+M) → HNO_3(+M)	220	3.2×10^{-12}	15 km, M = N_2
	222	1.6×10^{-12}	25 km
	235	5.5×10^{-13}	35 km
	250	2.7×10^{-13}	40 km
	260	1.4×10^{-13}	45 km
$_2$+NO+H_2O → 2HNO_2	300	$<1.1 \times 10^{-55}$	k defined by $\dfrac{-d[NO_2]}{dt} = k[NO][NO_2][H_2O]^2$ Main reaction probably heterogeneous
$_3$+hv → NO+O_2	300	$10^{-2}\ \text{s}^{-1}$ (daylight)	Strong absorption spectrum 600–700 nm
$_3$+M → NO_2+O+M	295	$\sim 8 \times 10^{-42}$	M = N_2 Value based on reverse rate
$_3$+O_2 → NO_2+O_3	300	7×10^{-34}	Value based on reverse rate
$_3$+NO → 2NO_2	300	8.7×10^{-12}	

We do not agree with this comment (M.J.M., L.F.P.).

Reactions of neutral species in their ground electronic states   (*continued*)

	Temp Range (K)	Rate Constant	Comments
$NO_3 + NO_2 \rightarrow NO_2 + O_2 + NO$	300–850	$2.3 \times 10^{-13} \exp(-1000/T)$	
$NO_3 + NO_2(+M) \rightarrow N_2O_5(+M)$	220	1.9×10^{-12}	15 km
	217	1.1×10^{-12}	20 km
	227	4.5×10^{-13}	30 km
	250	1.5×10^{-13}	40 km
	260	7.1×10^{-14}	45 km
$NO_3 + NO_3 \rightarrow 2NO_2 + O_2$	600–1100	$4.3 \times 10^{-12} \exp(-3850/T)$	$\dfrac{-d[NO_3]}{dt} = 2k[NO_3]^2$
	293–309	$5 \times 10^{-12} \exp(-3000/T)$	
$NO_3 + H_2O \rightarrow HNO_3 + OH$	300	2.3×10^{-26}	
$N_2O + O \rightarrow N_2 + O_2$	1200–2000	$1.7 \times 10^{-10} \exp(-14\,100/T)$	
$N_2O + O \rightarrow NO + NO$	1200–2000	$1.7 \times 10^{-10} \exp(-14\,100/T)$	
$N_2O + H \rightarrow N_2 + OH$	700–2500	$1.26 \times 10^{-10} \exp(-7600/T)$	
$NH_2 + H_2O \rightarrow NH_3 + OH$			No recommendation

$N_2O_5 + h\nu \rightarrow$ Products			
	λ (nm)	*Absorption cross section* $(cm^2\ molecule^{-1})$, *base e*	
	285	4.6×10^{-20}	
	300	2.3×10^{-20}	
	310	1.2×10^{-20}	
	320	0.69×10^{-20}	Reliability unknown
	330	0.39×10^{-20}	
	340	0.24×10^{-20}	
	360	0.095×10^{-20}	

$N_2O_5(+M) \rightarrow NO_2 + NO_3(+M)$	220	5.0×10^{-7}	15 km, k (s^{-1})
	217	1.8×10^{-7}	20 km
	227	5.5×10^{-7}	30 km
	250	1.6×10^{-5}	40 km
	260	4.4×10^{-5}	45 km
$N_2O_5 + H_2O \rightarrow 2HNO_3$	300	$<1 \times 10^{-20}$	
$NH_2 + O \rightarrow HNO + H(a)$ $\rightarrow OH + NH(b)$	300	3.5×10^{-12}	
$NH_2 + H_2 \rightarrow NH_3 + H$	800	$<10^{-16}$	approximate
$NH_2 + H + M \rightarrow NH_3 + M$	2000–3000	$1.3 \times 10^{-33} \exp(11\,200/T)$	M = Ar
$NH_2 + OH \rightarrow NH_3 + O$	300–1000	$1 \times 10^{-13} \exp(-2500/T)$	
$NH_3 + M \rightarrow NH_2 + H + M$	2000–3000	$1.5 \times 10^{-8} \exp(-42\,400/T)$	M = Ar P(M) < 4 atm
$NH_3 + O \rightarrow OH + NH_2$	300–1000	$2.5 \times 10^{-12} \exp(-3020/T)$	
$NH_3 + H \rightarrow NH_2 + H_2$	800	$<10^{-16}$	approximate
$NH_3 + OH \rightarrow NH_2 + H_2O$	298	$1.5 \times 10^{-13} – 4.2 \times 10^{-14}$	
$HNO + M \rightarrow H + NO + M$	230–700	$5 \times 10^{-8} \exp(-24\,500/T)$	M = H$_2$ Value based on reverse Estimated
$HNO + O_2 \rightarrow NO + HO_2$	300	$<2.1 \times 10^{-20}$	
$HNO + H \rightarrow H_2 + NO$	2000	7×10^{-12}	
	211–703	$>5 \times 10^{-14}$	
$HNO + OH \rightarrow H_2O + NO$	2000	6×10^{-11}	
$HNO + HNO \rightarrow H_2O + N_2O$	300	4×10^{-15}	
$HNO_2 + h\nu \rightarrow OH + NO$		6.45×10^{-4} s^{-1}(daylight)	
$HNO_2 + O \rightarrow OH + NO_2$			No data. Probably fas than $O + HNO_3$ since is 94 kJ mole^{-1} more e thermic
$HNO_2 + H \rightarrow$ Products			No data
$HNO_2 + OH \rightarrow H_2O + NO_2$	300	6.8×10^{-12}	Estimated, no data
$HNO_3 + h\nu \rightarrow OH + NO_2$	220	5.1×10^{-7}	15 km, k at noon (s^{-1})
	217	7.7×10^{-7}	20 km
	222	2.8×10^{-6}	25 km

actions of neutral species in their ground electronic states *(continued)*

	Temp Range (K)	Rate Constant	Comments
$O_3 + hv \rightarrow OH + NO_2$	227	1.4×10^{-5}	30 km
Contd.)	250	7.6×10^{-5}	40 km
	260	1.1×10^{-4}	45 km
$O_3 + O \rightarrow OH + NO_3$	300	$< 1.5 \times 10^{-14}$	
$O_3 + H \rightarrow$ Products	300	$< 1 \times 10^{-13}$	
$O_3 + OH \rightarrow H_2O + NO_3$	220–270	1.3×10^{-13}	
$+ O_2 \rightarrow SO_2 + O$	300	$< 8 \times 10^{-17}$	
	400–2500	$3.0 \times 10^{-13} \exp(-2800/T)$	
$+ SO \rightarrow SO_2 + S$	300	$< 3 \times 10^{-15}$	
	1000	$< 2 \times 10^{-13}$	
$+ O_3 \rightarrow O_2 + SO_2$	220–300	2.5×10^{-12}	
$_2 + O + M \rightarrow SO_3 + M$	250–1000	$1 \times 10^{-33} \exp(+500/T)$	$M = O_2$
$_2 + HO_2 \rightarrow SO_3 + OH$	300	9×10^{-16}	
$_2 + OH + M \rightarrow HSO_3 + M$	300	2×10^{-32}	$M = He$
$_3 + CH_3 + M) \rightarrow CH_3SO_2(+M)$	300	3×10^{-13}	
	296	$< 4 \times 10^{-25}$	
$+ OH \rightarrow CO_2 + H$	200–400	1.4×10^{-13}	
$+ HO_2 \rightarrow CO_2 + OH$	300	$< 10^{-19}$	
$_3 + O_2 \rightarrow CH_2O + OH$	295	3×10^{-16}	estimated
$_3 + NO_2 \rightarrow CH_3O + NO$	300–1400	3.3×10^{-11}	
$_3 + O_2 + M \rightarrow CH_3O_2 + M$	295	2.6×10^{-31}	$M = N_2$ low pressure limit
$_3 + O_2(+M) \rightarrow CH_3O_2$	295	4.3×10^{-13}	$M = N_2$, 2nd order high pressure limit
$_4 + O \rightarrow$ Products	350–1000	$3.5 \times 10^{-11} \exp(-4550/T)$	
$_4 + O_3 \rightarrow$ Products	310–340	$2.7 \times 10^{-13} \exp(-7700/T)$	
$_4 + OH \rightarrow CH_3 + H_2O$	240–370	$2.95 \times 10^{-12} \exp(-1770/T)$	
$O + O \rightarrow CO_2 + H\}$ $\rightarrow CO + OH\}$	297	2.1×10^{-10}	
$O + O_2 \rightarrow CO + HO_2$	300	5.7×10^{-12}	
$_2O + hv \rightarrow CHO + H$ (a) $\rightarrow CO + H_2$ (b)		$\varphi_a + \varphi_b = 1, 290 < \lambda < 360$ nm	

λ (nm)	φ_a	φ_b	Absorption cross section $(cm^2 \ molecule^{-1})$, base e	
290	0.81	0.19	31.8×10^{-21}	averaged for 10 nm bands
300	0.66	0.34	32.5×10^{-21}	
310	0.52	0.48	31.4×10^{-21}	
320	0.40	0.60	23.4×10^{-21}	
330	0.29	0.71	23.6×10^{-21}	
340	0.18	0.82	19.7×10^{-21}	
350	0.09	0.91	8.37×10^{-21}	
360	0.01	0.99	1.77×10^{-21}	

	Temp Range (K)	Rate Constant	Comments
$_2O + H \rightarrow H_2 + CHO$	297	5.4×10^{-14}	
$_2O + O \rightarrow CHO + OH$	300	1.6×10^{13}	
$_2O + OH \rightarrow CHO + H_2O$	300	1.4×10^{-11}	
$_2O + HO_2 \rightarrow CHO + H_2O_2$	200–1000	$1.7 \times 10^{-12} \exp(-4000/T)$	
$_3O + O_2 \rightarrow CH_2O + HO_2$ (1a)	298	$\sim 3 \times 10^{-18}$ $\sim 1.6 \times 10^{-13} \exp(-3300/T)$	
$_3O + NO \rightarrow CH_3ONO$ (2a)	298	$k_1/k_2 = 4.7 \times 10^{-5} \pm 20\%$ $k_{2(a,b)} \approx 8 \times 10^{-14}$	
$\rightarrow CH_2O + HNO$ (2b)	298	$k_{2b}/k_{2(a,b)} = 0.17$	
$_3O + NO_2 \rightarrow CH_3ONO_2$ (3a)	298	$k_2/k_3 = 1.2 \pm 0.1$ $k_{2a}/k_{3a} = 1.1$	
$\rightarrow CH_2O + HNO_2$ (3b)		$k_{3a}/k_3 = 0.9 \pm 0.1$ $k_{3b}/k_3 = 0.1 \pm 0.01$	
$_3O + CO \rightarrow$ Products (4)	298–423	$k_4/k_2 = 5 \times 10^{-4}$	
$_3O_2 + HO_2 \rightarrow CH_3OOH + O_2$	300	6.7×10^{-14}	estimate

Reactions of neutral species in their ground electronic states (*continued*)

	Temp Range (K)	Rate Constant	Comments
$CH_3O_2 + CH_3O_2$	300	6.8×10^{-14}	estimate
$\quad \to CH_3OOH + CH_2O_2$ (a)			
$\quad \to 2CH_3O + O_2$ (b)	300	6.8×10^{-14}	estimate, $k_a = k_b$
$CH_3O_2 + NO \to CH_3O_2NO$ (a)	298	$k_a/k = 0.6 \pm 0.1$	estimate, $k = k_a + k_b$
$\quad \to CH_2O + HONO$ (b)		$k_b/k = 0.4 \pm 0.1$	
$\quad \to CH_3O + NO_2$ (c)		$k_c/k < 0.02$	
$CH_3O_2 + NO_2 \to CH_3O_2NO_2$ (a)	298	$k_a/k = 0.75 \pm 0.05$	
$\quad \to CH_2O + HONO_2$ (b)		$k_b/k = 0.25 \pm 0.1$	
$\quad \to CH_3O + NO_3$ (c)		$k_c/k < 0.1$	
$CH_3ONO + h\nu \to CH_3O^* + NO$ (a)	298	$k_a/k = 0.76 \pm 0.02$	$\lambda = 366$ nm
$\quad \to$ Isomer (b)		$k_b/k = 0.24 \pm 0.04$	$\lambda = 366$ nm
$\quad \to CH_2O + HNO$ (c)		$(k_c + k_d)/k < 0.02$	$\lambda = 366$ nm
$\quad \to CH_2O + H + NO$ (d)			$k = k_a + k_b + k_c + k_d$
$C_2H_4 + O \to$ Products	200–500	$5.5 \times 10^{-12} \exp(-565/T)$	
$C_2H_4 + OH \to$ Products	298	3×10^{-12}	
$C_2H_4 + O \to$ Products	200–300	$6 \times 10^{-15} \exp(-2400/T)$	
$C_2H_4 + HO_2 \to$ Products	300	$\sim 1.7 \times 10^{-17}$	
$C_2H_6 + O \to$ Products	300–650	$4.1 \times 10^{-11} \exp(-3200/T)$	
$C_3H_6 + O \to$ Products	200–500	$4.1 \times 10^{-12} \exp(-38/T)$	
$C_3H_6 + OH \to$ Products	298	1.45×10^{-11}	
$C_3H_6 + O_3 \to C_3H_6O_3$	200–300	$7 \times 10^{-15} \exp(-1900/T)$	
$C_4H_{10} + OH \to$ Products	298	2.35×10^{-12}	

Reactions of neutral metastable species

	Temp Range (K)	Rate Constant	Comments
$O(^1D) + O_2 \rightarrow O_2(^1\Sigma_g^+) + O(^3P)$	298	7.4×10^{-11}	
$O(^1D) + O_3 \rightarrow O_2(^3\Sigma_u^-) + O_2(?)$ (a)	298	5.3×10^{-10}	$(k_a + k_b)$
$\rightarrow O_2 + 2O(^3P)$ (b)	298	$k_a/k_b \approx 1$	
$O(^1D) + CO \rightarrow CO + O(^3P)$	298	7.7×10^{-11}	
$O(^1D) + CO_2 \rightarrow CO_2 + O(^3P)$	298	1.8×10^{-10}	
$O(^1D) + N_2 \rightarrow N_2 + O(^3P)$	298	5.5×10^{-11}	
$O(^1D) + N_2 + M \rightarrow N_2O + M$	298	2.8×10^{-36}	
$O(^1D) + N_2O \rightarrow N_2 + O_2$ (a)	298	1.1×10^{-10}	
$\rightarrow NO + NO$ (b)	298	1.1×10^{-10}	
$O(^1D) + NO \rightarrow NO + O(^3P)$	298	1.7×10^{-10}	
$O(^1D) + NO_2 \rightarrow NO + O_2$	298	2.8×10^{-10}	
$O(^1D) + H_2 \rightarrow OH + H$	298	2.9×10^{-10}	
$O(^1D) + H_2O \rightarrow 2OH$	298	3.5×10^{-10}	
$O(^1D) + CH_4 \rightarrow CH_3 + OH$ (a)	298	4.0×10^{-10}	$(k_a + k_b)$
$\rightarrow CH_2O + H_2$ (b)		$k_a/k_b = 10$	
$O(^1D) + C_2H_6 \rightarrow C_2H_5 + OH$ (a)⎱	298	4.8×10^{-10}	$(k_a + k_b)$
$\rightarrow CH_3 + CH_2O$ (b)⎰			
$O(^1D) + NH_3 \rightarrow NH_2 + OH$	298	$\sim 3 \times 10^{-10}$	estimate
$O(^1D) + H_2O_2 \rightarrow OH + HO_2$	298	$> 3 \times 10^{-10}$	estimate
$O(^1S) + O(^3P) \rightarrow$ Products	300	7.5×10^{-12}	
$O(^1S) + O_2 \rightarrow$ Products	200–377	$4.3 \times 10^{-12} \exp(-850/T)$	
$O(^1S) + O_3 \rightarrow$ Products	300	5.8×10^{-10}	
$O(^1S) + CO_2 \rightarrow$ Products	200–450	$3.1 \times 10^{-11} \exp(-1320/T)$	
$O(^1S) + N_2 \rightarrow$ Products	200–380	$< 5 \times 10^{-17}$	
$O(^1S) + N_2O \rightarrow$ Products	300	1.4×10^{-11}	
$O(^1S) + NH_3 \rightarrow$ Products	300	5×10^{-10}	
$O(^1S) + NO \rightarrow$ Products	200–291	$3.2 \times 10^{-11} (T)^{0.5}$	
$O(^1S) + NO_2 \rightarrow$ Products	300	5×10^{-10}	
$O(^1S) + H_2O \rightarrow$ Products	300	$> 10^{-10}$	
$O(^1S) + CH_4 \rightarrow$ Products	300	2×10^{-14}	
$O_2(^1\Delta) + M \rightarrow O_2 + M$	285–322	$2.2 \times 10^{-18} (T/300)^{0.8}$	$M = O_2$
	300	$< 2 \times 10^{-15}$	$M = SO_2$
	300	$< 3 \times 10^{-16}$	$M = CO$
	300	$< 2 \times 10^{-20}$	$M = N_2$
$O_2(^1\Delta) + O_3 \rightarrow 2O_2 + O$	283–321	$4.5 \times 10^{-11} \exp(-2830/T)$	
$O_2(^1\Delta) + SO \rightarrow O_2 + SO(^1\Delta)$	300	$3.5(\pm 0.36) \times 10^{-13}$	
$O_2(^1\Sigma) + M \rightarrow O_2 + M$	300	1.5×10^{-16}	$M = O_2$
		2.0×10^{-15}	$M = N_2$
		4×10^{-12}	$M = H_2O$

Reactions of vibrationally excited species

	Temp Range (K)	Rate Constant	Comments
$N_2(v = 1) + M \rightarrow N_2(v = 0) + M$	1000–5000	$8.53 \times 10^{-7} \exp(-273.10/T^{1/3})$	$M = N_2, O_2, CO$
$N_2(v = 1) + H_2O(000) \rightarrow N_2(v = 0) + H_2O(010)$	200–2000	$3.48 \times 10^{-9} \exp(-95.94/T^{1/3})$	
$N_2(v = 1) + O_2(v = 0) \rightarrow N_2(v = 0) + O_2(v = 1)$	200–5000	$1.74 \times 10^{-10} \exp(-124.00/T^{1/3})$	
$N_2(v = 1) + CO(v = 0) \rightarrow N_2(v = 0) + CO(v = 1)$	200–2000	$1.78 \times 10^{-6} \exp(-209.90/T^{1/3})$ $+ 6.98 \times 10^{-13} \exp(-25.60/T^{1/3})$	
$N_2(v = 1) + O \rightarrow N_2(v = 0) + O$	200–3000	$1.07 \times 10^{-10} \exp(-69.9/T^{1/3})$	
$O_2(v = 1) + M \rightarrow O_2(v = 0) + M$	200–5000	$4.81 \times 10^{-8} \exp(-169.60/T^{1/3})$	$M = N_2, O_2, CO$
		$< 3.60 \times 10^{-10} \exp(-60.69/T^{1/3})$	$M = H_2O$
$O_2(v = 1) + O \rightarrow O_2(v = 0) + O$	200–2000	$6.88 \times 10^{-9} \exp(-76.75/T^{1/3})$	
$OH(v = 1) + M \rightarrow OH(v = 0) + M$	300	1×10^{-15}	$M = O_2$
		3.6×10^{-16}	$M = N_2$
		1.5×10^{-14}	$M = NO$
		4.8×10^{-15}	$M = N_2O$
		2.4×10^{-15}	$M = CO_2$
		2.0×10^{-14}	$M = H_2O$
$CO(v = 1) + M \rightarrow CO(v = 0) + M$	200–5000	$6.67 \times 10^{-8} \exp(-208.30/T^{1/3})$	$M = CO, N_2, O$
	1000–3000	$3.12 \times 10^{-10} \exp(-64.99/T^{1/3})$	$M = H_2O$
$CO(v = 1) + O_2(v = 0) \rightarrow CO(v = 0) + O_2(v = 1)$	1000–3000	$3.5 \times 10^{-10} \exp(-124.00/T^{1/3})$	
$CO(v = 1) + O \rightarrow CO(v = 0) + O$	200–3000	$9.9 \times 10^{-8} \exp(-118.1/T^{1/3})$	
$CO_2(010) + M \rightarrow CO_2(000) + M$	200–2000	$4.64 \times 10^{-10} \exp(-76.7/T^{1/3})$	$M = CO_2$
		$6.69 \times 10^{-10} \exp(-84.07/T^{1/3})$	$M = N_2, O_2$
	200–700	$3.22 \times 10^{-13} \exp(22.91/T^{1/3})$	$M = H_2O$
$CO_2(001) + N_2(v = 0) \rightarrow CO_2(000) + N_2(v = 1)$	200–2000	$1.71 \times 10^{-6} \exp(-175.30/T^{1/3})$ $+ 6.07 \times 10^{-14} \exp(15.27/T^{1/3})$	
$CO_2(001) + CO(v = 0) \rightarrow CO_2(000) + CO(v = 1)$	1000–3000	$1.56 \times 10^{-11} \exp(-30.12/T^{1/3})$	
$CO_2(001) + M \rightarrow CO_2(030) + M$	200–2000	$1.0 \times 10^{-15} + 5.2 \times 10^{-11}$ $\times \exp(-76.75/T^{1/3})$	$M = N_2, O_2$
		$3.0 \times 10^{-15} + 1.72 \times 10^{-10}$ $\times \exp(-76.75/T^{1/3})$	$M = CO_2$
	200–1000	4.0×10^{-13}	$M = H_2O$
$CO_2(100) + M \rightarrow CO_2(020) + M$	200–400	$\geqslant 3 \times 10^{-11}$	$M = N_2, O_2,$
$H_2O(010) + M \rightarrow H_2O(000) + M$	200–600	$5.37 \times 10^{-10} \exp(-70.00/T^{1/3})$	$M = N_2, CO_2$
$H_2O(010) + O_2(v = 0) \rightarrow H_2O(000) + O_2(v = 1)$	200–400	1.0×10^{-12}	
$O_3(001) + NO \rightarrow NO_2 + O_2$	300	2.2×10^{-13}	Rate ~ 20 times faster with $O_3(000)$

nary positive-ion reactions

eaction	k (cm^3 molecule^{-1} sec^{-1})	*Uncertainty*	*Remarks*
$^+ + O \rightarrow O^+ + H$	$3.8(-10)^*$	$\pm 50\%$	
$^+ + Fe \rightarrow Fe^+ + H$	$7.4(-9)$		
$^+ + NO \rightarrow NO^+ + H$	$1.9(-9)$	$\pm 30\%$	
$^+ + CO_2 \rightarrow COH^+ + O$	$3.0(-9)$	$\pm 30\%$	
$e^+ + H_2 \rightarrow$ products	$<1(-13)$		
$e^+ + N_2 \rightarrow N^+ + N + He$	$1.2(-9)$	$\pm 30\%$	$k_a/k_b = 1.5$
$\rightarrow N_2^+ + He$			
	$1.0(-9)$	$+0.3, -0.2$	$k_a/k_b = 1.2$
	$1.85(-9)$	$\pm 15\%$	
	$1.2(-9)$	$\pm 20\%$	
	$1.5(-9)$	$\pm 30\%$	$k_a/k_b = 1.1$
			$k_a/k_b = 2.2$
	$1.45(-9)$	$\pm 15\%$	
$e^+ + O_2 \rightarrow O^+ + O + He$	$1.0(-9)$	$\pm 30\%$	$k_a/k_b = 1.6$
$\rightarrow O_2^+ + He$			
	$8.5(-10)$	$+2.5; -2.0$	$k_a/k_b = 4$
	$1.5(-9)$	$\pm 25\%$	
	$1.2(-9)$	$\pm 30\%$	
	$1.10(-9)$	$\pm 15\%$.	
$e^+ + CO \rightarrow C^+ + O + He$	$1.7(-9)$	$\pm 30\%$	
	$2.0(-9)$	$\pm 15\%$	
	$1.6(-9)$	$\pm 30\%$	
$e^+ + NO \rightarrow N^+ + O + He$	$1.7(-9)$	$\pm 30\%$	
	$2.1(-9)$	$\pm 30\%$	
	$2.0(-9)$	$\pm 15\%$	
$e^+ + H_2O \rightarrow$ products	$5.6(-10)$	± 0.5	
	$4.5(-10)$	$\pm 30\%$	
$e^+ + CO_2 \rightarrow O^+ + CO + He$	$1.2(-9)$	$\pm 30\%$	
$\rightarrow CO^+ + O + He$			
	$1.6(-9)$	$\pm 15\%$	
$e^+ + CH_4 \rightarrow$ products	$1.5(-9)$	$\pm 15\%$	
$e^+ + C_2H_6 \rightarrow$ products	$2.3(-9)$	$\pm 15\%$	
$^+ + O_2 \rightarrow CO^+ + O$	$1.1(-9)$	$\pm 30\%$	
	$9.0(-10)$	$\pm 30\%$	
$^+ + CO_2 \rightarrow CO^+ + CO$	$1.9(-9)$	$\pm 30\%$	
	$1.6(-9)$	$\pm 30\%$	
$^+ + H_2O \rightarrow COH^+ + H$	$2.0(-9)$	$\pm 30\%$	
$^+ + Na \rightarrow Na^+ + N$	very small		
$^+ + Mg \rightarrow Mg^+ + N$	$1.2(-9)$		
$^+ + Ca \rightarrow Ca^+ + N$	$1.1(-9)$		
$^+ + Fe \rightarrow Fe^+ + N$	$1.5(-9)$		
$^+ + H_2 \rightarrow NH^+ + H$	$7(-10)$	$\pm 30\%$	
$^+ + CO \rightarrow CO^+ + N$	$5(-10)$	$\pm 30\%$	
$^+ + NO \rightarrow NO^+ + N$	$8(-10)$	$\pm 30\%$	
$^+ + O_2 \rightarrow O_2^+ + N$	$7(-10)$	$\pm 30\%$	
$\rightarrow NO^+ + O$			
	$5(-10)$	$\pm 30\%$	
	$6(-10)$	$\pm 30\%$	
	$6.1(-10)$	$\pm 30\%$	$k_a/k_b = 2.8$
			$k_a/k_b \sim 1$
			$k_a/k_b = 1.3$
	$7(-10)$		
$N^+ + H_2O \rightarrow H_2O^+ + N$	$2.6(-9)$	± 0.4	
	$2.6(-9)$	$\pm 30\%$	
$N^+ + CO_2 \rightarrow CO_2^+ + N$	$1.3(-9)$	$\pm 30\%$	
$)^+ + H \rightarrow H^+ + O$	$6.8(-10)$	$\pm 50\%$	
$)^+ + Na \rightarrow Na^+ + O$	small		
$)^+ + Mg \rightarrow Mg^+ + O$	small		
$)^+ + Ca \rightarrow Ca^+ + O$	$7.6(-10)$		
$)^+ + Fe \rightarrow Fe^+ + O$	$2.9(-9)$		
$)^+ + H_2 \rightarrow OH^+ + H$	$2.0(-9)$	$\pm 30\%$	

* $3.8(-10) = 3.8 \times 10^{-10}$.

Binary positive-ion reactions (*continued*)

Reaction	k (cm^3 molecule^{-1} sec^{-1})	Uncertainty	Remarks
$O^+ + N_2 \rightarrow NO^+ + N$	$1.3(-12)$	$\pm 15\%$	
	$1.2(-12)$	$\pm 10\%$	
$O^+(^2D) + N_2 \rightarrow N_2^+ + O$	$\sim 1(-9)$		
$O^+ + NO \rightarrow NO^+ + O$	$< 1(-12)$		
$O^+ + O_2 \rightarrow O_2^+ + O$	$2.0(-11)$	$\pm 15\%$	
	$2.0(-11)$	$+0.4, -0.3$	
	$2.0(-11)$	± 0.5	
$O^+ + CO_2 \rightarrow O_2^+ + CO$	$1.2(-9)$	$\pm 30\%$	
	$1.0(-9)$	$\pm 30\%$	
$O^+ + H_2O \rightarrow H_2O^+ + O$	$2.3(-9)$	± 0.25	
	$2.4(-9)$	$\pm 30\%$	
$O^+ + N_2O \rightarrow N_2O^+ + O$	$2.2(-10)$	± 1.3	
$\rightarrow NO^+ + NO$	$2.3(-10)$	± 1.4	
	$k_a + k_b = 6.3(-10)$	$\pm 30\%$	
$O^+ + NO_2 \rightarrow NO_2^+ + O$	$1.6(-9)$	$\pm 30\%$	
$Na^+ + O_3 \rightarrow NaO^+ + O_2$	$< 1(-11)$		Probably endother
$Mg^+ + O_3 \rightarrow MgO^+ + O_2$	$2.3(-10)$	$\pm 50\%$	
$S^+ + NO \rightarrow NO^+ + S$	$4.2(-10)$	$\pm 20\%$	
$S^+ + O_2 \rightarrow SO^+ + O$	$1.6(-11)$	$\pm 20\%$	
$S^+ + CO_2 \rightarrow products$	$< 1(-12)$		
$Ar^+ + H_2 \rightarrow ArH^+ + H$	$6.8(-10)$	$\pm 20\%$	
	$7.0(-10)$	$\pm 30\%$	
$Ar^+ + N_2 \rightarrow N_2^+ + Ar$	$6.6(-11)$	?	
	$5(-11)$	$\pm 75\%$	
$Ar^+ + CO \rightarrow CO^+ + Ar$	$1.2(-10)$	$\pm 30\%$	
	$9.0(-11)$	$\pm 30\%$	
	$5(-11)$	$\pm 90\%$	
$Ar^+ + NO \rightarrow NO^+ + Ar$	$3.9(-10)$	$\pm 30\%$	
	$2.5(-10)$	$\pm 30\%$	
$Ar^+ + O_2 \rightarrow O_2^+ + Ar$	$1.1(-10)$	$\pm 30\%$	
	$5.0(-11)$	$\pm 30\%$	
	$7(-11)$	$\pm 50\%$	
	$5.0(-11)$	$\pm 25\%$	
$Ar^+ + H_2O \rightarrow H_2O^+ + Ar$			
$\left.\rightarrow ArH^+ + OH \right\}$	$1.4(-9)$	± 0.1	
	$1.6(-9)$	$\pm 30\%$	
$Ar^+ + CO_2 \rightarrow CO_2^+ + Ar$	$7.0(-10)$	$\pm 30\%$	
	$7.6(-10)$	$\pm 30\%$	
	$4.6(-10)$	$\pm 15\%$	
$Ar^+ + CH_4 \rightarrow CH_3^+ + H + Ar$	$6.5(-10)$	$\pm 30\%$	
$\rightarrow CH_2^+ + H_2 + Ar$	$1.4(-10)$	$\pm 30\%$	
$\rightarrow products$	$k_a + k_b = 9.0(-10)$	$\pm 30\%$	
	$1.3(-9)$	$\pm 15\%$	
$Ar^+ + C_2H_6 \rightarrow products$	$1.1(-9)$	$\pm 15\%$	
$K^+ + O_3 \rightarrow KO^+ + O_2$	$< 1(-11)$		Probably endother
$Cu^+ + O_3 \rightarrow CuO^+ + O_2$	$1.6(-10)$	$\pm 50\%$	
$Fe^+ + O_3 \rightarrow FeO^+ + O_2$	$1.5(-10)$	$+50\%$	
$H_2^+ + H_2 \rightarrow H_3^+ + H$	$2.0(-9)$	$+10\%$	
	$1.85(-9)$		
	$1.95(-9)$	$\pm 20\%$	
$H_2^+ + N_2 \rightarrow N_2H^+ + H$	$1.95(-9)$	$\pm 20\%$	
$H_2^+ + Ar \rightarrow ArH^+ + H$	$1.2(-9)$	$\pm 20\%$	
$N_2^+ + N \rightarrow N^+ + N_2$	$< 1(-11)$		
$N_2^+ + O \rightarrow NO^+ + N$	$1.4(-10)$	$\times 2$	
$\rightarrow O^+ + N_2$	$< 1(-11)$		
$N_2^+ + H_2 \rightarrow N_2H^+ + H$	$1.7(-9)$	$\pm 30\%$	
	$1.4(-9)$	$\pm 20\%$	
$N_2^+ + O_2 \rightarrow O_2^+ + N_2$	$6.6(-11)$	$\pm 30\%$	
	$6(-11)$		
	$5(-11)$	$\pm 30\%$	
	$6.5(-11)$	± 1	

inary positive-ion reactions (*continued*)

eaction	k(cm³ molecule⁻¹ sec⁻¹)	Uncertainty	Remarks
$_2{}^+ + O_2 \to O_2{}^+ + N_2$ (*Contd.*)			
$\quad\quad \to NO^+ + NO$	$<3(-14)$		
$_2{}^+ + CO \to CO^+ + N_2$	$7(-11)$	$\pm 30\%$	
$_2{}^+ + NO \to NO^+ + N_2$	$3.3(-10)$	$\pm 30\%$	
	$4.8(-10)$	$\pm 30\%$	
$_2{}^+ + H_2O \to H_2O^+ + N_2$ }	$2.2(-9)$	± 0.3	
$\quad\quad \to N_2H^+ + OH$ }	$2.0(-9)$	$\pm 30\%$	
$_2{}^+ + CO_2 \to CO_2{}^+ + N_2$	$9(-10)$	$\pm 30\%$	
$_2{}^+ + Na \to Na^+ + N_2$	$5.8(-10)$	$\pm 50\%$	
	$1.9(-9)$		
$_2{}^+ + Mg \to Mg^+ + N_2$	$7.2(-10)$		
$_2{}^+ + Ca \to Ca^+ + N_2$	$1.8(-9)$		
$_2{}^+ + Fe \to Fe^+ + N_2$	$4.3(-10)$		
$_2{}^+ + N \to NO^+ + O$	$1.8(-10)$	$\times 2$	
$_2{}^+ + N_2 \to NO^+ + NO$	$<1(-15)$		
$_2{}^+ + NO \to NO^+ + O_2$	$6.3(-10)$	$\pm 30\%$	
	$7.7(-10)$	$\pm 30\%$	
	$7.2(-10)$	± 1.5	
$_2{}^+ + NO_2 \to NO_2{}^+ + O_2$	$6.6(-10)$	$\pm 30\%$	
$_2{}^+ + NH_3 \to NH_3{}^+ + O_2$	$2.4(-9)$	$\pm 30\%$	
$_2{}^+ + Na \to Na^+ + O_2$	$6.7(-10)$	$\pm 50\%$	
	$1.4(-9)$		
$_2{}^+(a^4\Pi_u) + Na \to Na^+ + O_2$	$2.0(-9)$		
$_2{}^+ + Na \to NaO^+ + O$	$1.2(-10)$		
	$<7(-11)$		
$_2{}^+ + Mg \to Mg^+ + O_2$	$1.2(-9)$		
$_2{}^+(a^4\Pi_u) + Mg \to Mg^+ + O_2$	$>3(-9)$		
$_2{}^+ + Ca \to Ca^+ + O_2$	$1.8(-9)$		
$_2{}^+(a^4\Pi_u) + Ca \to Ca^+ + O_2$	$3.5(-9)$		
$_2{}^+ + Fe \to Fe^+ + O_2$	$1.1(-9)$		
$O^+ + Na \to Na^+ + NO$	$7.7(-11)$	$\pm 50\%$	
$O^+ + Mg \to Mg^+ + NO$	$8.1(-10)$		
$O^+ + Ca \to Ca^+ + NO$	$4.0(-9)$		
$O^+ + Fe \to Fe^+ + NO$	$9.2(-10)$		
$O^+ + O_3 \to NO_2{}^+ + O_2$	$<1(-14)$		
$O^+ + O \to O^+ + CO$	$1.4(-10)$	$\pm 50\%$	
$O^+ + N \to$ products	$<2(-11)$		
$O^+ + NO \to NO^+ + CO$	$3.3(-10)$	$\pm 30\%$	
$O^+ + H_2 \to COH^+ + H$	$2.0(-9)$	$\pm 30\%$	
$O^+ + O_2 \to O_2{}^+ + CO$	$2.0(-10)$	$\pm 30\%$	
$O^+ + CO_2 \to CO_2{}^+ + CO$	$1.1(-9)$	$\pm 30\%$	
$O^+ + H_2O \to$ products	$2.2(-9)$	$\pm 30\%$	
$iO^+ + O \to Si^+ + O_2$	$\sim 2(-10)$		
$iO^+ + N \to Si^+ + NO$	$\sim 2(-10)$		
$\quad\quad \to NO^+ + Si$	$\sim 1(-10)$		
$MgO^+ + O \to Mg^+ + O_2$	$\sim 1(-10)$		
$O^+ + CO \to S^+ + CO_2$	$<1(-12)$		
$_3{}^+ + N_2 \to N_2H^+ + H_2$	$1.5(-9)$	$\pm 30\%$	
	$1.0(-9)$	$\pm 30\%$	
$_3{}^+ + CO \to COH^+ + H_2$	$1.4(-9)$	$\pm 30\%$	
$_3{}^+ + NO \to NOH^+ + H_2$	$1.4(-9)$	$\pm 30\%$	
$_3{}^+ + CO_2 \to CO_2H^+ + H_2$	$1.9(-9)$	$\pm 30\%$	
$_3{}^+ + NO_2 \to NO^+ + OH + H_2$	$7(-10)$	$\pm 30\%$	
$_3{}^+ + N_2O \to N_2OH^+ + H_2$	$1.8(-9)$	$\pm 30\%$	
$_3{}^+ + CH_4 \to CH_5{}^+ + H_2$	$1.6(-9)$	$\pm 30\%$	
	$7.5(-10)$	$\pm 30\%$	
$_3{}^+ + C_2H_4 \to C_2H_5{}^+ + H_2$	$1.9(-9)$	$\pm 30\%$	
$\quad\quad \to C_2H_3{}^+ + 2H_2$	$1.2(-9)$	$\pm 30\%$	
$_3{}^+ + C_2H_2 \to C_2H_3{}^+ + H_2$	$1.9(-10)$	$\pm 30\%$	
$_3{}^+ + H_2O \to H_3O^+ + H_2$	$\sim 3.0(-9)$		
$_3{}^+ + NH_3 \to NH_4{}^+ + H_2$	$\sim 3.6(-9)$		

Binary positive-ion reactions (*continued*)

Reaction	k (cm^3 molecule^{-1} sec^{-1})	Uncertainty	Remarks
$N_2H^+ + CO_2 \rightarrow CO_2H^+ + N_2$	9.2(−10)	±30%	
$N_2H^+ + N_2O \rightarrow N_2OH^+ + N_2$	7.9(−10)	±30%	
$N_2H^+ + CH_4 \rightarrow CH_5^+ + N_2$	8.9(−10)	±30%	
$N_2H^+ + H_2O \rightarrow H_3O^+ + N_2$	~5(−10)		
$H_2O^+ + H_2O \rightarrow H_3O^+ + OH$	1.7(−9)	±30%	
	1.8(−9)	±30%	
	1.6(−9)	±20%	
$H_2O^+ + Na \rightarrow Na^+ + H_2O$	1.9(−9)		
$H_2O^+ + Ca \rightarrow Ca^+ + H_2O$	4.0(−9)		
$H_2O^+ + Fe \rightarrow Fe^+ + H_2O$	1.5(−9)		
$CO_2^+ + H \rightarrow HCO^+ + O$ ⎫ 　　　　$\rightarrow H^+ + CO_2$ ⎬	6(−10)	±50%	$k_a/k_b \sim 5$
$CO_2^+ + O \rightarrow O_2^+ + CO$ ⎫ 　　　　$\rightarrow O^+ + CO_2$ ⎬	2.6(−10)	×2	$k_a/k_b \sim 1.7$
$CO_2^+ + N \rightarrow$ products	<1(−11)		
$CO_2^+ + H_2 \rightarrow CO_2H^+ + H$	1.4(−9)	±30%	
$CO_2^+ + NO \rightarrow NO^+ + CO_2$	1.2(−10)	±30%	
$CO_2^+ + O_2 \rightarrow O_2^+ + CO_2$	5.0(−11)	±30%	
$NO_2^+ + NO \rightarrow NO^+ + NO_2$	2.9(−10)	±30%	
$SO_2^+ + CO \rightarrow SO^+ + CO_2$	3.0(−10)	±20%	
$SO_2^+ + O_2 \rightarrow O_2^+ + SO_2$	2.8(−10)	±20%	
$H_3O^+ + Ca \rightarrow Ca^+ + H_2O + H$	4.4(−9)		
$N_4^+ + O_2 \rightarrow O_2^+ + 2N_2$	4(−10)	±30%	
$O_4^+ + O \rightarrow O_2^+ + O_3$	3(−10)	±2	
$O_4^+ + H_2O \rightarrow O_2^+.H_2O + O_2$	1.5(−9)	±0.5	
	1.3(−9)	±50%	
	2.2(−9)	±50%	
$O_2^+.N_2 + H_2O \rightarrow O_2^+.H_2O + N_2$	4(−9)	±2	
$O_2^+.N_2 + O_2 \rightarrow O_4^+ + N_2$	>5(−11)		80 K
$H_3O^+.OH + H_2O \rightarrow H_3O^+.H_2O + OH$	1.4(−9)	±0.5	
	>1(−9)		
	~3.2(−9)		
$O_2^+.H_2O + H_2O \rightarrow H_3O^+.OH + O_2$	1.0(−9)	±0.4	
	9(−10)	±50%	
	1.9(−9)	±50%	
$\rightarrow H_3O^+ + OH + O_2$	2(−10)	±1	
	3(−10)	±1	
	⩽3(−10)		
$NO^+.CO_2 + H_2O \rightarrow NO^+.H_2O + CO_2$	~1(−9)		
$NO^+(H_2O)_3 + H_2O \rightarrow H_3O^+(H_2O)_2 + HNO_2$	7(−11)	±2	
	8(−11)	±50%	
	7(−11)	±30%	
$NO^+.NO + H_2O \rightarrow NO^+.H_2O + NO$	1.4(−9)	±0.3	
$NO^+.NO + NH_3 \rightarrow NO^+.NH_3 + NO$	1.3(−9)	±0.2	
$NO^+.H_2O + NO \rightarrow NO^+.NO + H_2O$	9(−14)	±2	
$NO^+.H_2O + NH_3 \rightarrow NH_4^+ + HNO_2$	1.0(−9)	±30%	
$NO_2^+.H_2O + NH_3 \rightarrow NH_4^+ + HNO_3$	1.1(−9)	±30%	
$NO^+.NH_3 + NH_3 \rightarrow NH_4^+ + ONNH_2$	9.1(−10)	±30%	
$NH_3^+ + NH_3 \rightarrow NH_4^+ + NH_2$	1.9(−9)	±0.2	
	1.5(−9)	±30%	
	1.7(−9)	±30%	
$H_3O^+ + NH_3 \rightarrow NH_4^+ + H_2O$	2.1(−9)	±30%	
$H_3O^+(H_2O) + NH_3 \rightarrow$ products	2.6(−9)	±30%	
$H_3O^+(H_2O)_2 + NH_3 \rightarrow$ products	1.6(−9)	±30%	
$H_3O^+(H_2O)_3 + NH_3 \rightarrow$ products	2.1(−9)	±30%	
$NH_4^+(H_2O) + NH_3 \rightarrow NH_4^+(NH_3) + H_2O$	1.2(−9)	±30%	
$NH_4^+(H_2O)_2 + NH_3 \rightarrow NH_4^+(NH_3)(H_2O) + H_2O$	⩾9(−10)	±30%	
$H^- + H \rightarrow H_2 + e$	1.3(−9)	±50%	
	1.8(−9)	×2	
$H^- + CO \rightarrow HCO + e$	~5(−11)		

nary positive-ion reactions (*continued*)

action	k (cm³ molecule⁻¹ sec⁻¹)	Uncertainty	Remarks
+ NO → HNO + e	4.6(−10)	±30%	
+ O₂ → HO₂ + e	1.2(−9)	±0.2	
→ O⁻ + OH	<1(−11)		
→ O₂⁻ + H	<1(−11)		
→ OH⁻ + O	<1(−11)		
+ N₂O → OH⁻ + N₂	1.1(−9)	±0.3	
+ NO₂ → NO₂⁻ + H	2.9(−9)	±30%	
+ H₂O → OH⁻ + H₂	3.8(−9)	±30%	
+ H₂ → products	<1(−13)		
+ CO → C₂O + e	4.1(−10)	±30%	
+ O₂ → O⁻ + CO	4.0(−10)	±30%	
+ N₂O → CO + N₂ + e	9.0(−10)	±30%	
+ CO₂ → 2CO + e	4.7(−11)	±30%	
+ O → O₂ + e	1.9(−10)	×2	
+ N → NO + e	2.2(−10)	×2	
+ H₂ → H₂O + e	7.0(−10)	±0.5	
	6.0(−10)	±30%	
	7.2(−10)	±1	
→ OH⁻ + H	3.3(−11)	±0.5	
+ CO → CO₂ + e	6.5(−10)	±1	
	4.4(−10)	±30%	
	7.3(−10)	±0.7	
+ NO → NO₂ + e	2.2(−10)	±0.5	
	1.6(−10)	±30%	
+ N₂ → N₂O + e	<5(−13)		
	<1(−12)		
+ O₂(¹Δg) → O₃ + e	~3(−10)		
+ H₂O → OH⁻ + OH	1.4(−9)	±30%	
+ N₂O → NO⁻ + NO	2.2(−10)	±0.4	
	2.5(−10)	±0.5	
	1.95(−10)	±0.06	
+ NO₂ → NO₂⁻ + O	1.2(−9)	±30%	
+ O₃ → O₃⁻ + O	5.3(−10)	×2	
+ CH₄ → OH⁻ + CH₃	1.1(−10)	±0.1	
	1.0(−10)	±20%	
+ C₂H₄ → C₂H₄O + e	4.05(−10)	±0.5	
→ C₂H₂⁻ + H₂O	1.9(−10)	±0.3	
+ C₂H₂ → C₂H₂O + e	1.3(−9)	±0.09	
→ C₂H⁻ + OH	8.0(−10)	±0.5	
→ C₂OH + H	8(−11)	±1	
+ C₂H₆ → OH⁻ + C₂H₅	7.0(−10)	±20%	
+ C₃H₈ → OH⁻ + C₃H₇	9.3(−10)	±20%	
+ n-C₄H₁₀ → OH⁻ + C₄H₉	1.2(−9)	±20%	
+ H → HF + c	1.6(−9)	×2	
+ H₂ → H₂S + e	<1(−15)		
+ O₂ → SO₂ + e	3.0(−11)	±30%	
+ H → HCl + e	9.0(−10)	×2	
	1.0(−9)	×2	
H⁻ + H → H₂O + e	1.0(−9)	×2	
	1.8(−9)	×2	
H⁻ + O → HO₂ + e	2.0(−10)	±50%	
H⁻ + N → HNO + e	<1(−11)		
H⁻ + NO₂ → NO₂⁻ + OH	1.9(−9)	±30%	
S⁻ + H → H₂S + e	1.3(−9)	×2	
N⁻ + H → HCN + e	8.0(−10)	×2	
O⁻ + O₂ → O₂⁻ + NO	5.0(−10)	±30%	
O⁻ + N₂O → NO₂⁻ + N₂	2.8(−14)	±0.2	
O⁻ + NO₂ → NO₂⁻ + NO	7.4(−10)	±30%	
₂⁻ + H → products	1.5(−9)	×2	
₂⁻ + N → NO₂ + e	4.0(−10)	±50%	

Binary positive-ion reactions (*continued*)

Reaction	k (cm^3 molecule^{-1} sec^{-1})	Uncertainty	Remarks
$O_2^- + O \rightarrow O_3 + e$	$3.3(-10)$	$\pm 50\%$	
$O_2^- + H_2 \rightarrow$ products	$<1(-12)$		
$O_2^- + O_2(^1\Delta_g) \rightarrow 2O_2 + e$	$\sim 2(-10)$		
$O_2^- + N_2O \rightarrow O_3^- + N_2$	$<1(-12)$		
$O_2^- + NO_2 \rightarrow NO_2^- + O_2$	$8.0(-10)$	$\pm 30\%$	
$O_2^- + O_3 \rightarrow O_3^- + O_2$	$3.0(-10)$	$\pm 30\%$	
$O_2^- + SO_2 \rightarrow SO_2^- + O_2$	$4.8(-10)$	$\pm 30\%$	
$NO_2^- + H \rightarrow OH^- + NO$	$3.0(-10)$	$\times 2$	
	$4.0(-10)$	$\times 2$	
$NO_2^- + O \rightarrow$ products	$<1(-11)$		
$NO_2^- + N \rightarrow$ products	$<1(-11)$		
$NO_2^- + NO_2 \rightarrow NO_3^- + NO$	$\sim 4(-12)$		
$NO_2^- + O_3 \rightarrow NO_3^- + O_2$	$1.8(-10)$	$\pm 50\%$	
$O_3^- + H \rightarrow OH^- + O_2$	$8.4(-10)$	$\times 2$	
$O_3^- + N_2 \rightarrow$ products	$<1(-15)$		
$O_3^- + NO \rightarrow$ products	$1.0(-11)$	$\pm 50\%$	
$O_3^- + CO_2 \rightarrow CO_3^- + O_2$	$4.0(-10)$	$\pm 30\%$	
	$5.5(-10)$	± 0.5	
$O_3^- + NO_2 \rightarrow$ products	$2.8(-10)$	$\pm 30\%$	
$O_3^- + SO_2 \rightarrow SO_3^- + O_2$	$1.7(-9)$	$\pm 30\%$	
$CO_3^- + O \rightarrow O_2^- + CO_2$	$8.0(-11)$	$\pm 50\%$	
$CO_3^- + NO \rightarrow NO_2^- + CO_2$	$9.0(-12)$	$\times 2$	
	$1.8(-11)$	$\pm 30\%$	
$CO_3^- + NO_2 \rightarrow$ products	$2(-10)$	$\times 2$	
$CO_3^- + SO_2 \rightarrow SO_3^- + CO_2$	$2.3(-10)$	$\pm 30\%$	
$NO_3^- + N \rightarrow$ products	$<1(-11)$		
$NO_3^- + O \rightarrow$ products	$<1(-11)$		
$O_4^- + O \rightarrow O_3^- + O_2$ $\rightarrow O^- + 2O_2$	$4.0(-10)$	$\pm 50\%$	
$O_4^- + CO \rightarrow CO_3^- + O_2$	$<2(-11)$		
$O_4^- + NO \rightarrow NO_3^{-*} + O_2$	$2.5(-10)$	$\pm 30\%$	
$O_4^- + CO_2 \rightarrow CO_4^- + O_2$	$4.3(-10)$	$\pm 30\%$	
$O_4^- + H_2O \rightarrow O_2^-.H_2O + O_2$	$1.5(-9)$		
	$1.4(-9)$		
$CO_4^- + O \rightarrow CO_3^- + O_2$ $\rightarrow O_3^- + CO_2$	$1.5(-10)$	$\pm 50\%$	
$CO_4^- + NO \rightarrow NO_3^{-*} + CO_2$	$4.8(-11)$	$\pm 30\%$	
$O_2^-.H_2O + NO \rightarrow NO_3^- + H_2O$	$3.1(-10)$	$\pm 30\%$	
$O_2^-.H_2O + CO_2 \rightarrow CO_4^- + H_2O$	$5.8(-10)$	$\pm 30\%$	
$O_2^-.H_2O + O_3 \rightarrow$ products	$3(-10)$	$\pm 30\%$	
$O_2^-(H_2O)_2 + O_3 \rightarrow$ products	$3.4(-10)$	$\pm 30\%$	
$NO_3^{-*} + NO \rightarrow NO_2^- + NO_2$	$1.5(-11)$	$\times 2$	
$NO_3^- + NO \rightarrow NO_2^- + NO_2$	$\sim 3(-15)$		
$O_2^-.H_2O + O_2 \rightarrow O_4^- + H_2O$	$2.5(-15)$	± 1	
$O_3^-(H_2O) + CO_2 \rightarrow$ products	$3(-10)$	$\times 2$	
$O_3^-(H_2O)_2 + CO_2 \rightarrow$ products	$2(-10)$	$\times 2$	
$CO_3^-(H_2O) + NO \rightarrow$ products	$1.8(-11)$	$\times 2$	
$CO_3^-(H_2O) + NO_2 \rightarrow$ products	$1.5(-10)$	$\times 2$	
$NO_2^-(H_2O) + SO_2 \rightarrow NO_2^-(SO_2) + H_2O$	$1.5(-9)$	$\pm 30\%$	
$SO_4^- + NO_2 \rightarrow NO_2^- + SO_2 + O_2$ $\rightarrow NO_3^- + SO_3$	$2.5(-10)$ $1(-10)$	$\times 2$ $\times 2$	
$SO_3^- + H_2O \rightarrow H_2SO_4 + e$	$<1(-12)$		

General Index

Author Index